Intercellular and intracellular communication

General Editor: Professor B. Cinader

Interlocking ligand–receptor systems constitute an intercellular language in which molecules, e.g. factors and hormones, serve as words and convey signals – messages – through combination with membrane molecules, i.e. receptors. These signals can give rise to the production of other factors which can combine with receptors and thus form the sentences of the intercellular language. In the immune system, macromolecules of the *external* world cause distortions of the *internal* communication. The resulting change in the balance of molecular communication constitutes the immune response.

The analogy between verbal and molecular communication is supported by the existence of superfamilies of molecules, designed with a common two-chain structure and involved in recognition of different ligands. Members of these families show homologies in a variety of cell types among cells of different animals, from invertebrates to vertebrates.

Cell communication involves sequential molecular interactions between receptors and ligands, which are either cell-bound or are secreted by one cell and taken up by surface structures (receptors) of another cell. A succession of these interactions at the membranes of cells and organelles coordinate cell metabolism within the same and between different organs.

Receptors can be activated at a distance, or by receptor–ligand interaction between membranes of different cell types, i.e. via adhesion molecules which guide the structural development of organs, as exemplified by neural cell adhesion and embryological development under the influence of 'master' cells. Recognition and, thus, receptor–ligand interaction directs homing of cells in development, differentiation and cell migration.

Receptor–ligand communication plays various roles in the initiation of, and resistance to, infectious disease. Antigen recognition by B and T cells is one aspect of this process, the ability of a particular parasite to attach to a cell receptor is an example of another. In short, the interaction of the cellular milieu with external molecular changes occurs through receptors of the lymphoid system and through receptors of other cells, which control the ability of parasites to attach to special membrane sites, e.g. malarial merozoites to the N-terminal sugar N-acetyl glucosamine.

Interaction between ligand and receptor activates cells for a limited time. This limitation is achieved by endocytosis of receptor, or by dissociation of the complex in which the receptor is contained, resulting in decreased affinity for ligand.

Malfunction of a single step in cell communication results in disease and contributes to neoplastic transformation. Proliferation of neoplastic cells and spread of metastases may be promoted by the disappearance or blockage of receptors through which growth and metastatic spread are controlled.

The analysis of molecular language represents a major stream in biology in the twentieth century. This book is a part of a series in which we shall examine intercellular communication within and between organs, comparative as well as evolutionary aspects of this communication, and, finally, its effect on initiation and progression of disease.

Intercellular and intracellular communication

Hormones, receptors and cellular interactions in plants

EDITORS:

C. M. CHADWICK
Department of Biological Sciences, University of Pittsburgh

D. R. GARROD
Senior Lecturer, Cancer Research Campaign Medical Oncology Unit
University of Southampton

The right of the
University of Cambridge
to print and sell
all manner of books
was granted by
Henry VIII in 1534.
The University has printed
and published continuously
since 1584.

CAMBRIDGE UNIVERSITY PRESS

Cambridge

London New York New Rochelle

Melbourne Sydney

CAMBRIDGE UNIVERSITY PRESS
Cambridge, New York, Melbourne, Madrid, Cape Town, Singapore, São Paulo, Delhi

Cambridge University Press
The Edinburgh Building, Cambridge CB2 8RU, UK

Published in the United States of America by Cambridge University Press, New York

www.cambridge.org
Information on this title: www.cambridge.org/9780521117647

First published 1986
This digitally printed version 2009

A catalogue record for this publication is available from the British Library

Library of Congress Cataloguing in Publication data

Main entry under title:
Hormones, receptors, and cellular interactions in plants
(Intercellular and intracellular communication; 1)

  Includes bibliographical references and index.
  1. Plant hormones.   2. Hormone receptors.
3. Cell interaction.   I. Chadwick, C. M. (Christopher
Michael)   II. Garrod, D. R.
QK731.H65 1986   581.19'27   85–7853

ISBN 978-0-521-30426-9 hardback
ISBN 978-0-521-11764-7 paperback

CONTENTS

viii

CONTRIBUTORS

Clinton E. Ballou *Department of Biochemistry, University of California, Berkeley, California 94720, USA*

Eric G. Brown *Department of Biochemistry, University College of Swansea, Singleton Park, Swansea, SA2 8PP, UK*

C. M. Chadwick *Department of Biological Sciences, University of Pittsburgh, Pittsburgh, PA 15260, USA*

Adrienne E. Clarke *Plant Cell Biology Research Centre, School of Botany, University of Melbourne, Parkville, Victoria 3052, Australia*

Frank B. Dazzo *Department of Microbiology, Michigan State University, East Lansing, Michigan 48824, USA*

H. G. Dickinson *Department of Botany, Plant Science Laboratories, The University of Reading, Reading, UK*

J. E. Ellison *Cancer Research Campaign Medical Oncology Unit, Centre Block, Southampton General Hospital, Southampton SO9 4XY, UK*

L. C. Fowke *Department of Biology, University of Saskatchewan, Saskatoon, Sask., Canada, S7N 0WO*

D. R. Garrod *Cancer Research Campaign Medical Oncology Unit, Centre Block, Southampton General Hospital, Southampton SO9 4XY, UK*

M. A. Hall *Department of Botany and Microbiology, The University College of Wales, Aberystwyth, Dyfed, SY23 3DA, UK*

Mark A. Holden *Department of Botany, The University of Edinburgh, Edinburgh, EH9 3JH, UK*

Barbara J. Howlett *Plant Cell Biology Research Centre, School of Botany, University of Melbourne, Parkville, Victoria 3052, Australia*

K. R. Libbenga *Department of Plant Molecular Biology, Botanisch Laboratorium, Rijksuniversiteit Leiden, Leiden, The Netherlands*

P. C. G. v.d. Linde *Department of Plant Molecular Biology, Botanisch Laboratorium, Rijksuniversiteit Leiden, Leiden, The Netherlands*

A. C. Maan *Department of Plant Molecular Biology, Botanisch Laboratorium, Rijksuniversiteit Leiden, Leiden, The Netherlands*

A. M. Mennes *Department of Plant Molecular Biology, Botanisch Laboratorium, Rijksuniversiteit Leiden, Leiden, The Netherlands*

Peter C. Newell *Department of Biochemistry, University of Oxford, South Parks Road, Oxford, OX1 3QU, UK*

Russell P. Newton *Department of Biochemistry, University College of Swansea, Singleton Park, Swansea, SA2 8PP, UK*

Michael Pierce *Department of Biochemistry, University of California, Berkeley, California 94720, USA*

Julie E. Ralton *Plant Cell Biology Research Centre, School of Botany, University of Melbourne, Parkville, Victoria 3052, Australia*

I. N. Roberts *Department of Botany, Plant Science Laboratories, The University of Reading, Reading, UK*

John L. Stoddart *Plant Biochemistry Department, University College of Wales, Welsh Plant Breeding Station, Aberystwyth, Dyfed, SY23 3EB, UK*

Michael M. Yeoman *Department of Botany, The University of Edinburgh, Edinburgh, EH9 3JH, UK*

PREFACE

The complementary binding of a ligand and receptor is the basic language of intercellular communication. Processes of this type are generally categorized under the heading 'cellular recognition' which includes the binding between a cell and (a) another cell (b) a bacterium or foreign particle or (c) a molecule. Such behaviour requires that a cell should anticipate its interaction with the environment by the provision of specific molecules (receptors) which can receive environmental stimuli. These stimuli are as diverse as, for example, the appearance of hormone molecules or contact with a bacterial food organism. The molecular components of the stimulus may be regarded as the ligand. The crucial aspect of ligand–receptor interaction is that the former fits into and binds to the latter in a chemically specific and exclusive manner.

The study of ligand–receptor interactions is well advanced in animal systems, especially in relation to hormones, growth factors, endocytosis and the immune system. By contrast, studies with plants have met with varied success. The aim of this book is to take a broad look at the current status of ligand–receptor interactions in a variety of higher and lower plant systems. It is hoped that this may provide a cross-fertilization of ideas and a stimulus to research in some of the more difficult areas.

One of the most crucial factors affecting the rate of progress of this type of research appears to be the location of the receptors. Plant cells are surrounded by a tough cell wall and this appears to be a barrier which must be penetrated in order to permit identification of receptors. It is probably fair to say that this has hampered research on plant hormone receptors. However, progress is now beginning to be made. The current status of some hormone receptors of higher plants is reviewed in Chapter 1 (auxins), Chapter 2 (ethylene) and Chapter 3 (gibberellin).

We have begun the book with discussion of the major unsolved problems of receptors for these hormones because they stress the difficulties of this type of research. Problems relating to the role of cyclic AMP in plants (Chapter 4) seem to fall into the same category and may be contrasted with the degree of sophistication reached in the study of cyclic AMP receptors in *Dictyostelium* (Chapter 5). Here, some receptors are readily accessible on the exposed plasma membrane, facilitating their study.

In many plant interactions receptors on the extreme outer surface are involved in the initial stages, and in some instances study of receptor mechanisms are well advanced. Examples of this type of interaction considered here are the mating of yeast (Chapter 8), pollen–stigma interactions (Chapter 9), host–pathogen interactions (Chapter 10) and the infection of root hairs by nitrogen-fixing bacteria (Chapter 11).

Adhesion in the cellular slime moulds (Chapter 6), of which *Dictyostelium discoideum* is the most studied example, represents a system in which interaction occurs directly between surface membranes of adjacent cells. The methods used to study adhesion have been applied with great effect to the cells of vertebrate animals and may be useful in plants. However, it must be said that even here the details of the molecular mechanism of cell–cell adhesion have not been elucidated.

One way to study the plasma membrane of plant cells is to remove the cell wall and isolate the living protoplast. The technology for doing this has been available for some time and is now giving rise to the studies of membrane-associated receptor mechanisms (Chapter 7). Finally, a paradox is that most living systems, probably all, possess carbohydrate-binding proteins known as lectins. These possess, *par excellence*, the properties demanded of receptors, yet in the vast majority of cases their functions are unknown. Many plants are particularly richly supplied with lectins. The intriguing problem of their function is discussed in Chapter 12.

The idea for this book was generated by the series editor Dr Cinader. We plunged into it as something of a challenge, having little detailed knowledge of the systems involved. We have learned much during the editorial process and hope that this volume will provide a similar stimulus to its readers.

C. M. Chadwick
D. R. Garrod
January 1985

1

Auxin receptors

K. R. Libbenga, A. C. Maan, P. C. G. v. d. Linde and
A. M. Mennes

1.1   Introduction

It is generally accepted that growth and development of higher
plants are somehow controlled by low molecular weight compounds,
called growth substances or plant hormones. Up to now five classes
of growth substances have been identified, namely auxins, cytokinins,
gibberellins, abscissins and ethylene. However, we cannot exclude that
other classes of growth substances will be discovered in the near future.
For example, evidence is accumulating that endogenous steroids may
play a role in both vegetative and generative development of higher
plants (Geuns, 1982).

Growth substances are produced to a greater or lesser extent in all
parts of the plant and circulate via the vascular system and ground-tissue
through the whole plant body. Synthesis, degradation, inactivation,
interconversion, transport of the growth substances and the sensitivity
of tissues to them are not only influenced by internal factors (state of
tissue differentiation, age of tissues, etc.) but also by external ones (light,
temperature, gravity, mechanical stress, etc.). In general, growth substances have a wide action spectrum and most processes appear to be
controlled by more than one class of growth substances. The picture
is very complicated and in spite of more than 50 years of research we
still do not know the exact role of growth substances in overall and
local morphogenesis (Trewavas, 1981, 1982; Hanson & Trewavas, 1982).
Moreover, the literature comprises numerous reports describing effects
of growth substances, but virtually nothing is known about their molecular mechanism of primary action. However, increasing use of modern
concepts and techniques from cell biology, biochemistry and molecular
genetics during the past decade gives us hope that in the near future
major progress will be made in the unravelling of the molecular mecha-

nism of action of growth substances. The introduction of the receptor concept and the use of modern techniques to identify receptors for plant growth substances are examples. This chapter deals with auxin receptors, or merely with auxin-binding sites, since no receptor function has yet been identified unambiguously.

In addition to the naturally occurring auxins (Fig. 1.1), many synthetic auxin analogues have been synthesized and structure–activity rules have been formulated (Katekar, 1979; Kaethner, 1977; Farrimond, Elliott & Clack, 1978). It is general practice to call a compound an auxin if it stimulates growth by cell extension in particular bioassays, such as coleoptile and stem segments from etiolated monocotyledonous seedlings (mostly grasses such as *Avena sativa, Zea mays*) and stem-internode segments from etiolated dicotyledonous seedlings (mostly leguminous species such as *Pisum sativum, Glycine max, Phaseolus vulgaris*) (Larsen, 1961). A substance is called an anti-auxin if it reversibly inhibits growth induced by an auxin (Housley, 1961).

Major production centres of endogenous auxins are the shoot apex, leaves, developing seeds and fruits. Auxin produced by these tissues is translocated through the whole plant via the vascular system and unidirectionally (polar transport) via ground-tissue. Auxins appear somehow to control elongation and branching of shoots and roots, formation and activity of cambia, cell division and formation of adventitious shoots and roots, flower and fruit development, development and abscission of leaves, etc.

From all this, we must conclude that auxin receptors may be present in a great variety of tissues along with receptors for other growth sub-

Fig. 1.1. A. Structure of the naturally occurring auxin indole-3-acetic acid (IAA) and three synthetic auxins, indole-3-butyric acid (IBA), naphthalene-1-acetic acid (NAA) and 2,4-dichlorophenoxyacetic acid (2,4-D).
B. Structure of the phytotoxin fusicoccin (FC).

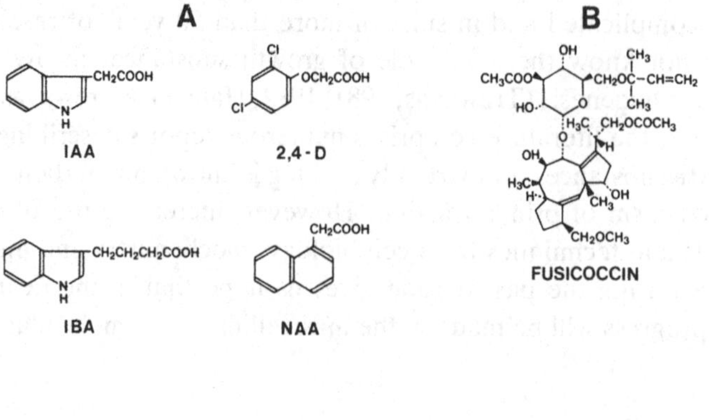

stances. To study exclusively auxin receptors, the experimental systems most widely used are those in which auxin seems to be a major limiting factor, i.e., the classical auxin bioassays. In this chapter we will try to evaluate critically the present state of auxin-receptor research. Previous reviews on growth-substance receptors are those by Kende & Gardner (1976), Venis (1977a), Stoddart & Venis (1980) and Rubery (1981), whereas short evaluations have been given by Libbenga (1978), Lamb (1978), Bogers & Libbenga (1981) and Venis (1981). For general information about plant growth substances, including auxins, the reader is referred to Letham, Goodwin & Higgins (1978).

1.2   Membrane-bound binding sites
1.2.1   *Characterization of auxin-binding sites*
During the last ten years auxin-binding studies have been performed with membrane preparations from various tissues and species. We have listed the main binding characteristics in Table 1.1. As can be seen from this table, membranes from maize coleoptiles have been studied most extensively and, consequently, this system will be discussed first.

(1) *Zea mays* coleoptiles
The first report on high-affinity auxin-binding sites was that by Hertel, Thomson & Russo (1972). They successfully introduced the centrifugation method for separation of bound and free ligand, experimental determinations of non-specific binding and Scatchard analysis (see section 1.4.4). Their report initiated a number of very interesting auxin-binding studies in particular on coleoptile membranes.

In discussing these studies we shall confine ourselves to the major reports by Ray, Dohrmann & Hertel (1977a,b), Ray (1977), Batt, Wilkins & Venis (1976), Batt & Venis (1976), Murphy (1980a), and Dohrmann, Hertel & Kowalik (1978). Solubilized binding sites will be treated separately in section 1.2.2.

Ray *et al.* (1977a), using crude microsomal preparations, demonstrated high-affinity, low-capacity binding for NAA (K_d, $0°C = 5–7 \times 10^{-7}$ M, $n = 30–50$ pmol/g fresh wt) from straight Scatchard plots obtained after correcting for experimentally determined non-specific binding. They found that rapid binding (half exchange time for naphthylacetic acid (NAA) < 15 min) occurs at 0°C and shows a sharp optimum at pH 5.5. The binding site is rapidly inactivated at temperatures above about 30°C and is sensitive to pronase, in particular after pretreatment of the membrane preparations with phospholipase.

Table 1.1. *Characteristics of membrane-bound binding sites*

Material	Fraction[a]	Ligand	Binding sites K_d (μM)	Binding sites conc.[b]	Specificity	pH	Temp. (°C)	Time[c]	Reference
Zea									
coleoptiles	CM	IAA	3–4	30	—	6	0	<30	Hertel et al. 1972
	CM	IAA	1.7	51	+	5.5[d]	0	<15	Batt et al.[e] 1976
	CM	IAA	5.8	96	+	5.5[d]	0	<15	Normand et al. 1975
	CM	IAA	—	—	+[g]	6	0	5	Hertel et al. 1972
	CM	NAA	1–2	40	+	6	0	<30	Ray et al. 1977a,b
	CM	NAA	0.5–0.7	30–50	+	5.5[d]	0	<15	Moloney et al. 1981
	CM	NAA	0.5	38[h]	—	5.5	20	10	Normand et al. 1977
	CM	NAA	1.9	620[i]	—	6	4	5	Batt et al.[e] 1976
	CM	NAA	0.15	38	+	5.5[d]	0	<15	Murphy 1980a
	CM	NAA	1.6	96	+	5.5[d]	0	<15	Batt et al. 1976f
	CM	NAA	0.15	100[i]	+	5.5	0	<15	Murphy 1980a
	H	NAA	1.16	32	—	5.5[d]	0	<15	Batt et al. 1976f
	H	NAA	0.5	—	—	5.5	0	<15	Murphy 1980a
	L	NAA	0.39	24	—	5.5[d]	0	<15	Batt et al. 1976f
	L	NAA	0.44	—	—	5.5	0	<15	Murphy 1980a
	ER	NAA	1.7	15	+	6	0	5	Normand et al. 1977
	ER	NAA	0.4	40	—[k]	5.5[d]	0	<30	Ray 1977
	TP	NAA	1.3	20	—	5.5	0	<30	Dohrmann et al. 1978
	PM	2,4-D	5	40	—	5.5	0	<30	Normand et al. 1977
mesocotyls	CM	NAA	0.75	52	—	6	4	5	Walton et al. 1981
	ER	NAA	0.2	1.9[j]	—[k]	5.5[d]	—	—	Moloney et al. 1981
roots	PM	NAA	0.8	16	+[k]	5.5	20	10	Moloney et al. 1981
	PM	NAA	1.2	19		5.5	20	10	Moloney et al. 1981

Avena									
roots	CM	IAA	12	346	—	5.5^d	37	30	Bhattacharyya et al. 1978
	CM	IAA	52	989	—	5.5^d	37	30	Bhattacharyya et al. 1982
coleoptiles	CM	IAA	0.68	15^i	—	5.5	0	<15	Bhattacharyya et al. 1982
	CM	IAA	7	57^i	—	5.5	0	<15	Bhattacharyya et al. 1982
	CM	IAA	0.2	12^i	—	5.5	0	<15	
	CM	IAA	7	81^i	—	5.5	0	<15	
Glycine									
hypocotyls	PM	IAA	—	—	—	6	4^d	i	Williamson et al. 1977
Phaseolus									
hypocotyls	PM	IAA	—	—	—	6.5^l	4	20	Kasamo et al. 1976
Pisum									
cotyledonary buds	CM	IAA	0.1	—	—	6.5^d	25^d	30	Jablonović et al. 1974
epicotyls	PM	IAA	1^m	0.1^i	—	$6–6.5^d$	4	f	Döllstädt et al. 1976
roots	PM	IAA	1^m	0.1^i	—	$6–6.5^d$	4	f	Döllstädt et al. 1976
epicotyls	PM	MCPA	1^m	0.1^i	—	$6–6.5$	4	f	Döllstädt et al. 1976
roots	PM	MCPA	1^m	0.1^i	—	$6–6.5$	4	f	Döllstädt et al. 1976
Cucurbita									
hypocotyls	PM	IAA	1.5	20	+	5^d	2–4	<15	Jacobs et al. 1978
Helianthus									
tuber explants	CM	IAA	1	22^h	—	5.5	4	<30	Trewavas 1980
Nicotiana									
callus	CM	NAA	0.3	58	+	5^d	36^d	30^d	Vreugdenhil et al. 1979
stem pith	CM	NAA	0.3	32	+	5^d	36^d	30^d	Vreugdenhil et al. 1979
leaves	CM	NAA	0.5	18	—	5	36	30	Vreugdenhil et al. 1980
leaf protopl.	CM	NAA	0.4	36	—	5	36	30	Vreugdenhil et al. 1980
cell suspension	CM	NAA	0.2	$—^n$	—	—	36	30	Vreugdenhil et al. 1981

Table 1.1 (Contd.)

Material	Fraction[a]	Ligand	Binding sites		Specificity	pH	Temp. (°C)	Time[c]	Reference
			K_d(μM)	conc.[b]					
Cucumis young fruits	CM	NAA	10–20	1250	+	3.75[d]	0	10	Narayanan *et al.* 1981*a*
Fragaria receptacles of developing fruits	120K	NAA	1.1	100	+[o]	4[d]	0	10	Narayanan *et al.* 1981*b*

See Appendix 1 for definitions of abbreviations.
[a] CM = crude membrane, PM = plasma membrane, ER = endoplasmic reticulum, TP = tonoplast, H = heavy band, L = light band.
[b] Concentration expressed as pmol/g fresh weight, unless otherwise indicated.
[c] Time expressed in minutes or: i = immediate, f = fast.
[d] Optimal conditions determined.
[e] Batt, Wilkins & Venis (1976).
[f] Batt & Venis (1976).
[g] No K_d determined.
[h] Not given by authors but determined by us from their Scatchard plots.
[i] pmol per mg protein.
[j] Maximum value.
[k] Related to elongation test or curvature test.
[l] Optimum binding at pH 8.
[m] Not from Scatchard plots.
[n] Varying, see text.
[o] Related to receptacle enlargement.

The specificity of the site was described by Ray *et al.* (1977*b*), who determined K_d values for many auxins, auxin analogues (agonists and antagonists) and non-auxins. These K_d values were compared with the C_{50} values (the concentration at which a half-maximal effect is reached) obtained from classical maize coleoptile elongation tests (see section 4). They found large discrepancies between the K_d and C_{50} values, especially for the phenoxyacetic acids ($K_d/C_{50} \gg 1$, indicating a higher efficiency than would be expected from binding studies) and for phenyl derivatives ($K_d/C_{50} \ll 1$).

This is no surprise because:

> Binding of different analogues do not necessarily all lead to the same maximal (quantitative) response (see e.g., Ariens, Beld, Rodrigues de Miranda & Simonis, 1979).

> The analogues tested are possibly subject to differences in *in vivo* uptake (Evans & Hokanson, 1969), or in sequestration and metabolic alteration (Klämbt, 1961).

> Ray *et al.* (1977*b*) found that the C_{50} values may depend upon whether growth is measured after a relatively short or after a long period. A reasonable explanation for this phenomenon is that different, unsynchronized processes, initiated by auxin application, are treated as one process, i.e., growth.

> The binding experiments were performed at 0°C while elongation tests are performed at higher temperatures (23°C).

> The elongation of maize coleoptiles may not be affected by the binding site under consideration.

The only conclusion that can be drawn from the experiments of Ray *et al.* (1977*b*) is that, in general, auxins and anti-auxins (those compounds that inhibit elongation reversibly) bind to the same binding site competitively while non-auxins do not, or do so with very low affinity. In this respect the binding sites can be considered as specific.

In a previous report Ray, *et al.* (1977*a*) had already shown that the supernatant from microsome pellets contains a factor that lowers the affinity of the binding sites for NAA but does not affect their number. This so-called supernatant factor (SF) was later identified by Venis & Watson (1978) as a mixture of 6-methoxy-2-benzoxazolinone (MBOA) and 6,7-dimethyl-2-benzoxazolinone (DMBOA). Ray *et al.* (1977*b*) tested the influence of crude SF on the K_d values of most analogues studied. They found that SF increases the K_d values for some analogues (e.g., 1-NAA, 2-NAA, 3-indole-3'-propionic acid (IPA)), whereas K_d values for some other compounds (e.g., 2,4-dichlorophenoxyacetic acid

(2,4-D), 3,4,5-triiodobenzoic acid (3,4,5-TIBA) and 2-naphthoxyacetic acid (2-NOA)) are decreased. However, these effects of SF did not generally improve the correlation between K_d and C_{50} values.

Of course we cannot conclude from these experiments that SF is an endogenous modifier of binding affinity of auxin receptors. SF is found only in a limited number of grasses and only after the cells have been damaged (Venis & Watson, 1978 and references therein). Cross & Briggs (1979) have tested whether supernatant isolated from other species (onion, carrot, potato, apple) influences binding of NAA to binding sites in microsomes from maize coleoptiles and solubilized binding sites from the same material. They found that all species tested show some inhibitory effect, and that the largest effect is caused by supernatant isolated from maize coleoptiles.

With respect to the localization of the binding sites, Ray (1977) found that after isopycnic centrifugation of microsomes, the majority of high-affinity NAA binding sites coincide with endoplasmic reticulum (ER) markers (NADPH: cytochrome c reductase and rotenone- and antimycin-insensitive NADH-cyt-c reductase). At high Mg^{2+} concentrations, which prevent the ribosomes from being stripped off the ER, the majority of NAA binding shifts to higher densities. This agrees with a localization of the binding sites at the ER. This was quite unexpected since, with regard to the acid-growth theory of auxin action (see section 1.2.4) it was plausible to assume that auxin receptors would be localized at the plasma membrane. However, evidence indicating the presence of two classes of auxin-binding sites in maize coleoptile membranes – one localized at the ER (site 1) and another one at the plasma membrane (PM) (site 2) – is reported in Batt et al. (1976) and in Batt & Venis (1976). They concluded from biphasic Scatchard plots that two binding sites – one with a high affinity (site 1) and one with a lower affinity (site 2) – are present in a membrane preparation, obtained by centrifugation at 4000–38 000 g (Batt et al. 1976). Further experiments by Batt & Venis (1976) showed that a light and a heavy microsome fraction can be obtained after sucrose density gradient centrifugation. With each of these fractions straight Scatchard plots with different slopes were found. The K_d values found for the heavy and light fractions corresponded with the K_d values found for sites 1 and 2, respectively, as obtained previously by Batt et al. (1976) from biphasic Scatchard plots. Although it is tempting to assume the presence of two different classes of binding sites from these data, there are a few problems. In the first place, the presence of a binding site of high affinity (site 1) can indeed be concluded from the biphasic Scatchard plot shown by Batt et al. (1976), but their conclusion that another site with lower affinity is also present was too hasty. They analysed the plot by means of the extrapolation method without taking into account that the flat part of the curve may have been caused by non-specific binding. In the second place, the straight Scatchard plots for the light and the heavy microsome fractions were obtained using NAA in the range of 2×10^{-7} to 1×10^{-6} M. The danger of using narrow concentration ranges is pointed out in section 1.4.2 and, indeed, when the experiments of Batt & Venis (1976) were repeated by Murphy (1980a), he found comparable plots in the same concentration range, but after extending the range towards higher concentrations, the straight lines turned into curves. He then analysed the curves with the computer program MLP (see section 1.4.2) and concluded that the best fits are obtained with a 3-parameter model, i.e., one high-affinity binding site plus non-specific binding. This analysis showed that K_d values of high-affinity binding in the light and heavy fractions are almost identical, but that the concentrations of the binding sites are different. When we re-examined his calculations we found something different. Using the equation

$$\frac{B}{F} = \frac{K_1 R_1}{1 + K_1 F} + N \tag{1}$$

where B = concentration of bound hormone; F = concentration of free hormone; K_1 = association constant; R_1 = concentration of receptor; N = amount of non-specific binding, the values of B and B/F can be computed and a theoretical Scatchard plot can be constructed, when K_1, R_1 and N are known, for any hormone concentration H. We have tried to construct such plots for binding to both the heavy and light bands, in order to compare them with the experimental plots. We used K_1 and R_1 as given by Murphy (1980a). As the amount of non-specific binding was omitted in his paper, we tested a range of values for N (by means of a short computer program based upon eq. (1). A value for N that would give a plot that could be superimposed upon the experimental plot was readily found for the heavy band, but for the light band such a value could not be found. This would mean that either the model used in Murphy's paper (1980a) was incorrect, or that K_1, R_1 and/or N were calculated wrongly. We tested both a 3-parameter model and a 4-parameter model using the programs described in section 1.4.2. We found no significant difference between the models, but we did find a difference between K_d values for the light and the heavy bands (respectively 2.3×10^{-6} and 8.0×10^{-6} M), thus indicating the presence of different classes of binding sites.

Further supporting evidence for two different classes of binding sites was found by Batt *et al.* (1976), in that some compounds, which were inactive in the coleoptile elongation test, such as benzoic acid, 2,6-dichlorobenzoic acid (2,6-D) and 2,4-dichlorobenzoic acid (2,4-D), compete with NAA for site 1 but not for site 2. This would indicate that site 1, which has the highest affinity for strong auxins like NAA, nevertheless binds some inactive compounds, whereas site 2, which has a lower affinity, is more selective in this respect.

Comparison of the different binding sites in sucrose density gradients with enzyme markers showed that the position of site 1 coincides with ER markers and that of site 2 with PM markers. Batt & Venis (1976) rightly draw attention to the fact that the use of enzyme markers to determine PM fractions is not unequivocal (see also Cross & Briggs, 1976, 1979, and references therein). However, they obtained additional evidence that site 2 is located at the PM from specific staining methods and the determination of sterol : phospholipid ratios.

Binding parameters and localization of site 1 are comparable with the NAA-binding site described by Ray and coworkers, but Ray *et al.* (1977*b*) found hardly any competition with benzoic acid, and a much lower affinity for 2,4-benzoic acid and 2,6-dichlorophenoxyacetic acid.

Finally, Dohrmann *et al.* (1978) reported that possibly three classes of binding sites are present in maize coleoptile membranes. On sucrose density gradients the major amount of the binding sites (site I) coincides with ER markers, another site (site II) is associated with acid phosphatase, which is possibly located at the tonoplast, and a third site (site III) appears at 30–38% sucrose coinciding with a plasma membrane marker (NPA-binding). The sites show a difference in affinity for various analogues. Thus, the affinity of site I, for instance, NAA, indoleacetic acid (IAA) and the anti-auxin phenylacetic acid (PAA) is substantially higher than the affinity of site II for these analogues. Apparently PAA can be used to discriminate between site I and site II in unfractionated microsome preparations. Both site I and site II have a relatively low affinity for 2,4-D, whereas site III has a relatively high affinity for 2,4-D and a relatively low affinity for NAA and IAA. SF has no effect upon binding affinities of site II, but in the presence of SF, binding affinities of site I towards auxin analogues are altered in such a way that they resemble binding affinities of site II. In accordance with Ray (1977), the majority of the binding sites (site I), (also in the presence of SF) does not bind benzoic acid, nor does site II. In respect to site III, we must notice that the amount of 2,4-D bound was very low (75 000 cpm added, 9000 non-specifically bound and only 200 specifically) and the test was performed by two-point analysis, which cannot distinguish between low- and high-affinity binding (see section 1.4.2).

In conclusion, it seems fairly well established that ER in maize coleoptiles contains high-affinity binding sites. The presence of other classes of binding sites located at the plasma membrane and possibly at the tonoplast seems not unlikely, but these binding sites need further characterization before a definite conclusion can be reached.

(2) Other tissues
(a) Mesocotyls and roots of *Zea mays*
Ray *et al.* (1977*b*) found that microsome preparations from coleoptiles, primary leaves and mesocotyls show relatively much high-affinity NAA-binding, whereas microsomes from node and shoot tip, roots and seed endosperm exhibit only relatively little binding per mg of membrane protein. The presence of NAA-binding sites in mesocotyls was already reported by Normand, Schuber, Benveniste & Beauvais (1977) and these sites were further studied by Walton & Ray (1981). The latter workers found that, as in coleoptiles, most of the NAA-binding sites are located at the ER, but have a somewhat higher affinity towards NAA. Interestingly, the amount of these binding sites, whether expressed per mg of membrane protein or per g fresh weight, shows a positive correlation with the maximum auxin-induced growth response of excised segments, which were taken from different parts of either etiolated or irradiated mesocotyls. The implication of this finding for a possible receptor function for the binding sites will be discussed in section 1.2.4.

In microsomes from maize roots high-affinity binding of auxin seems to be associated with a 10 000 g rather than with a 48 000 g pellet, indicating that the binding sites are not associated with the endoplasmic reticulum, but presumably with the plasma membrane (Moloney & Pilet, 1981, 1982). The affinity of this site for IAA is somewhat higher than that for NAA and, as judged from displacement experiments, much higher than that for 2,4-D.

(b) Roots of *Avena sativa*
Bhattacharyya & Biswas (1978) have demonstrated high-affinity IAA-binding to crude membrane fractions of oat roots. They did not determine an optimum incubation time nor an optimum temperature, but their experiments were performed at 37°C during 30 min. From a biphasic Scatchard plot they determined K_d values of 1.2×10^{-5} and 5.2×10^{-5} M, unfortunately using the extrapolation method (the disadvantages are discussed in section 1.4.2). Furthermore, the concentrations of IAA tested were high (1.6×10^{-6} to 10^{-3} M), so it is possible that sites of

higher affinity were overlooked. In this range neither NAA nor indole-3-butyric acid and tryptophan compete with IAA-binding. According to Bhattacharyya & Biswas (1978) this suggests that auxin-binding to oat root membranes is specific for IAA only. Bhattacharyya & Biswas (1982) later detected a binding site of higher affinity after lowering the auxin concentrations and pretreatment of the roots with 10^{-5} M IAA. The induced high-affinity IAA-binding site in roots occurs in comparable amounts to those in coleoptiles, and K_d values lie within the same order of magnitude. A low-affinity binding site is also present in control roots and in pretreated roots but the affinity of these sites (K_d values of 8.4×10^{-7} and 7×10^{-7} M, respectively) is still higher than the 'high-affinity' site from their previous paper (K_d 1.2×10^{-5} M, Bhattacharyya & Biswas, 1978).

(c) Hypocotyls of *Glycine max*

The report by Williamson, Morré & Hess (1977) was directed towards the localization rather than the kinetic characterization of auxin-binding sites. Two-point analysis showed that IAA-binding is located in a fraction enriched in plasma membrane, but other fractions bound IAA as well (on a cpm/mg basis, the lowest amount of bound IAA was 43, the highest was 93 cpm/mg). This result, combined with the inability of the method used to distinguish low-affinity binding from high-affinity binding, does not seem very convincing to us.

(d) Hypocotyls of *Phaseolus mungo*

Kasamo & Yamaki (1976) also reported high-affinity binding of IAA to plasma-membrane-enriched microsome fractions. No kinetic data were given, but from the presented figures we calculated a K_d of roughly 3×10^{-6} M. Surprisingly, routine binding tests were performed at pH 6.5 while high-affinity binding at pH 8 was significantly greater. From competition experiments it can be inferred that the binding sites had a lower affinity for NAA and a much lower affinity for 2,4-D than for IAA. Indole-3-propionic acid, indole-3-*n*-butyric acid and indole-3-acetonitrile did not interfere with high-affinity IAA-binding. Since Kasamo & Yamaki (1976) were able to demonstrate IAA-sensitive ATPase activity in plasma-membrane-enriched fractions from mung bean hypocotyls (see also section 1.2.4), it is a pity that the above-mentioned binding sites were only roughly characterized by these workers.

(e) *Pisum sativum*

(1) Cotyledonary buds. Jablonović & Noodén (1974) described binding of IAA to microsome preparations from cotyledonary buds of pea. Although no kinetic data were given, some remarkable results were

presented. After incubation of membranes with labelled IAA, only 0.5–1.7% of the total amount of bound radioactivity could be replaced by excess IAA or NAA. After washing (resuspension) of the incubated membranes, most (85%) of the label had disappeared but now more (percentwise) binding was competable. We would suggest that most of the non-specific binding was removed by washing while high-affinity binding remained. The dissociation rate constant must thus be very low.

Furthermore, it was found that high-affinity binding increased after leaving the homogenate at 4°C for 1 h, before addition of labelled IAA. No explanation was given, nor can we think of any easy explanation. Optimum pH was found to be 6.5 and temperature optimum 25°C. Although not indicated, we assume that equilibrium was reached after the incubation time of 30 min. High-affinity IAA-binding was enhanced by Ca^{2+}. Not much information was given about the specificity of the binding site. It was found that not only IAA, NAA and the strong anti-auxin and auxin-transport inhibitor 2,3,5-TIBA compete with bound radioactive IAA, but also the cytokinin benzyladenine. Jablonović & Noodén (1974) suggest that the ability of benyladenine to compete for IAA-binding sites in this system might reflect the ability of this cytokinin to counteract the inhibitory effect of auxin in apical dominance. A possible role of these binding sites in apical dominance is suggested by experiments in which release of apical dominance by decapitating seedlings just above the cotyledonary node caused rapid outgrowth and a large decrease of high-affinity IAA-binding sites in the cotyledonary buds. This correlates with a decreased ability of auxin to inhibit released buds, as was previously found by Sachs & Thimann (1964).

(2) Epicotyls and roots. Döllstädt, Hirschberg, Winkler & Hübner (1976) studied the binding of IAA and a few phenoxyacetic-acid derivatives (methylchlorophenoxyacetic acid (MCPA), 4-Cl-PA and 2-Cl-PA) to microsome fractions from pea epicotyls and roots. In auxin bioassays MCPA is a strong auxin, whereas 4-Cl-PA and 2-Cl-PA have only 50% and 10% respectively of the activity of MCPA. From displacement plots it was found that IAA and MCPA bind with a K_d of less than 10^{-6} M, whereas, in accordance with their lower biological activity, the affinity for 4-Cl-PA is lower, while 2-Cl-PA is inactive as a competitor (with 2-Cl-[^{14}C]-PA only non-specific binding was found). In cross-competition experiments it was found that MCPA is a better competitor for IAA-binding than IAA itself in roots; this was less pronounced in epicotyls, while IAA had no effect upon MCPA-binding in microsomes from either tissue. These results cannot be explained by assuming differences in

K_d (both are about the same and 4-Cl-PA, which has a lower affinity than IAA, has the same effect), but possibly the MCPA binds to another receptor site, thereby decreasing the affinity towards IAA.

The localization of the binding sites was tentatively given as the plasma membrane. This finding resulted from differential centrifugation and no further tests are reported. A preliminary paper by Slone & Bilderback (1983) reports the presence of an auxin-binding protein in Alaska pea stems with an optimum pH of 4. The soluble fraction of the cells is claimed to stabilize this binding.

(f) Hypocotyls of *Cucurbita pepo*

A study by M. Jacobs & Hertel (1978) on zucchini membranes indicated the presence of a plasma membrane-bound auxin-binding site. Later it was shown that this 'receptor' was nothing but uptake of auxin into sealed vesicles (Hertel, Lomax & Briggs, 1983).

(g) Tuber explants from *Helianthus tuberosus*

Upon culturing Jerusalem artichoke tuber explants in the presence and in the absence of 2,4-D, Trewavas (1980) found an IAA-binding site present in microsome preparations, which site can only be detected if the explants are cultured in the presence of 10^{-5} to 10^{-7} M 2,4-D. Maximal high-affinity IAA-binding was found after about 2 days of culture. Specificity was not determined other than the demonstration that neither gibberellic acid nor abscisic acid competed with IAA-binding.

(h) *Nicotiana tabacum*

In our laboratory, high-affinity binding of auxins to tobacco membranes from different tissues (leaf mesophyll, stem pith, cell and tissue cultures) has been demonstrated. Binding of NAA to membranes has a fairly constant affinity, ranging from 0.5×10^{-6} M for callus to 0.2×10^{-6} M for cell suspensions. Binding experiments were performed at 36°C for 30 min (Vreugdenhil, Burgers & Libbenga, 1979; Vreugdenhil, Harkes & Libbenga, 1980); longer incubation resulted in loss of binding activity. By decreasing the temperature to 25°C equilibrium was reached after *c.* 2 h and binding remained constant for a few more hours (Vreugdenhil, 1981; Maan, Vreugdenhil, Bogers & Libbenga, 1983). The decrease of the temperature had some effect on the affinity of binding of NAA (K_d, callus 36°C is *c.* 0.5×10^{-6} M while K_d, callus 25°C is *c.* 0.2×10^{-6} M), as is to be expected on thermodynamic grounds. Binding at 0°C was not detectable. The specificity of the binding was determined by testing a number of analogues (Vreugdenhil *et al.* 1979); although these binding data were compared with coleoptile straight-growth tests, we think that specificity was demonstrated, at least according to the criterion discussed in section 1.2.3.).

In a study to determine the concentration of binding sites during a growth cycle of tobacco cells in batch culture Vreugdenhil, Burgers, Harkes & Libbenga (1981) found that both cell number and high-affinity NAA-binding increased exponentially, but at different rates and for different periods. This caused a characteristic modulation of binding sites per cell during the growth cycle: during the first day of the lag phase it decreased, in the exponential phase it markedly increased till the sixth day, then decreased again while it remained constant in the stationary phase (Fig. 1.2).

Because of the slow attainment of equilibrium it is possible to construct time courses of association and dissociation. Time courses of dissociation in semilogarithmic plots are non-linear, while one would expect linearity from simple binding kinetics. Moreover, the dependence of initial rates of complex formation on NAA concentrations shows saturation kinetics; in most experiments these curves are biphasic (Fig. 1.3). We can rule out retention of NAA by carrier-mediated uptake into membrane vesicles, since even after an incubation of 18 h at 25°C, binding is a hyperbola-like function of the hormone concentration: there is no linear relationship as one would expect for uptake. These data also indicate a more complex binding, since in simple binding, initial rates of complex formation would have shown a linear dependency on the NAA concentration (Vreugdenhil, 1981; Libbenga, Maan & Bogers, 1982; Maan *et al.* 1983).

Fig. 1.2. Changes in the amount of specific NAA binding per cell to isolated membranes during the growth cycle of tobacco cell-suspension cultures grown on medium containing NAA and kinetin. From Vreugdenhil (1981).

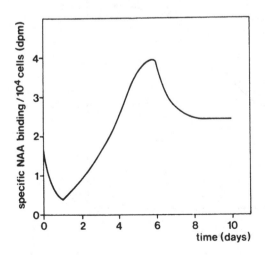

Attempts to localize the binding by sucrose density gradient centrifugation failed because at 35% sucrose NAA-binding, glucan-synthetase-II activity (a presumed PM marker) and NADPH-cytochrome-c reductase (ER marker) coincide (Vreugdenhil, 1981). However, a tentative localization at the plasma membrane was proposed (Vreugdenhil *et al.* 1980). We found that treatment of viable plasmolysed leaf cells with pronase drastically reduced the number of binding sites and it seemed unlikely that pronase was taken up by viable cells. We also found that freshly prepared leaf protoplasts had lost NAA-binding sites; these reappear after culturing the protoplasts for 4–5 days. The loss of binding sites during preparation is mostly due to protease contamination of the enzymes (macerozyme and cellulase Onozuka R-10) used to dissolve the cell walls.

(i) Young fruits of *Cucumis sativus*

As mentioned in the Introduction, auxin regulates processes in many tissues, so it is no surprise that an auxin-binding site has been found in fruits (see also (j)). Narayanan, Mudge & Poovaiah (1981*a*) reported the existence in membrane preparations from cucumber fruits of a binding site for NAA with a K_d of *c.* 10^{-5} M and an optimum pH of 3.75–4. To determine the time needed to reach equilibrium between binding

Fig. 1.3. Initial velocity of NAA binding (V_0) to membrane fractions isolated from tobacco callus as a function of the NAA concentration.

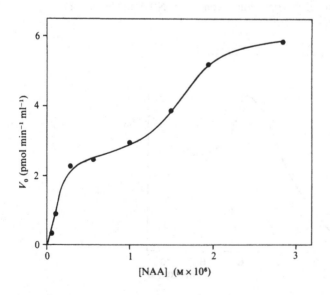

V_0 (pmol min^{-1} ml^{-1})

[NAA] (M × 10^6)

site, NAA and [^{14}C]-NAA, they incubated membranes with [^{14}C]-NAA followed by addition of [^{12}C]-NAA. After 2 min at 0°C, 1.3% [^{14}C]-NAA was bound, after 10, 30 and 60 min, 0.8%, indicating that equilibrium had been reached after *c.* 10 min. However, when hormone was added in the sequence [^{12}C]-NAA and [^{14}C]-NAA, a similar figure was obtained, although in this case one would expect the [^{14}C]-NAA binding to increase. Possibly the figure they presented is a wrong one. The presence of two rather dubious points in a displacement plot of high-affinity NAA-binding is the basis of their interpretation that possibly positive coopera-tive binding is present. Transformation of this plot into a Scatchard plot results in a very steep upward part at low NAA concentrations, again interpreted as positive cooperative binding without checking for possible artifacts (see, for instance, Chamness & McGuire, 1975; De Lean & Rodbard, 1979). Further, it is evident from their Scatchard plot that, apart from using the extrapolation method, they calculated the K_d wrongly because they did not determine the intercept with the abscissa correctly. If we use the (wrong) intercept for K_d determination, then, indeed, $K_d = 1.7 \times 10^{-5}$ M, the value given by Narayanan *et al.* (1981*a*). From displacement experiments it was found that the affinity of the binding sites for analogues decreased in the order NAA, 2,4-D, TIBA > IAA, 2-NOA > 2-NAA, IPA. The first two groups of com-pounds are reported to be very effective inducers of parthenocarpy in cucumbers. Further data and correct analytical methods are needed before any conclusions with respect to this binding site can be drawn.

(j) Developing fruits of *Fragaria anassa*

High-affinity binding of NAA to membranes from receptacles of develop-ing Strawberry fruits was demonstrated by Narayanan, Mudge & Poovaiah (1981*b*). Interestingly, the optimum pH of binding was relatively low as was also found for cucumber fruits (Narayanan *et al.* 1981*a*). Also, the displacement plot of NAA-binding showed an upward curvature at low NAA concentrations, which can indicate positive coopera-tive binding. Again, the authors did not check for possible artifacts such as breakdown of unoccupied binding sites (see also (i)). The displace-ment plot (which shows half-maximal binding – a rough indication for the K_d – at 10^{-5} M) can be transformed into a straight Scatchard plot (neglecting the upward part at low NAA concentrations) from which a K_d of *c.* 10^{-6} M can be calculated. The relative stimulation of *in situ* fruit growth by 12 auxin analogues at 10^{-4} M (stimulation of 10^{-4} M NAA was set at 100%) was compared with their ability to displace *in vitro* 5×10^{-7} M [^{14}C]-NAA from the binding sites, also at 10^{-4} M (competition by 10^{-4} M NAA was set at 100%). By linear regression of the data a correlation of $r = 0.74$ was calculated, which is a rather good correlation under these circumstances.

(3) Conclusions

Membrane-bound auxin-binding sites have been found in cells of many tissues. As we have pointed out in section 1.1, auxin receptors are indeed expected to occur in a variety of cell types. The properties of the auxin-binding sites described thus far (affinity, pH optimum, temperature dependence, cellular localization, analogue specificity) seem to differ quite significantly. Therefore, the question arises whether cells contain various types of auxin receptors (including receptors that are not membrane-bound, see section 3) of which the relative proportions may differ among different cell types. However, only binding sites have been identified and some of them may be artifacts (see also Murphy, 1979). The crucial point is the identification of receptor functions and one should focus on that, rather than describe even more binding sites for even more tissues and species. In section 1.2.4 we shall discuss some possible receptor functions for the membrane-bound auxin-binding sites.

1.2.2   Characterization of solubilized auxin-binding sites

For a better understanding of the mode of action of a membrane-bound receptor, ideally all components involved in the transduction of the signal of auxin binding to the receptor should be isolated, purified and re-incorporated into an artificial membrane.

The first step, solubilization of binding sites, has been described by a number of workers, all using *Zea mays* coleoptiles and primary leaves in their experiments (see Table 1.2).

An early report by Dohrmann & Ray (1976) was concerned mainly with detection methods (see also 1.4.4). Using a gel-filtration technique, they showed that upon incubation of membranes with Triton X-100, radioactive hormone was bound to a fraction that was eluted in the void volume, indicating a high molecular weight; they also showed that unlabelled hormone displaced the labelled hormone.

Batt *et al.* (1976) also described solubilization using Triton X-100 giving 87% solubilized sites. Although the concentration of Triton X-100 was lowered before the binding assay, it cannot be excluded that part of the binding was caused by uptake of hormone into Triton X-100 micelles.

Another method involving the use of acetone, was used by Venis (1977b) to characterize binding of NAA to the solubilized sites. The binding was found to be hetrogeneous with apparent K_d values of 1.7×10^{-7} and 1.4×10^{-6} M; these values are comparable with values

Table 1.2. *Characteristics of solubilized auxin-binding sites from Zea mays*

Fraction[a]	Binding site K_d (μM)[b]	conc.[b]	Specificity	pH	Mol. wt (K)	Purification[c]	Method[d]	Temp. (°C)	Time (h)[e]	Reference
133	0.5[f]	—	—	5.5	—	—	TX-100	0–4	—	Dohrmann et al. 1976
4–38	—	—	—	5.5	—	—	TX-100	4	23	Batt et al. 1976[g]
4–80	0.17	14.3	—	5.5	47.3	100×	acetone	4	5	Venis 1977
	1.4	51	—	5.5	40.3					
145	—	—	—	5	90	+	TX-100	0–4	f	Cross et al. 1978[h]
48	—	—	—	5	80	+	TX-100	0–4	f	Cross et al. 1978[i]
48	0.046	25–40	+	5	80	+	acetone	0–4	f	Cross et al. 1978[i]
4–80	—	—	—	5.5	40	87%	acetone	4	5	Venis 1979
ER[j]	—	—	—	5	—	—	acetone	0–4	f	Cross et al. 1979
	0.12	750[k]	—	5.5	<45	100–150×	acetone	0–4	f	Tappeser et al. 1981

[a] From differential centrifugation. Values × 10^{-3} **g**.
[b] pmol/g fresh weight, unless indicated otherwise.
[c] Purification in %, times (×), or + when no amount was given.
[d] TX-100 = Triton X-100.
[e] A fast reaction is denoted by: f.
[f] About half-saturation of the binding sites.
[g] Batt, Wilkins & Venis (1976).
[h] Cross, Briggs, Dohrmann & Ray (1978).
[i] Cross & Briggs (1978).
[j] Endoplasmic-reticulum fraction.
[k] pmol/mg protein.

found for intact membranes (Batt *et al.* 1976). In this case, solubilization of the binding sites does not significantly alter binding parameters. The presence of two binding sites may be inferred from the fact that upon solubilization and gel filtration on Sephadex-G-100 a peak at *c.* 47 kilodaltons and a shoulder at *c.* 40 kilodaltons are found (Venis, 1977*b*). Another interpretation, as given by Cross, Briggs, Dohrmann & Ray (1978) is the possible degradation of binding sites by proteolytic enzymes. They found a molecular weight of about 90 kilodaltons upon solubilization of the binding sites in the presence of PMSF (an inhibitor of proteolytic enzymes), and a molecular weight of *c.* 70 kilodaltons in the absence of the inhibitor. Their methods of solubilization, detection of binding, and determination of molecular weight (Triton X-100, gel filtration and Biogel gel filtration, respectively) were different from those employed by Venis (1977*b*) (acetone, equilibrium dialysis and Sephadex-G-100 gel filtration, respectively). Later (Cross & Briggs, 1978) a comparison between Triton X-100 and acetone was made, but even then a mol. wt of 80 kilodaltons was found for both treatments. Venis (1979) showed that in the system he used, PMSF had no influence on the determination of the molecular weight. Two possibilities for the discrepancies in the mol. wt determinations are: (a) differences in varieties of the material used, and (b) aggregation of two 40-kilodalton proteins, resulting in a mol. wt of 80 kilodaltons.

Another, more interesting discrepancy between the results of Venis (1977*b*, 1979), Cross & Briggs (1978), and Cross *et al.* (1978) is that in the former experiments the affinities of the binding sites were more or less unaltered, while in the latter one a large decrease was found in K_d: from *c.* 8×10^{-7} M for intact membranes to *c.* 4.6×10^{-8} M for solubilized preparations. however, the specificity of the solubilized sites was roughly equal to that of the intact membranes (Ray *et al.* 1977*b*). Finally, this solubilized binding was shown to have been derived from ER membranes (Cross & Briggs, 1979).

Tappeser, Wellnitz & Klämbt (1981) have solubilized NAA-binding sites from maize coleoptiles and purified them, using affinity-chromatography techniques. Surprisingly, the high-affinity NAA-binding proteins ($K_d = 1.2 \times 10^{-7}$ M) could be eluted more easily from the affinity absorbant than the non-specific NAA-binding proteins. Competition experiments showed that unlabelled NAA, 2,3,5-TIBA and 2,4-D displaced bound, labelled NAA; the 50% displacement values ($3–4 \times 10^{-7}$, 10^{-6} and 2×10^{-2} M, respectively) cannot be used to calculate K_d values because the concentration of labelled NAA was too high. Upon Sephadex-G-100 gel filtration, most NAA-binding occurs at a mol. wt of *c.* 45 kilodaltons. This finding was based upon single-point analysis (one concentration of [^{14}C]-NAA per fraction) but gel electrophoresis of these fractions showed no indications of the presence of large amounts of proteins (that could bind NAA with low affinity and high capacity).

Recently, Löbler (1984) solubilized membrane proteins and partially purified an auxin-binding fraction by means of 2-OH-3,5-DIBA-Sepharose affinity chromatography and AcA54 gel filtration. This fraction was used as antigen for the preparation of IgG fractions that were coupled to CNBr-activated Sepharose. It contained antibodies against the auxin-binding site as well as contaminating proteins. Almost all contaminating proteins could be isolated by affinity chromatography on a Sepharose matrix without 2-OH-3,5-DIBA. These proteins were immobilized and used to isolate the contaminating antibodies from the IgG fraction. These antibodies were then used to obtain a highly purified auxin-binding protein with a rather high affinity for NAA, K_d 5.7×10^{-8} M. Subsequent isolation of monospecific antibodies and immunohistochemical experiments showed that the binding site is predominantly located in the epidermal cells of the coleoptiles.

1.2.3   Characterization of fusicoccin-binding sites

Fusicoccin (FC, Fig. 1.1) is a vivotoxin produced by the phytopathogenic fungus *Fusicoccum amygdali* (Ballio *et al.* 1964, 1976). In auxin-responsive tissues FC mimics auxin in many respects and there is some evidence indicating that in higher-plant cells an endogenous FC-like ligand exists (Aducci, Crosetti, Federico & Ballio, 1980). Therefore, it is not unlikely that FC and its putative endogenous equivalent modulate effector systems which are similar, or at least related, to effector systems that are controlled by auxin. Thus, since knowledge of FC-receptor function may be very useful to identify receptor functions for membrane-bound auxin-binding sites, we shall discuss briefly some major reports on FC-binding sites. The possible functions of FC- and auxin-binding sites will be discussed elsewhere (section 1.2.4).

In general, dose-response curves for FC lie in the range of *c.* 10^{-8} to 10^{-6} M. This range has been used as a first indication for the detection of FC-binding sites. In these studies, [^{3}H]dihydro-FC was used as labelled ligand. Dihydro-FC is known to have the same effects and activities upon plant tissues as FC (Ballio *et al.* 1971; De Michelis, Lado & Ballio, 1975).

The first report on FC-binding sites was by Dohrmann *et al.* (1977). They demonstrated the presence of a high-affinity binding site in microsome preparations from maize coleoptiles (see Table 1.3 for details). The affinity ($K_d = c.$ 10^{-9} M) was remarkably high, at least when com-

Table 1.3. *Characteristics of fusicoccin-binding sites*

Material	Binding sites[a]		Binding sites[b]		pH	Temp. (°C)	Time (h)	Reference
	K_d(nM)	conc.[c]	K_d(nM)	conc.[c]				
Zea								
	2	0.7	2 or 1	1.4 / 0.7	5.5	26	1	Dohrmann et al. 1977
	1.2	2.4	0.75	2.4[d]	5.5	26	2	Pesci et al. 1979
	0.68	0.17	0.2	0.092[e]	5.5	26	2	Ballio et al. 1980
	6	0.57	6.0	0.57[e]	5.5	26	2	
Spinacea								
	5.7	1.41	1.4	0.5[e]	5.5	26	1.5	Ballio et al. 1980
	22	3.96	57	6.9[e]	5.5	26	1.5	

[a] Values as given.
[b] Values recalculated by us (analysed with computer program Scatpack, see pp. 53–4).
[c] pmol/g fresh weight.
[d] Recalculated assuming one binding site.
[e] Recalculated assuming two binding sites.

pared with membrane-bound auxin-binding sites. Upon sucrose-density centrifugation, the binding activity coincided with plasma-membrane markers. The bound [³H]dihydro-FC could be displaced by non-labelled FC but no cross-competition with NAA was observed. Later experiments carried out by Pesci, Cocucci & Randazzo (1979) confirmed the existence of FC-binding sites. They presented a biphasic Scatchard plot, but their conclusion that a second binding site with lower affinity was also present is not very convincing (the biphasic plot might be caused by small experimental errors of residual non-specific binding).

Ballio, Federico, Pessi & Scalorbi (1980) found FC-binding sites in microsome preparations from spinach leaves (biphasic Scatchard plot) and also in microsome fractions from maize coleoptiles. The latter binding sites bind FC with a *c.* 10-fold higher affinity than the binding sites from spinach (see Table 1.3). Here again no cross-competition between FC and auxin (in this case IAA) can be observed.

In both maize and spinach microsomes binding of FC is, even at elevated temperatures, a relatively slow process and, as was already found for maize by Dohrmann *et al.* (1977), this corresponds to a very slow exchange of bound [³H]dihydro-FC for unlabelled FC. Using the exchange data from Dohrmann *et al.* (1977, Table II) for maize, and from Ballio *et al.* (1980) for spinach, we estimated the k_{off} values for FC binding to have orders of magnitude of 10^{-3} min^{-1} and 10^{-2} min^{-1}, respectively, at 26°C. Thus, the difference in K_d values between FC-binding sites from maize and spinach microsomes is largely due to differences in k_{off}. The order of magnitude of k_{off} is such that the dissociation of FC from maize binding sites is very slow, even at 26°C (*c.* 0.1% per min). Pesci, Tognoli, Beffagna & Marrè (1979) used this property to mark FC-binding sites with radioactive ligand prior to solubilization and fractionation at low temperature. They found that in maize microsomes bound radioactive FC can be partially solubilized by treatment with Triton X-100, deoxycholate, sodium dodecyl sulphate, urea, sodium perchlorate or trypsin. By Sephadex-G-200 gel filtration of a perchlorate-solubilized FC-macromolecule complex, a molecular weight of *c.* 80 kilodaltons was determined.

Using essentially the same procedure described by Venis (1977*b*), Aducci, Ballio & Federico (1982) were able to detect solubilized [³H]dihydro-FC-binding-site complexes in buffer extracts from acetone powders of microsomal preparations from spinach leaves. Biogel-P100 gel filtration revealed a molecular weight of *c.* 80 kilodaltons. The biphasic Scatchard plot of FC-binding to the solubilized sites suggests, according to the authors, the presence of two binding sites. But upon isoelectric focussing only one sharp binding peak could be detected. Therefore, Aducci, Ballio & Federico (1982) do not exclude the possible existence of negative cooperativity. In contrast to membrane-bound binding sites, the binding of FC to solubilized binding sites is independent of pH in the range 5.5–9.

Ballio, Aducci & Frederico (1982) reported that purification of this solubilized FC-binding site on a large scale, as required for biochemical studies, was hampered by the fact that binding sites progressively lost their affinity upon purification. Results obtained so far suggest that FC-binding sites are phosphorylated glycoproteins, a structure consistent with their localization at the outer surface of the plasma membrane. Evidence for this localization was obtained by Aducci, Federico & Ballio (1980), who used a conjugate of bovine serum albumin (BSA) with the dialdehyde of dideacetyl FC. This conjugate is as effective as FC to compete with [^{3}H]dihydro-FC for binding sites in microsome preparations. Tobacco-leaf protoplasts bound [^{3}H]dihydro-FC, and the binding was strongly reduced by simultaneous addition of the conjugate, which was not taken up by the protoplasts. When a suspension of protoplasts was incubated with the conjugate and then treated with antibodies against the conjugate, complete aggregation of the protoplasts took place within 5 min, while appropriate controls were negative.

It is generally believed that FC stimulates particular classes of plasma-membrane-bound, divalent-cation-stimulated, proton-translocating ATPases (see section 1.2.4). A much-studied ATPase component found in a number of higher plant species has the following characteristics:

it is probably located at the plasma membrane

it is activated by Mg^{2+} and further stimulated by K^+

there is no synergistic effect of $Na^+ + K^+$

it shows little sensitivity towards the inhibitors ouabain and oligomycin, but is inhibited by N,N'-dicyclohexyl carbodiimide (DCCD), diethyl stilbestrol (DES), and orthovanadate

pH optimum is between 6 and 7, K_m for Mg-ATP is 1–2 mM

it is sensitive to *p*-chloromercuribenzoate (Poole, 1978).

Therefore, a number of workers paid attention to possible relationships between FC-binding sites and ATPase activity. Tognoli, Beffagna, Pesci & Marrè (1979) reported that both FC-binding and DCCD-sensitive ATPase activities in microsome preparations from maize coleoptiles were inhibited by DES. Upon solubilization FC-binding and ATPase activity could be separated by conventional biochemical methods. In subsequent reports from the same laboratory (Beffagna, Pesci & Marrè, 1981; Tognoli & Marrè, 1981; Pesci & Beffagna, 1982) it was shown that microsome preparations from subapical internode segments of pea seedlings contains an ATPase that shows most of the characteristics summarized above. In the absence of Mg^{2+} during homogenization, a substantial amount of the ATPase activity was solubilized from the microsomes. Saturable re-incorporation of the solubilized ATPase into the 'low Mg^{2+}' microsomes occurred in the presence of 2 mM Mg^{2+}. Sucrose density-gradient centrifugations of 'low Mg^{2+}' and reconstituted microsomes revealed that ATPase activities in both coincide. Morover, specific [³H]dihydro-FC binding showed a parallel distribution in the gradient, with a single peak of bound radioactivity in the same (plasma-membrane-enriched) sucrose fractions (41–44%). FC did not bind to the solubilized enzyme, and solubilization of the ATPases did not affect the binding capability of FC to the microsomes. Additionally, sonication of 'high Mg^{2+}' microsomes removed more than 40% of the ATPase activity, while FC-binding capacity was not affected. DES inhibited both membrane-bound (and solubilized) ATPase activity and FC-binding, as was previously found for maize microsomes (Tognoli *et al.* 1979). Pesci & Beffagna (1982) concluded that ATPase and FC receptors are located on the same membrane, the plasmalemma, but that they are different proteins.

In vivo activity and *in vitro* binding of a number of FC analogues were studied by Ballio, De Michelis, Lado & Randazzo (1981) and Ballio, Federico & Scalorbi (1981). They found that active FC analogues have a relatively high affinity for FC-binding sites in microsomal preparations from maize coleoptiles, while non-active analogues do not compete with [³H]dihydro-FC in the concentration ranges tested. According to the criterion discussed in section 1.4.3, binding of FC to the microsomal preparations is specific.

Finally, we want to mention the report by Stout & Cleland (1980). Using roots from oat seedlings they found that the largest amount of specific [³H]dihydro-FC-binding was present in plasma-membrane-enriched microsome fractions. Trypsin treatment of these fractions reduced [³H]dihydro-FC-binding as well as Mg^{2+}/K^+-ATPase activity by 75 and 80%, respectively. Upon solubilization and gel filtration, the binding activity and the ATPase activity could be separated. The molecular weight for the ATPase activity was estimated to be *c*. 100–200 kilodaltons, for the FC-binding sites 60–100 kilodaltons. In contrast to the findings by Tognoli *et al.* (1979) the Mg^{2+}/K^+-ATPase activity is insensitive to DES (but sensitive to DCCD).

Conclusions
High-affinity FC-binding sites are probably phosphorylated glycoproteins (mol. wt 60–100 kilodaltons) that are located at the outer surface of the plasma membrane. Both unoccupied and occupied Fc-binding sites can be separated easily from a certain class of divalent-cation-dependent ATPases which are also located at the plasma membrane. However, in a number of reports it has been shown that both FC-binding and ATPase activity are affected by inhibitors such as DES. There is no conclusive evidence whether more than one specific binding site exists. Auxin and FC do not show cross-competition, indicating that auxin- and FC-binding sites are different entities.

1.2.4   Possible functions for membrane-bound auxin-binding sites
(1) Proton-translocating ATPases
We shall describe the rapid response of tissues to auxins (proton efflux), how this can lead to cell enlargement (acid-growth theory), the primary enzymes possibly responsible for this acidification (ATPases), and, finally, the influence of auxins upon these ATPases.

The most rapid effects of auxin added to excised responsive tissues include: net proton efflux, hyperpolarization of the cell's transmembrane potential and subsequent increase in net influx of K^+ ions. These processes are inhibited by drugs that uncouple oxidative phosphorylation and that make the plasma membrane more permeable to protons, and by inhibitors of protein synthesis (e.g., cycloheximide) (Marrè, 1977).

According to the acid-growth theory, independently proposed by Rayle & Cleland (1970) and Hager, Menzel & Krauss (1971), auxin-induced acidification of the cell wall's free space activates enzymes that are capable of cleaving structural bonds in the wall matrix, thus causing a rapid increase in growth rate of the tissues. This initial increase is independent of RNA synthesis, but undisturbed protein and RNA synthesis are required to prolong a high growth rate (Vanderhoef, 1979). [For reviews and discussions of the acid-growth theory the reader is referred to Cleland, 1976, 1979; Zeroni & Hall, 1980; and Penny & Penny, 1978.]

How this acidification is brought about is not known exactly, but one possibility is that in this process (which requires energy) an ATPase, capable of translocating protons, is involved. Indeed, during the past decade many indications have been found that electrogenic plasma-membrane-bound, proton-translocating ATPases are present in higher plants. They probably constitute a mechanism whereby the electrochemical potential difference for protons may be responsible for driving fluxes of major ions like K^+, Na^+ and Cl^- via symports or antiports and of metabolites like sugars and amino acids. Part of this evidence consists of the identification of ATPases in microsome preparations enriched in plasma membrane (Spanswick, 1981; Poole, 1978).

The correlations between kinetic data of K^+ stimulation of Mg^{2+}-dependent ATPase activity *in vitro* and the kinetics of short-term K^+ influx *in vitro* suggest that this enzyme plays a part in K^+ uptake (Leonard & Hotchkiss, 1976). Hence, this enzyme may be the energy-generating part of an electrogenic H^+/K^+-exchange system. Experiments with fusicoccin (FC) support the presumed occurrence of these systems in plasma membranes (see section 1.2.3). FC brings about a rapid increase in H^+ extrusion in a variety of shoot and root tissues, growing or non-growing. The increase in H^+ extrusion is coupled with an increase in cation uptake, especially of K^+, and a hyperpolarization of the plasma membrane. FC-stimulation of these processes is affected by the same agents that affect Mg^{2+}/K^+-ATPase activity (Mg^{2+}/K^+-ATPase activity is the activity that is stimulated by Mg^{2+} and further stimulated by K^+). In many aspects FC mimics auxin but generally speaking it acts much more strongly and is much less sensitive to cycloheximide. A lag phase can be observed only when the applied concentration of FC is below $1 \mu M$ (shown for oat coleoptiles by Rubinstein & Cleland, 1981) and only at these sub-optimal concentrations are the rates of acidification similar to the rates induced by optimal auxin concentrations. There are more differences between auxin and FC effects. FC also induces H^+ extrusion and associated effects in tissues that do not respond to auxins, and coupling of H^+ extrusion and K^+ uptake in oat coleoptiles is much stronger for FC than for auxin (Cleland & Lomax, 1977). This led the latter authors to conclude that FC may stimulate an electrogenic H^+/K^+-exchange system, whereas auxin may stimulate an electrogenic H^+ pump coupled with passive K^+ uptake. (For a comparison between auxin and FC effects see further Marrè, 1977.)

In spite of all the evidence, a more rigorous biochemical approach is needed to prove:

> the existence of (plasma-)membrane-bound ATP-driven electrogenic H^+ pumps or H^+/K^+-exchange mechanisms,
> the role of auxin- and FC-binding sites in the regulation of these systems.

Attempts to demonstrate ATPase-driven electrogenic proton-pumping *in vitro* (in microsome preparations) failed until recently because the vesicles used were leaky (Sze & Hodges, 1977). However, Sze (1981) reported how a population of 'sealed' vesicles can be prepared from tobacco callus. Since the appearance of this report it has been demonstrated by workers from different laboratories that Mg^{2+}/K^+-ATPase activity is associated with electrogenic proton-pumping in membrane vesicles from either plasma-membrane or tonoplast origin (Mettler, Mandala & Taiz, 1982; Stout & Cleland, 1982; Dupont, Bennett & Spanswick, 1982; Macri, Vianello & Dell'Antone, 1982).

In conclusion, recent evidence indicates that ATP-driven proton pumps do occur in membranes from higher plants.

Now that this fact is better established, we have to search for the possible role of auxins and FC in the modulation of these pumps. It is plausible to assume that auxin and/or FC affect plasma-membrane-bound pumps, although we cannot exclude a role in controlling proton pumping into the vacuole.

There are a number of problems, however, that require some attention. Cross et al. (1978) showed that auxin-binding sites solubilized from microsome fractions of maize coleoptiles are clearly separated from basal Mg^{2+}- and Mg^{2+}/K^+-ATPase activity upon gel filtration. Similar findings were reported by Venis (1979). This can be expected because unoccupied receptors and effectors may be physically separated (Jacobs & Cuatrecasas, 1976, 1977). Although Kasamo & Yamaki (1974) reported that Mg^{2+}-stimulated ATPase can be stimulated by IAA in vitro, a maximal stimulation is reached already at a concentration of 10^{-13} M. Making a few reasonable assumptions, Venis (1977a) calculated that at a concentration of 10^{-13} M only one molecule of IAA is present for every 250 ATPase molecules, which seems incredibly low. Later Kasamo & Yamaki (1976) used more purified plasma membrane-enriched fractions and showed a stimulation of c. 15% at 10^{-10} M and 10^{-6} M IAA. This report has no follow up and attempts to reproduce these results have failed (see Cleland & Lomax, 1977). According to Scherer & Morré (1978) irreproducibility of in vitro auxin stimulation of ATPase is due to classical membrane isolation techniques, e.g., the one described by Hodges & Leonard (1974), with which marked degradation of phospholipids takes place (see also remarks on in vitro proton pumping, page 27). The inclusion of choline, ethanolamine and nupercain in the isolation buffer to prevent phospholipid degradation apparently enabled Scherer & Morré (1978) to reproduce a stable in vitro auxin-sensitive membrane system from etiolated soybean hypocotyls. At pH 5.5, in the presence of Mg^{2+} but in the absence of K^+, maximal stimulation of ATPase activity (20–30%) in plasma-membrane-enriched fractions is obtained at 10^{-9} M 2,4-D. The inactive structural analogue 2,3-D has no effect. At higher pH levels (6–6.5) in either the presence or in the absence of K^+, there is no stimulation by 2,4-D in the concentration range tested, $0-10^{-3}$ M. This *in vitro* auxin stimulation of Mg^{2+}-ATPase may be mediated by the auxin-binding site described by Williamson *et al.* (1977).

Recently, Scherer (1984*a*,*b*) studied the influence of auxin on ATPase activity in sealed vesicles from zucchini hypocotyls by measuring the liberation of $[^{32}Pi]$ from $[\gamma\text{-}^{32}P]$-ATP. At suboptimal ATP concentrations only he found that auxin apparently lowers the K_m of the ATPase for its substrate by a factor of 2.2. The stimulation of ATPase activity with 2,4-D was five times higher than with the weak auxin 2,3-D.

Thompson, Krull & Venis (1983) have prepared artificial membranes into which they incorporated partially purified, solubilized, membrane proteins from *Zea mays*. These proteins exhibited both auxin binding and ATPase activity. ATP, proteins and NAA were added to the membranes in different sequences and changes in the transmembrane current were monitored. There was an increase in transmembrane current only after all components had been added. This effect is specific for NAA, because it was not observed with BA. The optimal pH for stimulation is 5.3, which is in agreement with the optimum pH of binding (5.5). Also the K_d of the binding, which is *c.* 10^{-7} M, is close to the detection limit of the NAA effect. Unfortunately, no experiment is shown in which the proteins and NAA were added before the ATP.

A possible function of the ER-bound auxin-binding site in maize coleoptiles has been proposed by Ray (1977) as follows: 'Combination of auxin with its receptor on the ER could induce H^+ transport from the cytoplasm into the ER cisternal space. This acid contained therein would be transported along with secretory proteins contained in the ER space to the cell exterior, probably via the Golgi system'. According to Ray, this hypothesis explains the widely observed absolute lag phase of 10–12 min in auxin action on elongation and acid secretion, since this is about the time needed for transport to the cell exterior via the Golgi system. It offers an explanation for the rapid inhibition of auxin-controlled proton extrusion by inhibitors of protein synthesis, since secretory flow of proteins from the ER space outwards would then be stopped. Cleland (1979, 1982) has pointed out that this 'bucket-brigade' system is not without its problems:

(i) In order to transport sufficient protons the pH within the vesicles would have to be less than 2.
(ii) Inhibition of protein synthesis in pea stems does not prevent wall matrix synthesis, which presumably involves exocytosis of Golgi vesicles.

(iii) Auxin-induced proton extrusion cannot be detected by an external pH of 4.5 or less and it is not obvious how the external pH could regulate the rate at which vesicles fuse with the plasma membrane or open to the cell exterior.

Circumstantial evidence for some function of the ER-bound binding sites has been reported by Walton & Ray (1981). In mesocotyls of etiolated maize seedlings, they found that red light considerably reduced both the amount of NAA-binding sites (30–50%) and the sensitivity to auxin induced elongation. The red-light treatment did not alter the affinity of the binding site. Dose-response curves for auxin obtained with mesocotyl segments isolated from either irradiated or etiolated seedlings did not show a difference in C_{50} but only in the maximum response value, which can be expected if the response is proportional to the amount of receptors occupied. The picture is somewhat complicated by the finding that red light apparently not only reduces the concentration of ER-bound receptors along with ER-bound enzymes such as glucan-I synthetase, but also the total amount of ER material in the cells. As the authors have pointed out, other explanations are possible. However, at the present stage, the hypothesis that reduction in maximum elongation response is due to the reduction in the number of ER-bound receptors is at least as good as any other explanation. It is not necessarily contradictory to the interesting suggestion made by Venis (1977a), that the ER binding sites may represent 'proreceptors', which undergo a maturation process to a final form in the plasmamembrane.

Root elongation is stimulated at low auxin concentrations (c. 10^{-11}–10^{-9} M), while higher concentrations (c. 10^{-8}–10^{-6} M), which stimulate stem tissues, are inhibitory (Larsen, 1961). Like stem growth, root growth is promoted by low pH (3–4) and inhibited by high pH (7) (Edwards & Scott, 1974; Evans, 1976). FC stimulates proton extrusion in roots, while for low auxin concentrations this has never been demonstrated unambiguously (Pitman, Anderson & Schaefer, 1977). However, Evans, Mulkey & Vesper (1980) found that inhibition of root growth at high auxin concentrations is associated with an increase in apparent uptake of protons, increasing the pH of the cell wall's free space. Thus, although root tissues apparently are more complex in this respect, auxin seems to play some part in regulating transmembrane transport of protons in roots also. In addition, the presence of Mg^{2+}/K^+- or Ca^{2+}/K^+-ATPases and ATPase-driven proton pumping in root membranes has been demonstrated (Leonard & Hotchkiss, 1976; Stout & Cleland, 1982; Dupont et al. 1982). Erdei, Toth & Zsoldos (1979) demonstrated the presence of a Ca^{2+}/K^+-stimulated ATPase in crude microsome fractions from rice roots. ATPase activity measured in the presence of K^+ and Ca^{2+} increased when the microsomes were preincubated with 2,4-D or IAA. With both auxins maximum stimulation was found at 10^{-10} M (resp. 27 and 36%). Longer pretreatment (*in vivo*, for five days with either 10^{-10} M 2,4-D or 10^{-8} M IAA) enhanced the *in vitro* stimulation (resp. 52 and 58%). Hence, as was found by Kasamo & Yamaki (1974) for mung-bean hypocotyls, pretreatment with auxin apparently made the enzyme more sensitive to auxin. Extraction of lipids from microsomal fractions by acetone reduced the basal ATPase activity, whereas addition of Ca^{2+} and K^+ no longer had any stimulative effect. Addition of phosphatidyl choline had no effect upon the basal activity, but completely restored the stimulative effect of Ca^{2+} and K^+. This suggests that the Ca^{2+}/K^+-ATPase requires phosphatidyl choline for activity. (The basal activity must need different, crucial lipid components since this activity was restored upon the addition of a total polar-lipid extract.) The effect of IAA upon the Ca^{2+}/K^+ activity was also restored by addition of phosphatidyl choline. Erdei *et al.* (1979) did not report the presence of auxin-binding sites in this system.

As we have discussed earlier, Moloney & Pilet (1981) demonstrated the presence of a high-affinity auxin-binding site in plasma membrane-enriched preparations from pea roots. Moloney, Henri & Pilet (1983) found, that H^+ uptake by sealed vesicles in these preparations was dependent on the presence of ATP. Other nucleoside triphosphates and ADP had no effect. Kinetic studies of the H^+ uptake showed that IAA apparently increased the affinity of the proton pump for their substrate ATP. The maximum effect was found with 10^{-6} M IAA. Interestingly, these data are comparable with those obtained by Scherer (1984*a,b*) for zucchini membranes.

As we have shown in section 1.2.3, high-affinity FC-binding sites are localized at the plasma membrane and probably at the external surface, at least in the tissues examined thus far. The FC-binding sites coincide with Mg^{2+}/K^+-ATPase activity in microsome preparations enriched in plasma membrane, but solubilized binding sites, both occupied and unoccupied, can be separated from this ATPase activity by gel permeation.

Beffagna, Coccuci & Marrè (1977) have shown that plasma membrane-enriched preparations from maize coleoptiles and from spinach leaves contain an FC-sensitive Mg^{2+}/K^+-ATPase. Addition of FC to such preparations in the presence of Mg^{2+} and increasing amounts of K^+ caused a stimulation in total ATPase activity with a maximum of *c.* 29% at the highest K^+ concentrations (50 mM). In the presence of DCCD (an ATPase inhibitor) no effect of FC was observed; this was interpreted as further evidence that FC stimulates an Mg^{2+}/K^+-ATPase. FC stimulation results obtained with plasma membrane-enriched fractions from spinach leaves were rather unreproducible due to an apparent loss of sensitivity to FC with increasing times of storage. More reproducible results were obtained with crude preparations. In these fractions FC stimulated total ATPase activity by *c.* 23%. Here again the FC effect was enhanced by increasing amounts of K^+. Absence of Mg^{2+} or replacing it with Ca^{2+} markedly reduced the ATPase activity and completely abolished the FC effect. The inactive derivative of FC, 8-oxo-9-epidideacetyl fusicoccin (DAK), had no effect upon ATPase activity of these preparations.

A major follow-up of these investigations has not yet been published and it appears also that *in vitro* FC stimulation of ATPase activity is hampered by irreproducible results (Marrè, 1979). Thus, although all the available evidence indicates that FC is a modulator of plasma membrane-bound Mg^{2+}/K^+-ATPase-driven electrogenic H^+ pumps, to date no conclusive evidence exists that FC-binding sites activate such pumps.

Conclusions

At present the theory that auxin and FC modulate particular, not necessarily the same, classes of plasma-membrane-bound, ATPase-driven proton pumps via coupling to their respective receptors can neither be accepted nor rejected. The *in vitro* stimulation of the catalytic sites (the ATPases) of these pumps has been frustrated by negative and irreproducible positive results.

Further investigations should proceed along the following lines:

Isolation and fractionation of membranes may induce all kinds of alterations resulting in, for instance, altered lipid and protein composition and subsequent changes in membrane-bound functions and permeability characteristics. As a minimum requirement one should try to isolate from the appropriate membrane system microsome preparations, enriched in vesicles, which have not lost their capability to drive ATPase-dependent proton pumps. We have shown that such membrane vesicles have recently been prepared.

Full characterization of auxin- and FC-binding sites located in such microsomal fractions.

Exploration of conditions that might allow *in vitro* auxin and/or FC modulation of proton pumping.

Solubilization and purification of binding sites and ATPase-driven proton pumps.

Reconstitution of auxin- and/or FC-sensitive proton-pumping systems by incorporation of purified auxin- and/or FC-binding sites together with purified ATPase H^+ pumps into an artificial membrane. In this connection it is interesting to note that incorporation of purified Mg^{2+}/K^+-ATPase H^+ pumps from plasma membranes of yeast into artificial membranes results in electrogenically proton-pumping liposomes (Villalobo, Dufour & Goffeau, 1982).

The authors realize that this flow scheme does not pave the way to immediate success in elucidating the receptor function of auxin- and FC-binding sites. It should rather be considered as a list of requirements for the verification of an attractive hypothesis.

It is encouraging that a few recent attempts along this line have produced very interesting results (Scherer, 1984*a,b*; Thompson *et al.* 1983; Moloney *et al.* 1983).

(2) Other functions

A completely different biological function for a membrane-bound auxin-binding site is proposed by Maan, Van der Linde, Harkes & Libbenga (1985). They discovered that the membrane-bound binding site from tobacco cells disappears when the cells are grown on medium containing only 2,4-D as phytohormone. This does not affect the growth rate of the culture, but the apparent absence of the binding sites is correlated with inability of the cells to regenerate roots. After replacement of 2,4-D by NAA and kinetin the membrane-bound receptor reappears after about eight weeks along with the ability of the cells to regenerate roots. Consequently, it is suggested that this binding site is involved in root regeneration. The high external concentration of auxin required to induce this regeneration (Skoog & Miller, 1957) is entirely consistent with the kinetic behaviour of the binding site, i.e. the affinity of the binding site for IAA is rather low (K_d c. 2×10^{-5} M). Since no competition of cytokinins is observed with the membrane-bound receptor, it seems likely that the influence of cytokinins (i.e. inhibition of root regeneration) occurs via a different receptor.

1.3   Receptors that are not membrane-bound

1.3.1   *Characterization*

In trying to evaluate the data about soluble receptors we want to limit ourselves to those auxin-binding proteins for which at least the criteria of high affinity and specificity are fulfilled. For complete reviews on all auxin-mediator proteins we refer to Kende & Gardner (1976),

Venis (1977*a*), Stoddart & Venis (1980), and Rubery (1981). The number of publications in which these criteria are met is almost surpassed by the number of criticisms dedicated to them. This illustrates the progress made in this field during the last decade.

Since 1972 only eight research groups have published data on soluble auxin-binding sites. The characteristics of these binding sites are listed in Table 1.4. We shall treat them separately.

(1) The receptors from nuclei from immature coconut endosperm

The group of Biswas and co-workers detected two IAA-binding proteins (= IRP) in isolated nuclei from immature coconut endosperm. One is present in the nucleoplasm (n-IRP) and can be detected by equilibirum dialysis at pH 8.0 for 18 h (Mondal, Mandal & Biswas, 1972) as well as by the dextran-coated-charcoal (DCC) method (Roy & Biswas, 1977). The protein was purified to apparent homogeneity by conventional methods and affinity chromatography on carboxyl-linked 2,4-D-poly(lysyl)-Sepharose. The molecular weight of this protein, estimated by means of SDS-polyacrylamide-gel electrophoresis (SDS-PAGE), is 94 kilodaltons and the K_d for IAA is 7.5 μM. The other IAA-binding protein, which was detected by the DCC method after incubation at 25°C for 30 min, is located on the chromatin (c-IRP) and can be extracted with 2 M NaCl. This protein was also purified to apparent homogeneity with conventional methods and the above-mentioned affinity matrix. The molecular weight, determined by SDS-PAGE, is 70 kilodaltons. The c-IRP exhibits high- ($K_d = 0.058$ μM) and low- ($K_d = 8.2$ μM) affinity binding for IAA. The number of IAA-binding sites they found for purified c-IRP is 0.4 pmol/μg protein and 5.1 pmol/μg protein for the high- and low-affinity sites, respectively (Roy & Biswas, 1977).

Some critical remarks should be made concerning these reports. In order to demonstrate the IAA-binding capacity of the purified n-IRP, equilibrium dialysis was used in the presence of one concentration of [^{14}C]-IAA (sp.act. 8×10^5 cpm/nmol) only, namely 0.91 μM. From these values one would expect to detect 7.3×10^5 cpm/ml outside the dialysis bag in case no IAA-binding is present. Mondal *et al.* (1972) reported an average number of 1.3×10^4 cpm/ml, which differs by a factor 56.

Roy & Biswas (1977) determined the affinity of IAA-binding to c-IRP using a DCC assay with various concentrations of [^{14}C]-IAA. According to them, two IAA-binding sites were detected with K_d values of 5.8×10^{-8} M and 8.15×10^{-6} M, respectively. In contrast we calculated K_d values of 8.15×10^{-6} M and 5.8×10^{-4} M, respectively, by means of extrapolation of their Fig. 2, which represents the Scatchard analysis of the results of this binding assay. Furthermore, in their Fig. 1, which shows both the elution profile (absorbance at 280 nm) of c-IRP upon purification on the affinity column and the specific activity of IAA-binding to the fractions obtained, one would expect a plateau for the specific activity of the fractions containing the c-IRP. However, the specific activity in this peak increased with the protein concentration and reached a maximal value of 75 pmol/μg protein; this means that c-IRP (mol. wt = 70 kilodaltons) can bind 5 molecules of IAA, whereas the number of binding sites found for purified c-IRP (namely 0.4 pmol/μg protein and 5.1 pmol/μg protein for the high- and low-affinity sites, respectively) indicates a stoichiometry of binding of one IAA/35 c-IRP for the high-affinity site and *c.* one IAA/3 c-IRP for the low-affinity site.

(2) The soluble receptor from tobacco pith explants

Oostrom, Van Loopik-Detmers & Libbenga (1975) and Oostrom, Kulescha, Van Vliet & Libbenga (1980) detected a soluble IAA-binding protein in Sepharose-CL-6B elution fractions of concentrated cytosol, prepared from cultured tobacco pith explants. This protein exhibits a molecular weight of *c.* 300 kilodaltons calculated from the Sepharose-CL-6B elution profile, a K_d for IAA of 0.01 μM and structure-affinity relations for hormone analogues, which correlate well with the structure-activity relations of these compounds in physiological-response tests. Optimal binding was observed at pH 7.5–7.8 using the DCC method after incubation of the preparation at 24°–30°C for 25–30 min. The total number of binding sites was strongly modulated during culture of the calluses. The protein was not present in freshly excised stem pith, but it was detectable after one day of culture on medium with or without IAA and kinetin (Bogers, Kulescha, Quint, Van Vliet & Libbenga, 1980). Van der Linde, Bouman, Mennes & Libbenga (1984) improved the purification of this binding protein. When boric acid buffers at pH 6.8 were used for homogenization and gel filtration, contamination of the cytosol by polyphenols was substantially reduced. The IAA binding was now found in fractions representing proteins with a molecular weight of 150–200 kilodaltons. Similar high-affinity binding was demonstrated in soluble fractions from cell-suspension cultures from tobaccoo and sycamore and in 0.5 M KCl extracts from nuclei isolated from sycamore cell-suspension cultures.

The progress of this research is hampered by the rather poor reproducibility of the detection of this IAA-binding protein and by its very low concentration (less than 200 fmol mg^{-1} protein) in cultured tissues. However, recently Van der Linde, Maan, Mennes & Libbenga (1984) found

Table 1.4. *Characteristics of auxin-binding sites that are not membrane-bound*

Material	Fraction	Ligand	Binding sites K_d (μM)	Binding sites conc.	Specificity	pH	Temp (°C)	Time (min)	Mol. weight (K)	Method[a]	Reference
Acer											
cell suspension	gel-filtrated cytosol	IAA	0.0024	0.044[c]	—	7.5	25	30	—	DCC	Van der Linde et al. 1984
	nuclei	IAA	0.008	0.590[c]	—	7.5	25	30	—	DCC	Van der Linde et al. 1984
Cocos											
immature endosperm	n-IRP purified	IAA	—	—	—	8.0	—	980	—	ED	Mondal et al. 1972
	n-IRP purified	IAA	7.5	—	—	—	—	—	94	—	Roy et al. 1977
	c-IRP purified	IAA	0.058[b] 8.2[b]	0.4[c] 5.1[c]	—	8.0	25	30	70	DCC	Roy et al. 1977
Glycine											
cotyledons	vv[d]G-75	IAA	—	1300[c]	+[e]	7.2	0	120	> 50	DCC	Ihl 1976
Nicotiana											
pith callus	gel-filtrated cytosol	IAA	0.01	0.01[c]	—	7.8	30	20	—	DCC	Oostrom et al. 1975
	gel-filtrated cytosol	IAA	0.01	—	+	7.5–7.8[f]	24–30[f]	25–30[f]	300	DCC	Oostrom et al. 1980
	gel-filtrated cytosol	IAA	0.01	0.018[g]	—	7.8	24–30	25–30	—	DCC	Bogers et al. 1980
	gel-filtrated cytosol	IAA	0.006	<0.02[c]	—	7.5	25	30	150–200	DCC	Van der Linde et al. 1984
cell suspension	gel-filtrated cytosol	IAA	0.004	0.03[c]	—	7.5	25	30	—	DCC	Van der Linde et al. 1984
Phaseolus											
primary leaves	gel-filtrated cytosol	IAA	0.09[h]	100[c]	+[i]	8.5[f]	0	5	315	ASP	Wardrop et al. 1977
	RuBPCase purified	IAA	0.8[h]	1.2[j]	+	8.0[f]	4	5	550	ASP	Wardrop et al. 1980a,b

	Preparation	Ligand			+/−					Method	Reference
seedlings	RuBPCase purified homogeneous protein	IAA	0.8[h]	0.7[i]	—	7.0	30	2520	550	ED	Wardrop et al. 1980a
	homogeneous protein	IAA	3.2	1.95[i]	—	7.6[f]	25	10	390	ASP	Sakai et al. 1983
	homogeneous protein	2,4-D / 2,4-D	9.3 / 9.3	1.79[g] / —	+	7.6	25	10	—	ASP	Sakai 1984
Pisum primary leaves	RuBPCase purified	IAA	1.3[h]	1[i]	+	8.0	4	5	550	ASP	Wardrop et al. 1980a,b
	RuBPCase purified	IAA	0.5[h]	0.4[i]	—	7.0	30	2520	550	ED	Wardrop et al. 1980a
epicotyls	vv[d]G-75	2,4-D	21[h]	230[c]	+[k]	8.5	0	5	>50	ASP	Jacobsen 1981
	vv[d]G-75	2,4-D / IAA / NAA	0.1–0.4	10–20	+	8.0[f]	0	5	>50	ASP	Jacobsen 1982
Triticum seedlings	cytosol	IAA	—	75[c]	—[l]	7.0	10–12	45	—	DCC	Likholat 1974
Zea coleoptiles	partly purified cytosol	NAA	3.6	11.6[c]	+[m]	4.75[f]	4	10	38.7	ASP	Murphy 1980b

[a] ASP = ammonium sulphate precipitation, DCC = dextran-coated charcoal, ED = equilibrium dialysis.
[b] Value is in contrast with the Scatchard analysis in their paper.
[c] Expressed as pmol/mg protein.
[d] vv = void volume.
[e] Determined by *in vivo* competition.
[f] Optimal condition determined.
[g] Expressed as fmol/g fresh weight.
[h] Per pure protein molecule.
[i] Calculated using the extrapolation method.
[j] Determined using 0.5 μM labelled IAA and 50 μM competitor.
[k] Competition with one analogue (NAA) was determined.
[l] Only one analogue was tested in a 50-fold excess over labelled ligand.
[m] Determined using 0.1 μM labelled NAA and 10 μM and 100 μM competitor.

that the amount of high-affinity IAA binding in crude cytosol preparations can be increased if Mg^{2+} and ATP and/or a phosphatase substrate
are included in the binding assay. These results suggest that the affinity
of the receptor towards IAA is modulated by phosphorylating and
dephosphorylating processes. At present, the research in our laboratory
is focused on this interesting phenomenon.

Using the improved isolation procedure as described by Van der Linde
et al. (1984), Bailey *et al.* (1984) confirmed the presence of essentially
the same auxin-binding protein in both cytosol preparations and salt
extracts of isolated nuclei from tobacco cell-suspension cultures. They
purified the binding protein 50–60-fold by affinity chromatography and
made the interesting observation that the number of binding sites varies
according to the growth stage of the culture. In the cytosol preparations
high-affinity binding could be detected only during the lag and stationary
phase of the culture, whereas salt extracts from nuclei contained detectable binding sites only in the exponential phase. It is concluded that
during the growth cycle, the binding protein is recycling between a soluble and a nuclear-bound state.

(3) The soluble receptor from primary leaves of dwarf-bean seedlings
Wardrop & Polya (1977) reported the presence of a soluble IAA-binding
protein in fractions obtained by gel filtration on agarose A-15m of the
redissolved 30–50% ammonium sulphate precipitate of cytosol, prepared
from 6–8-day-old seedlings of dwarf bean. Optimal binding occurs at
pH8.5 and was detected with the ammonium sulphate precipitation
method after incubation of the receptor preparation at 0°C for 5 min.
The molecular weight estimated by gel filtration was 315 kilodaltons
and the K_d for IAA is 0.09 μM. The number of binding sites did not
exceed 100 pmol/mg protein. In two subsequent publications Wardrop
& Polya (1980*a,b*) identified this protein as ribulose-1,5-biphosphate
carboxylase (RuBPCase), an enzyme abundantly present in chloroplasts,
involved in the response of plants to light and consisting of several subunits with an overall molecular weight of 550 kilodalton. Binding to
purified RuBPCase was detectable at pH 7.0 after equilibrium dialysis
for 42 h at 30°C, or with the ammonium sulphate precipitation method
after incubation of the preparation at pH 8.0 for a very short time at
4°C. The K_d for IAA was 0.8 μM and the stoichiometry of binding was
about one IAA/RuBPCase. RuBPCase isolated from 7-day-old pea
seedlings exhibited a K_d for IAA of either 1.3 μM when the binding
assay was performed with the ammonium sulphate precipitation method
or 0.5 μM when equilibrium dialysis was used. The stoichiometry of bind-

ing was the same as for RuBPCase isolated from dwarf-bean seedlings. An impressive number of analogues was tested to show the specificity of these binding sites.

Although the work of Wardrop & Polya (1977, 1980*a,b*) is one of the most extensive studies on auxin binding to a soluble protein, some critical notes should be made. As Venis (1981) has pointed out, the use of precipitation methods to separate bound from free ligand in binding assays must be considered with caution because by these methods ligand–protein interactions can be severely modified. Venis (1981) showed that in fractions which do not exhibit specific IAA-binding in equilibrium dialysis, specific IAA-binding is detectable when the precipitation method is used. This may also account for the results of Wardrop & Polya (1980*a*). They reported that equilibrium dialysis revealed specific IAA-binding in only 9 of the 20 preparations in which specific IAA binding was observed with the precipitation method. Secondly, it is strange that RuBPCase needs 30°C for 42 h to bind IAA in equilibrium dialysis, but binds IAA immediately at 4°C when the precipitation method is used.

There is also a marked discrepancy between the K_d of the receptor mentioned in their subsequent publications. In (1977) they calculated a K_d for IAA of 0.09 μM from a Scatchard analysis using the extrapolation method. We calculated a K_d of 0.015 μM by applying hyperbolical regression to their data. But in (1980*a*) the K_d for IAA had increased to 0.8 μM. The use of a relatively high basal concentration of radioactive IAA in the binding assays as described in the latter publication, makes it almost impossible to detect a binding protein with a K_d of 0.01 μM or less. Therefore, we think that in their second paper (1980*a*) they may have overlooked the protein described in their first publication (1977). This probably explains the difference in molecular weights reported in (1977) and in (1980*a*), 315 kilodaltons and 550 kilodaltons, respectively.

(4) The soluble receptor from pea epicotyls

In a short report, Jacobsen (1981) describes the presence of an auxin-binding protein in epicotyls of 7-day-old pea seedlings. This binding site is detected with the ammonium sulphate precipitation method in concentrated protein-containing fractions, obtained by Sephadex-G-75 fractionation of cytosol. The time required for binding is 5 min at 0°C. The K_d values for NAA and 2,4-D are 0.23 and 0.21 μM, respectively. Competition between these ligands was shown. The number of binding sites is *c.* 20 pmol/mg^{-1} protein.

In a subsequent paper (Jacobsen, 1982) additional information was given. The proteinaceous nature was proven by proteinase K digestion. The binding protein was sensitive to freezing and thawing and prolonged storage at 0°C. There was also a linear relationship between the total (specific plus non-specific) NAA binding at 5×10^{-8} M and the protein concentration in successive experiments. The pH dependency of the binding was shown by an increase in specific binding from c. 100 dpm at pH 7.0 to 200 dpm at pH 8.0, while lowering the pH to 5.5 hardly affected the specific binding. Moreover, the specific binding observed at 6×10^{-8} M NAA was displaced 137% by IBA, 100% by NAA, 83% by MCPA, 78% by 2,4-D, 74% by TIBA and 69% by IAA, while weak auxins displace only 31–37%. (The concentration of the competitors was 5×10^{-5} M.)

Although the presence of specific auxin binding by soluble proteins from pea epicotyls was undoubtedly shown, some of the results have to be discussed. The most important criticism certainly involves the deduction of the Scatchard plot. The binding data, which were presented as a displacement plot, reveal 2500 dpm binding at 5×10^{-8} M [^{14}C]-NAA, which decreases to 1500 dpm after increasing the NAA concentration to 10^{-5} M with unlabelled NAA. After substracting this background value a displacement plot for specific NAA binding was obtained from which we calculated a K_d value of 0.13×10^{-6} M. When we transformed this plot into a Scatchard plot a perfect straight line was found giving the same K_d value. After transformation Jacobsen (1982), however, found a Scatchard plot revealing both specific and non-specific binding. He obtained a straight line after a second correction for non-specific binding. To us the difference between the correction for background and for non-specific binding is not clear and we do not know the implications of the two-step mathematical correction procedure for the binding parameters he presented. Moreover, the Scatchard plot constructed for specific NAA binding lacks a point, which certainly will influence the K_d and the number of binding sites. Finally, some other remarks should be made:

(i) The pH dependence of specific NAA binding does not show an optimum and therefore it clearly differs from the results found for tobacco-pith callus (Oostrom, Kulescha, van Vliet & Libbenga, 1980) and for dwarf bean (Wardrop & Polya, 1980a).

(ii) The relevance of the linearity between protein concentration and auxin binding (which reflects total binding and from which a significant amount is non-specific) is not clear.

(5) The soluble receptor from mung bean seedlings

Sakai & Hanagata (1983) reported the purification of an auxin-binding protein from mung bean seedlings. By using several cycles of ammonium sulphate precipitation and filtration on Sepharose-4B, alternated by affinity chromatography on 2,4-D-Sepharose-4B, they obtained an apparently homogenous protein preparation. This preparation showed 2,4-D and IAA binding after incubation at 25°C for 10 min at an optimal pH of 7.0–7.6. Separation of bound and free ligand was performed with the ammonium sulphate precipitation method. The K_d for IAA was 3.2×10^{-6} M and for 2,4-D 9.3×10^{-6} M. One protein molecule is able to bind two hormone molecules. The protein consists of two subunits with molecular weights of 45 and 15 kilodaltons, respectively (ratio large over small unit is 0.83). It has an overall molecular weight of 390 kilodaltons as determined by gel filtration. The reversibility of the binding and the displacement of radiolabelled 2,4-D from the binding site by increasing concentrations of IAA, NAA, and PCIB, but not by BA, was described in a second article (Sakai, 1984).

The binding protein is not identical with RuBPCase, since RuBPCase activity is not retained by the auxin-affinity matrix. The subunit structure, stoichiometry and temperature dependence of the binding also differ from those of RuBPCase. The protein seems to be very stable, since it still binds auxin after several precipitation steps. This makes this system accessible for more detailed studies.

(6) Other binding sites that are not membrane-bound

Likholat, Pospelov, Morozova & Salganik (1974) demonstrated binding at pH 7.0 of radioactive IAA to cytosol proteins isolated from wheat coleoptides, using the DCC method after incubation of the sample at 10–12°C for 60 min. By addition of a 50-fold excess of unlabelled IPA or IAA to the incubation mixture, they found that IPA competes weakly, while IAA inhibits the binding of labelled auxin by 50%. The maximal number of binding sites is *c.* 73 pmol/mg protein (not corrected for non-specific binding). They observed a sharp- decrease in binding of [^{14}C]-IAA to the cytosol during the transition of the cells from the division phase (36-h-old seedlings) to the elongation phase (72-h-old seedlings). This decrease is correlated with a decline in auxin content in 72-h-old seedlings. Addition of 10 μM IAA to crude homogenates of 72-h-old seedlings increases the number of binding sites in the cytosol to the level observed for 36-h-old seedlings. The authors suggested that the soluble binding sites observed in the 36-h stage change from their free active state to a latent membrane-bound state at the 72-h stage; they would be released again by IAA. However, they did not report IAA binding to membranes, neither did they determine a K_d, which should have been in the order of 1–10 μM. No evidence in favour of a proteinaceous nature for the binding site was presented.

Ihl (1976) showed that part of the [^{14}C]-IAA taken up by soybean cotyledons at 25°C during 1 h is bound to proteins that are eluted in the void volume of Sephadex-G-25 and G-50 columns during fractionation of cytosol, prepared from these cotyledons. Recharging of these proteins with [^{14}C]-IAA at pH 7.2 is detectable with the DCC method after a 2-h incubation at 0°C. The IAA-binding is sensitive to pronase and heat. Incubation of the cotyledons for 60 min at 25°C with [^{14}C]-IAA in the presence of a 275-fold excess of unlabelled auxins reduces the radioactivity bound to the proteins in the void volume of Sephadex-G-75-fractionated cytosol by 64–80%. The highest number of binding sites is *c*. 1.3 nmol/mg protein. Neither K_d nor *in vitro* binding of auxin analogues was determined.

Murphy (1980*b*) demonstrated specific NAA binding at pH 4.75 to fractionated cytosol, prepared from 4–7-day old dark-grown maize coleoptiles. He used the ammonium sulphate precipitation method, after 10 min of incubation at 4°C. As for the receptor from tobacco pith explants, secondary plant metabolites had to be removed before conventional purification methods could be applied successfully. In this case extraction of cytosol with DCC was preferred. The binding site was further purified by precipitation with 80% ammonium sulphate, Bio-Gel-P-6 chromatography and Sephadex-G-100 gel filtration, the latter in the presence of 0.1 μM [^{14}C]-NAA. The molecular weight of the binding site was estimated to be 38.7 kilodaltons. The K_d for NAA was *c*. 3.5 μM and the concentration of binding sites *c*. 150 pmol/g fresh weight or 11.6 pmol/mg protein (for the Sephadex-G-100 purified fraction). Murphy (1980*b*) demonstrated that the binding site, partially purified on Sephadex-G-100, was heterogenous, by subsequently purifying it on DEAE-Sephacel and hydroxyl apatite. The binding site was distinct from the solubilized membrane-bound NAA-binding site (see Table 1.1 section 1.2.1) with regard to its molecular weight, its specificity for auxin analogues, its pH optimum of binding and its behaviour in the purification procedures.

Evidently this binding site behaves like a macromolecule, but Murphy made no attempts to investigate whether it has a proteinaceous nature. When this binding site represents a soluble cytoplasmic protein, then the optimal pH of binding is rather strange (pH 4.75), since the pH of the cytoplasm of plant cells is about 7 (Kurkdjian & Guern, 1978). At neutral pH no binding assays were reported, so we assume that no binding is detectable under this condition. As we have described above, the binding site, which was partially purified on Sephadex-G-100, was found to be heterogeneous after subsequent purification on DEAE-Sephacel and hydroxyl apatite in the presence of radioactive NAA. A close examination of the number of dpm in each fraction obtained from these gels shows that these values correlate well with the absorbances at 280 nm of these fractions, which is indicative for nonspecific binding. We do not understand why the same purification procedure was not performed in the presence of excess unlabelled NAA, to determine the specificity of the hormone binding in these fractions.

1.3.2   Possible receptor functions in gene expression
We have shown in section 1.2.4 that auxin induces a rapid increase in growth rate in excised coleoptile and stem sections. This rapid increase is preceded by a stimulation of proton extrusion and associated transmembrane transport processes. As a kind of secondary messengers the protons bring about alterations in cell-wall metabolism, leading to enhanced plasticity of the cell wall and thus to increased growth rates. However, this initial response is only transient if RNA and protein synthesis are blocked. In fact, auxin also induces alterations in RNA and protein metabolism in all kinds of responsive tissues. (For a review on RNA alterations see Jacobsen, 1977.) This is not surprising, since cell division, cell extension and cell differentiation, which are major processes controlled by auxin and other growth substances, are closely connected with RNA and protein metabolism. However, the problem is how directly auxin is involved in the control of these long-term processes, and at which level of the information stream DNA → RNA → protein it possibly acts.

The first indications of a direct action of auxin at the RNA and protein level may be given by relative short response times. Thus, Meyer & Chartier (1981) reported that already 18 h after application of auxin to tobacco mesophyl protoplasts one auxin-induced protein appeared in the 2D-PAGE analysis; it was followed by changes in the levels of several other proteins. However, much shorter response times have been reported. Melanson & Trewavas (1982) showed that a first change in tissue protein pattern occurred 3 h after *in vivo* auxin treatment of artichoke tuber slices, while Zurfluh & Guilfoyle (1980) reported that such an effect occurred in soybean hypocotyl sections after 5 h. More rapid responses can of course be expected by looking at changes in the level of *in vitro* translatable RNA in response to the addition of the hormone. Bevan & Northcote (1981) reported such a change within 2 h after hormone treatment of *Phaseolus vulgaris* L. cell suspension cultures.

Theologis & Ray (1982), working with isolated polyadenylylated RNA from etiolated pea stem tissue, found increased levels of three mRNA sequences within 20 min of exposure of the tissue to IAA, upon 2D-PAGE analysis of the *in vitro* translation products. Also, Zurfluh & Guilfoyle (1982), now using the *in vitro* translation technique, showed that the amount of at least one polypeptide increased within 15 min after hormone treatment of soybean hypocotyls.

These examples of a rapid response in gene expression show that one can be reasonably sure that this response is closely related to a primary hormone effect. To demonstrate that auxin really induces changes in the RNA population of auxin-responsive tissue one should, by means of c-DNA probes, titrate RNA species that code for auxin-induced sequences. In the near future such probes will undoubtedly become available. Baulcombe & Key (1980) used recombinant-DNA techniques to produce c-DNA probes, with which they could demonstrate changes in the concentration of a minor abundant class of poly(A^+)-RNA sequences following auxin treatment of soybean hypocotyls.

The question now arises how auxin might cause relatively rapid changes in translatable mRNA. One possibility is that these alterations are brought about indirectly via rapid modulations of membrane functions. Another possibility is that auxin exerts its influence upon RNA synthesis via binding to cytoplasmic or nuclear receptors, in a way that is similar to the way in which animal steroid hormones do this (O'Malley & Schrader, 1976). In that case the occupied receptors should not only increase total RNA-polymerase activity, but also induce the synthesis of specific RNA sequences, as both responses are observed upon auxin treatment *in vivo*. Some of the auxin-binding sites that are not membrane-bound, which are described in the preceding section, might have such receptor function. Publications in which an influence of these receptor proteins on transcription *in vitro* was reported will now be discussed separately.

(1) The receptors from nuclei from immature coconut endosperm
Considering the location of these receptors a function at the genomic level seems evident. In their first publication Mondal, Mandal & Biswas (1972) found a two-fold increase in transcription as measured by the incorporation of [^{32}P]-AMP or [^{3}H]-UMP into TCA-precipitable material, after addition of purified n-IRP and IAA to a homologous transcription system. However, essential control experiments, to check the influence of IAA on the transcription systems and on the endogenous nucleotide-polymerase activity of the n-IRP preparation, were lacking.

Roy & Biswas (1977) showed that the IAA-n-IRP complex is able to bind to isolated chromatin and to stimulate transcription. However, they did not explain why in this case they preferred to use *Esherichia coli* polymerase. Moreover, the binding of [^{14}C]-IAA-n-IRP to isolated chromatin reached 'saturation' at 90 μg IAA-n-IRP/42 μg DNA equivalent of chromatin. At this point 1600 cpm [^{14}C]-IAA (sp. act. 100 cpm/pmol) had been bound to the chromatin. This means that only 10% of the [^{14}C]-IAA-n-IRP complex was able to bind to chromatin. No control experiments to determine nonspecific binding of [^{14}C]-IAA-n-IRP to the filter were performed. This probably explains why a straight line, representing nonspecific, unsaturable binding, would fit their 'saturation curve' much better than a Michaelis-Menten-type curve representing specific, saturable binding.

There is also a marked discrepancy in the reported nature of the stimulated RNA, namely, 25 s and 12 s in 1972 (Mondal *et al.*) and 9–12 s in 1977, respectively (Roy & Biswas). According to Fig. 4 in Roy & Biswas (1977) and after calculating the total radioactivity in the gel slices, we see that the RNA synthesized in the presence of IAA-n-IRP contains only 20% more radioactivity than the control, instead of the 200% that would be expected from their Table 2. Furthermore, their stimulation experiments were performed in the presence of 0.1 μM IAA, 18 μg n-IRP and 40 μg DNA equivalent of coconut chromatin. From the binding experiments with IAA-n-IRP to chromatin they concluded that 90 μg IAA-n-IRP are required to reach saturation with the amount of chromatin used. Since the K_d for IAA of n-IRP is 7.6 μM, this means that under the experimental conditions used to determine the stimulation of RNA synthesis at most 0.4% of the available binding sites for IAA-n-IRP on the chromatin are occupied, while RNA synthesis is already stimulated for more than 100%.

When we also consider the questions raised to the IAA-binding ability of the nuclear receptors in coconut endosperm (see section 1.3.1), we are forced to doubt the significance of these results.

(2) The auxin-binding protein from wheat coleoptiles
Likholat *et al.* (1974) demonstrated the liberation of auxin-binding sites from membrane preparations, isolated from 72-h-old seedlings, by incubation with 10 μM IAA. This was not observed for membranes isolated from 36-h-old seedlings. Chromatin isolated from this stage in culture showed a template activity that was above twice as high as in chromatin isolated from the 72-h stage. Each membrane preparation was incubated separately with chromatin isolated either from the 36-h or the 72-h stage. They observed that only the 72-h membrane preparation increased the

template activity of 72-h chromatin in the presence of 1 μM IAA. However, in our opinion the most important control experiment (chromatin with IAA) was not reported.

(3) The auxin receptor from cultured tobacco pith explants

Stimulation of RNA synthesis *in vitro* by the soluble auxin-binding site from cultured tobacco tissue is described by Van der Linde *et al.* (1984). They used a transcription system consisting of nuclei isolated from tobacco callus tissue. This has been described in several reports (Mennes, Voogt & Libbenga, 1977; Mennes, Bouman, Van der Burg & Libbenga, 1978; Bouman, Mennes & Libbenga, 1979; Bouman, Van Paridon, Vogelaar, Mennes & Libbenga, 1981). Addition of partially purified receptor preparations from tobacco callus to these nuclei resulted in an IAA-dependent stimulation of RNA-polymerase activity, which was not observed with similar preparations that did not contain detectable amounts of specific IAA-binding sites. The average stimulation in the presence of 1 μM IAA was 42%; it was achieved by an increase in RNA-polymerase-II activity. By increasing the IAA concentration from 0 to 10^{-6} M a good correlation (correlation coefficient = 0.95) was found between the receptor occupancy and the relative stimulation (Fig. 1.4).

Fig. 1.4. Dose–response relationship between the IAA concentration and the stimulation of RNA synthesis in isolated tobacco-callus nuclei in the presence of a partially purified auxin-receptor preparation.

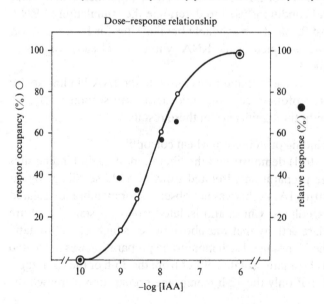

No stimulation occurred in the presence of 10^{-6} M tryptophan, an indole-like non-auxin, that does not bind to the receptor (Oostrom *et al.* 1980). Meanwhile Bailey *et al.* (1984) obtained similar results with partially purified receptor preparations and nuclei isolated from tobacco cell-suspension cultures. Although these results are promising, confirmation of involvement of this auxin-binding protein in nuclear transcription processes awaits purification of the receptor, and further characterization of the transcription products.

1.3.3 Concluding remarks

The presence of soluble hormone-binding proteins is often regarded as essential for a cell to respond to a hormonal signal especially when the response includes alterations in expression of the genome. From section 3 it is evident that soluble auxin-binding proteins can be isolated from a variety of tissues in which auxins induce changes in gene expression. These proteins were identified on basis of three criteria: (i) the binding is reversible; (ii) the binding is saturable within physiological concentrations of the hormone; and (iii) biologically active hormone analogues bind with high affinity. These criteria are often used to distinguish real receptors from non-functional binding proteins, but a mammalian protein like BSA also fulfils these criteria (Murphy, 1979) and we think a lot of other proteins will do, too. So we have to regard these criteria as a first step in the identification of a receptor. The second step is to clarify the biological function of the binding protein. In other words, which biochemical reaction or process is regulated by the receptor in the presence of physiological concentrations of the hormone? According to this criterion, only two IAA-binding proteins can be classified as putative receptors, both modulating RNA-polymerase activity in the presence of IAA. The first was isolated from immature coconut endosperm nuclei, but the many questions raised here and in other reviews and the lack of a follow-up decreases the significance of these results. The second one was isolated from cultured tobacco tissues, but because of the impurity of the fractions used to determine the function of the receptor, the results must be regarded as a first step to prove an involvement of this protein in the auxin-mediated regulation of transcription.

It is quite possible that the latter process is regulated by a chemical signal, which is liberated after binding of auxin to a membrane-bound receptor. The work of Hardin, Cherry, Morré & Lembi (1972) suggests this possibility, although they did not determine specific IAA-binding to the membrane preparation. Since their publication some membrane-bound auxin receptors have been characterized extensively, but at present no report is available that demonstrates the liberation of such a chemical signal upon auxin-binding to membrane-bound receptors. Since we do not know whether research has been performed to exclude this pathway for auxin action on gene expression, we think it should be given more attention.

In conclusion we can say that the research on soluble receptors as mediators in the action of auxin on gene expression is in a very juvenile phase and the progress is slow. The bulk of literature produced by similar research on animal hormones may provide us with clues, which can accelerate the developments in this field.

1.4   Theory and criteria

The receptor theory was originally developed in drug and animal hormone research, but it seems justified to presume that perception and transduction of chemical signals in higher plants are based on the same principle.

Although there exist a large number of treatises dealing with general theoretical and technical aspects of receptor research (O'Brien, 1979, and references therein), we shall formulate briefly the receptor concept for the plant growth substance auxin, as we feel it is used implicitly by many workers, and comment upon some of the applied techniques and analytical methods.

1.4.1   Receptor concept and criteria

Dose-response curves observed in classical bioassays were originally described by a number of workers using enzyme kinetics (Houseley, 1961, and references therein). Many of the observed curves could, indeed, be described by simple Michaelis and Menten kinetics, but the biochemical basis was not understood.

Dose-response curves can better be described using the receptor concept, which leads to the same mathematical formalism as enzyme kinetics; now the biochemical basis is formed by a particular class of regulatory proteins (receptors) that specifically and reversibly bind ligands but, unlike enzymes, do not convert them chemically. Upon binding, such receptor molecules are by conformational change transformed into activated phases by which other processes are stimulated or inhibited. Two possible mechanisms for signal transduction by receptor activation have been proposed:

(1) Occupation theory – The signal is transduced during the existence of a ligand–receptor complex.
(2) Rate theory – The signal is transduced each time an association between ligand and receptor takes place, but not when the receptor is occupied by the ligand (Paton, 1961; Van Haastert, 1980).

In the auxin–receptor literature it is implicitly assumed that the response is proportional to the number of occupied receptors, but we cannot exclude the occurrence of rate receptors.

In its most simple form the occupation theory proposes a linear relationship between the number of occupied receptors and the response:

$$H + R \underset{k_{-1}}{\overset{k_1}{\rightleftharpoons}} HR \overset{k_2}{\rightarrow} \text{Response} \tag{2}$$

Analogous to Michaelis and Menten kinetics it can be found that:

$$Y = k_e R_o H / (H + K_d) \tag{3}$$

with Y = effect, k_e = proportionality constant between occupied receptors and response, R_o = initial receptor concentration, H = hormone concentration $(H \gg R_o)$, K_d = dissociation constant of hormone–receptor complex.

It can easily be shown that in this case maximum response is $k_e R_o$, that half-maximal response is reached at $H(C_{50}) = K_d$, and that the range of 9 to 90% maximal response lies between $0.1 \times K_d$ and $10 \times K_d$. This gives an indication of the physiological range of the hormone.

Figure 1.5 shows a typical example of a dose–response curve obtained with the coleoptile system, which is most extensively used in auxin-

Fig. 1.5. Dose–response curve obtained in a classical coleoptile straight growth test. In the graph the response (f), expressed in percentage of the maximal response, is plotted against the initial external hormone concentration. The dotted line represents an approximation of the suboptimal part of the experimental curve based on the equation

$$f = \frac{[IAA]}{[IAA] + C_{50}} \times 100\%$$

The arrows, a, b, and c, indicate respectively, 9, 50 and 90% response found for the theoretical curve. (Data modified from Posthumus, 1967.)

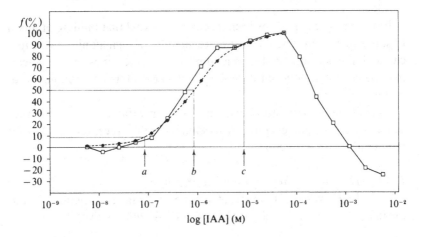

receptor research. The suboptimal part apparently satisfies equation (3), and in studying binding of auxin to subcellular components, one might expect that as a first approximation 9 to 90% occupation of binding sites is reached at auxin concentrations between 8.5×10^{-8} M and 8.5×10^{-6} M; $K_d = 8.5 \times 10^{-7}$ M and the concentration of binding sites is some orders of magnitude lower than K_d. Moreover, the binding sites should discriminate between auxins and unrelated compounds, and within the group of auxins and anti-auxins, differences in K_d values should correspond with differences in C_{50} values.

Although, indeed, binding of auxin to subcellular components should be of high affinity, low capacity and high specificity, one should be aware of the fact that dose–response curves give only (approximate) information about binding parameters and specificity. Auxin is rapidly taken up and metabolized by the cells and these processes differ among the various analogues. Thus, external concentrations may not reflect concentrations at possibly internal auxin-receptor sites. This may be a source of discrepancies between dose–response and binding parameters. Also, dose–response curves do not always provide unambiguous information. For example, Nissl & Zenk (1969) and Cleland (1972) found that coleoptile sections exhibit dose–response curves in which the apparently inhibitory effects at higher auxin concentrations are absent, if initial growth rates are taken rather than total growth over a prolonged period, as is customary in classical bioassays. The dose–response curves thus obtained still satisfy equation (3), but C_{50} values are now considerably lower than those generally obtained with classical bioassays (Nissl & Zenk, 1969, C_{50} c. 10^{-9} M; Cleland, 1972, C_{50} c. 6×10^{-8} M). The reason for these differences is unknown.

Finally, the coupling between receptor occupancy to response may be more complex than is assumed when deriving equation (3) (see section 1.4.3).

In summary, auxin dose–response curves reveal that binding of auxin to subcellular components must be reversible, of high affinity, low capacity and high specificity; due to a number of uncertainties, discrepancies may be expected between dose–response parameters and binding parameters.

Apart from such possible discrepancies, uncertainties about the presence of real auxin receptors will remain, until a receptor function of the binding sites has been found.

1.4.2   Detection of high-affinity binding sites

In a reversible reaction of a receptor R and a hormone H resulting in a complex C, the relation between H, R and C is given by:

$$K_a = \frac{[C]}{[H][R]} = \frac{1}{K_d} \qquad (4)$$

There are a number of ways to calculate K_a, all of which require determination of the concentration of the complex. In most methods, radioactively labelled hormone is used to bind to the receptor, and free hormone is separated from bound hormone (the complex C). However, certain complications can obscure the real amount of complex:

Hormone can be bound to non-receptor macromolecules with a lower affinity (low-affinity binding).

Inclusion of free hormone, for instance by membranes (nonspecific binding).

Incomplete recovery of the complex, e.g. by rapid dissociation during separation.

Although it is preferable to use a method that is immune to these complications, it is still possible to compute K_a and R correctly. Defining B (bound hormone) as complex concentration [C] and F (free hormone) as [H], equation (4) can be rewritten

$$B = K_a F [R] \qquad (5)$$

When the initial receptor concentration is denoted by R_0, then

$$[R] = R_0 - B$$

Substituting equation (6) into equation (5) and dividing by F yields:

$$B/F = K_a R_o - K_a B \qquad (7)$$

Determination of B at different hormone concentrations and plotting B against B/F yields a straight line with a slope of $-K_a$ and an intercept on the abscission of R_o. The line can be fitted by hand or by simple linear regression.

The different hormone concentrations are often made by diluting a fixed amount of labelled hormone with unlabelled hormone of a known concentration.

There are other methods for the determination of K_a but we only mention the following:

the Woolf plot (F/B vs F) which seems to be more reliable than the Scatchard plot, especially when the binding data contain outliers (i.e. points in the graph which do not fit well into the expected pattern) (Keightley & Cressie, 1980);

the direct linear plot as described by Eisenthal & Cornish-Bowden (1974);

and the displacement plot (see section 1.4.3).

In all the methods mentioned above, only the interaction of one hormone with one receptor is considered. unfortunately, in almost all hormone–receptor interactions in plants a considerable amount of nonspecific and/or low-affinity binding is present, due to the use of impure preparations. All linear plots now become non-linear and extracting correct K_a and R_o values from plots is somewhat harder.

Because of the many pitfalls we will pay some attention to one special case, the binding of hormone to two different binding sites with low and high-affinity, analysed by means of a Scatchard plot (Scatchard, 1949).

The high-affinity binding alone would result in a straight line P (Fig. 1.6), with a slope of $-K_1$ and intercept R_1. The low-affinity binding alone would result in a line Q with slope $-K_2$ and intercept R_2. Taken together they form a hyperbola C of which P and Q are asymptotes (Feldman, 1972).

As can be seen from Fig. 1.6 it is important to use a wide range of hormone concentrations. If concentrations which are too high are used, one will miss the first part of the curve in Fig. 1.6 and overlook the high-affinity part. If concentrations which are too low are used, one will miss the low-affinity part which – although it may be uninteresting in itself – influences the slope of the high-affinity part of the plot. A typical example of such a case is shown by Murphy (1980a).

There are a number of ways of determining the parameters K_1, K_2,

Fig. 1.6. Theoretical Scatchard plot of binding of a ligand to a receptor with high affinity (line P), and a receptor with low affinity (line Q). Line C is the resulting hyperbola. B (bound ligand) is plotted against B/F (F = free ligand).

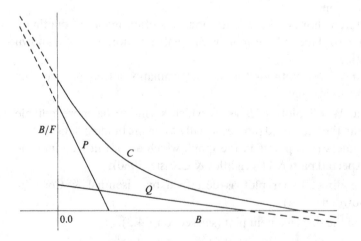

R_1 and R_2 (R_1 = concentration of high-affinity binding sites, R_2 = concentration of low-affinity binding sites). Furthermore, if non-specific binding is also present, an extra parameter N is introduced,

$$N = \lim_{B \to \infty} (B/F) \tag{8}$$

(Chamness & McGuire, 1975). A few methods of determining these parameters will be discussed.

(1) A graphical method, which is fast but often unreliable, consists of drawing tangents to the hyperbolic Scatchard plots at the intercepts with ordinate and abscissa (for instance, Buller, Schrader & O'Malley, 1976). This method is commonly known as the extrapolation method. The slopes, however, approximate the true values only when K_1 and K_2 differ by some orders of magnitude and when relatively little low-affinity binding proteins are present. We calculated that an error of less than 5% in K_1 and R_1 can be expected only when $K_1R_1/K_2R_2 > 100$. When only non-specific binding is present, the total amount of non-specifically bound hormone must be less than 5% of the total amount of hormone bound. These restrictions are, of course, very severe.

The reason we mention this extrapolation method is because it was frequently used wrongly. We would strongly advise only to use this method to get a very rough estimate of K_1 and R_1.

(2) When only non-specific binding is present, the high-affinity binding can be calculated as

$$B_{\text{high-affinity}} = B_{\text{total}} - F.(\lim_{B \to \infty} B/F) \tag{9}$$

(Chamness & McGuire, 1975). The corrected values of B, plotted against the corrected B/F values will result in a straight line.

(3) It is also possible to subtract the non-specific binding experimentally. For each point of the Scatchard plot a second determination is carried out with an excess of unlabelled hormone. Now all bound radioactivity is bound nonspecifically, but the plot can still be curved when low-affinity binding is present.

(4) A number of computer programs have been developed to compute the different parameters K_1, R_1 and N. Instead of giving a long, probably incomplete list, we would like to draw your attention to two programs.

– SCATPACK, written by the authors of LIGAND (Munson & Rodbard, 1980), is a BASIC program and can fit up to 18 parameters, including those for cooperative binding. The program can be adapted for use on microcomputers with at least 16 Kbytes of memory. The SCATPACK program is distributed by the Biomedical Computing Technology Information Center, R-1302 Vanderbilt Med. Center, Nashville, Tennessee 37232, USA.

– MLP, Maximum Likelihood Program is a FORTRAN program that runs on larger systems. This program can fit up to at least 8 parameters. An explanation of the operation of this program is given by Murphy (1980*a*). Both programs work by estimating the parameters, constructing a theoretical curve and comparing this curve with the experimental one. If needed the parameters are adjusted and the whole process is repeated, until a good fit is obtained.

A word of caution regarding computerized results is appropriate here. In all cases the user has to determine according to which model (number of binding sites, non-specific binding, if applicable) he wants to fit his data. Furthermore, the user has to supply the program with initial estimates of the parameters. In this way one can influence the outcome of the results. Therefore, the determination of the goodness of fit for a few possible models is very important. Sometimes the curve can be fitted with a 3-parameter model (1 binding site + non-specific binding), a 4-parameter model (2 binding sites) or even a 5-parameters model (2 binding sites + non-specific binding) etc. We suggest that a more complicated model should be used only when the fit is significantly better, since introducing an extra parameter should always improve the fit. Still, when one has found evidence for more than one binding site by computer analysis, other methods should be used to verify those results (e.g. biochemical separation of the sites).

Often, when one is interested only in the presence or absence of the receptor, a so-called 'two-point analysis' is carried out. Binding is determined at two hormone concentrations, one at a low (only labelled hormone) and one at a high concentration (the same concentration of labelled hormone plus a large additional amount of unlabelled hormone). This is in essence the determination of the first and the last points in a regular Scatchard plot, but since this plot can be curved it is impossible to see the difference between a low amount of receptor with high affinity and a high amount of receptor with low affinity for the hormone. We advise, when one insists on performing these 'two-point analyses', to use a smaller amount of unlabelled hormone for the determination of the second point, preferably a concentration not greater than the concentration at which the curvature of the Scatchard plot is most pronounced.

1.4.3   *Determination of binding specificity*

Specificity has already been mentioned briefly in relation to dose–response curves (1.4.1). When receptor occupancy is linearly related to physiological effect, then binding of an auxin to a receptor with affinity $K_d(1)$ should lead to the same value for half-maximal response $C_{50}(1)$, while binding of another auxin with a different $K_d(2)$ should lead to a value of $C_{50}(2)$ equal to $K_d(2)$ (eq. 3). But coupling between receptor occupation and response may be more complex than described in eq. (2); it could be described in a general way by the polynoma

$$Y = aHR + bHR^2 + cHR^3 + \ldots \tag{10}$$

Even now a dose–response curve could still have a shape as described by eq. (3), but a K_d value is not necessarily equal to C_{50}. The most reliable information about K_d values and specificity can be obtained using anti-auxins, i.e. compounds that, although they are not auxins themselves, can displace auxin competitively from its receptor (see also Cuatrecasas & Hollenberg, 1976; Hollenberg & Cuatrecasas, 1979). In the case where auxin and anti-auxin are both present

$$HR = R_0 \frac{H}{H + K_d(A) - [1 + I/K_d(I)]} \tag{11}$$

with HR = concentration of occupied receptor, R_0 = total receptor concentration, H = auxin concentration, $K_d(A)$ = dissociation constant of auxin–receptor complex, I = concentration of anti-auxin and $K_d(I)$ = dissociation constant of anti-auxin–receptor complex. Independently of the coupling of receptor occupation to response, equal responses are obtained at equal receptor occupancy by auxin. Thus, choosing an auxin concentration H that gives the same effect as a (higher) auxin concentration H^x in the presence of an anti-auxin

$$R_0 \frac{H^x}{H^x + K_d(A)[1 + I/K_d(I)]} = R_0 \frac{H}{HK_d(A)} \tag{12}$$

From eq. (12) it follows that

$$K_d(I) = \frac{I}{H^x/H - 1} \tag{13}$$

Thus true K_d values, at least for anti-auxins, can now in principle be obtained from dose-response curves, irrespective of the nature of the coupling.

When the coupling between occupancy and effect is as described by (10), and a,b,c etc. do not change for different auxins, then it is clear that the same response will be reached at the same receptor occupancy, at respective hormone concentrations H_1 (auxin 1) and H_2 (auxin 2). Consequently, H_1R (auxin 1) and H_2R (auxin 2) must be equal. Since

$$H_1R = R_0H_1/(H_1 + K_d(1)) \text{ and } H_2R = R_0H_2/(H_2 + K_d(2)),$$

it follows that

$$H_1/H_2 = K_d(1)/K_d(2) \tag{14}$$

i.e., the quotient of C_{50} values and the quotient of K_d values should be equal.

The determinations of parameters can be disturbed severely by, for instance, a difference between external (applied) concentration of hormone and internal concentration (the concentration the receptor 'feels') and metabolic conversion (see introduction to this section). Also, responses can differ due to allosteric effects or due to different parameters a,b,c etc. (from the polynomial eq. (10)); an indication is that maximal responses may differ, as was demonstrated already by McRae, Foster & Bonner (1953); they found different maximal responses for IAA, NAA and 2,4-D. In such cases, comparison of K_d and C_{50} values of different auxins yields only little, if any information about specificity.

To determine specificity two methods are used the most:
(a) Competition for the same binding site is measured by only adding one (excess) concentration of competitor to a mixture of receptor and labelled hormone. This method is only qualitative and gives hardly any useful information in the determination of specificity as described above. It can, however, be used as an indication of the ability of a receptor to bind the competitor. (b) Displacement plots, already mentioned in 1.4.2, are obtained in the following way: a receptor preparation is incubated with a fixed concentration of labelled auxin and a range of concentrations of unlabelled competitor (C) until equilibrium is reached. The concentration of C is then plotted against the amount of bound hormone. The concentration of competitor that is able to displace 50% of the bound hormone is often thought to equal K_d, but this is true only when a number of criteria have been met; for instance, both receptor and labelled hormone concentration must be far less than K_d (Jacobs, Chang & Cuatrecasas, 1975; Blondeau, Rocher & Robel, 1978).

For these displacement plots the influence of a possible second binding site with lower affinity was not taken into account but, as Fig. 1.7 shows, the presence of such low-affinity binding sites can influence the shape of the curve considerably.

In practice, a strict correlation between K_d and C_{50} is seldom observed, which is understandable in view of the facts mentioned above. In the discussion of the different receptors (sections 1.2.1, 1.2.2, 1.2.3 and 1.3.1) we have considered a binding site specific when it has at least the following characteristics: it should be saturable within a factor of ten around the range where auxin has its physiological effect; it should discriminate between auxins and non-auxins, while auxins and anti-auxins should show cross-competition.

1.4.4   Performing a binding assay
There are a number of pitfalls when one performs a binding assay.

(a) An incubation can be started by adding labelled hormone (H^x) as well as unlabelled hormone (H) to the receptor (R). When H^x and H are added sequentially this should be done as fast as possible or one should verify that equilibrium has been reached at the end of the incubation, especially when exchange of H^x and H is slow (low k_{off}).

(b) The amount of complex formed will be influenced by any inhomogeneity of the solution (of a soluble receptor) or of the suspension (for instance, of membranes) during incubation.

(c) Separation of bound and free hormone is often the biggest problem in the procedure. A few separation methods and their disadvantages will be discussed.

(1) Soluble receptors

(a) Removal of free hormone by dextran-coated charcoal (DCC). The DCC preferably binds small ions and molecules and will remove the free hormone. This process will take some time, e.g., 10–20 min at 0°C.

Fig. 1.7. Influence of the presence of a low-affinity receptor upon displacement of hormone by competitor (C). Maximal binding is defined as the total amount of hormone (H) bound at equilibrium to the high-affinity receptor (R) and to the low-affinity receptor (P). Initial concentrations are: $H = 10^{-8}$ M; $R = 10^{-10}$ M; $P = 0$ M (a), 10^{-9} M (b), 10^{-8} M (c,d). K_d values of different complexes are: K_d, H-R $= 10^{-7}$ M; K_d, H-P $= 10^{-5}$ M; K_d, H-C $= 10^{-7}$ M; K_d, P-C $= 10^{-5}$ M (a,b,c) or ∞ (d).

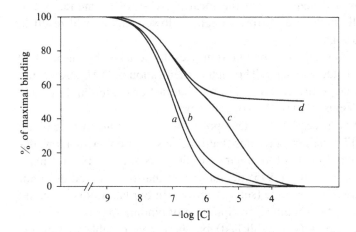

The DCC is then removed by centrifugation, leaving a solution containing the complex. The disadvantages are:

A high dissociation rate at 0°C will lead to loss of the complex.

Buffering capacity is lost upon addition of DCC.

The method cannot be used for viscous solutions.

DCC always binds minor amounts of protein, so the protein concentration should be high enough. (For practical aspects of this method see Poulsen, 1981; Ralet & Brombacher, 1981; De Hertogh, Van der Heyden & Ekka, 1975.)

(b) Equilibrium dialysis. Separation is no problem here, but it can take hours to reach equilibrium, so breakdown of the receptor could occur. And when only little binding is present it certainly will not rise above the background.

(c) Gel filtration. Sephadex G-25 is a good tool to separate complex from free hormone. The time needed for this procedure can be reduced if necessary by centrifugation of adapted columns (Helmerhorst & Stokes, 1980).

A different method is to equilibrate the column with labelled hormone, followed by the addition of the receptor.

(1) When the receptor is preincubated with labelled hormone, the amount of labelled complex will remain constant due to the presence of hormone in the column. A single peak will appear above the background in the void volume.

(2) When the receptor is not preincubated, it will bind labelled hormone, causing a local decrease in free hormone concentration. The complex will elute in the void volume, giving a peak as in (1), but it will be followed by a trough, due to the local decrease of hormone. A fast binding will result in a narrow trough, a slower binding will result in a broader trough.

(d) Ammonium sulphate precipitation. Results obtained by this method should be treated with care. The high concentration of $(NH_4)_2SO_4$ could very well change the protein–hormone interactions resulting in coprecipitation (Venis, 1981) or loss of complex.

(e) Protamine precipitation. This polycation was found to precipitate oestradiol-17 receptors. Proteins that bind this hormone with a low affinity have been found not to precipitate (Steggles & King, 1970). Chamness, Huff & McGuire (1975) demonstrated that this method does not coprecipitate free hormone (see (d) above). In our laboratory we could also demonstrate protamine precipitation of binding sites isolated from tobacco-pith callus (not published) but here considerable nonspecific binding was still present.

(2) Membrane-bound receptors

(a) Centrifugation

This is the most widely used method to separate membrane-bound receptors from free hormone. After centrifugation one should take care that the pellet is not contaminated by the supernatant. The comparatively long time needed to centrifuge samples is a disadvantage in non-equilibrium studies.

(b) Filtration

Filtration of membranes (preferably under reduced pressure) is a rapid method. Binding of free hormone to the filter should be avoided by pre-washing with unlabelled hormone and by washing after filtration. Thus, the method is only suitable for slow-dissociating complexes (it was applied successfully for the determination of association and dissociation of NAA and binding sites isolated from tobacco-pith callus, Maan *et al.* 1983).

(c) Equilibrium dialysis

The same applies as mentioned above for soluble receptors.

It is obvious that no method is always reliable. One should therefore check thoroughly the method one wants to use.

The authors are indebted to Dr R. J. Bogers for critically reading the manuscript.

1.5   References

Aducci, P., Ballio, A. & Federico, R. (1982). Solubilization of fusicoccin binding sites. In *Plasmalemma and Tonoplast: their Functions in the Plant Cell*, ed. D. Marmé, E. Marrè & R. Hertel, pp. 279–284. Amsterdam: Elsevier.

Aducci, P., Crosetti, G., Federico, R. & Ballio, A. (1980). Fusicoccin receptors. Evidence for endogenous ligand. *Planta* **148**: 208–10.

Aducci, P., Federico, R. & Ballio, A. (1980). Interaction of a high molecular weight derivative of fusicoccin with plant membranes. *Phytopath. Medit.* **19**: 187–8.

Ariens, E. J., Beld, A. J., Rodrigues de Miranda & Simonis, A. M. (1979). The pharmacon-receptor-effector concept. In *The Receptor*, ed. R. D. O'Brian, pp. 33–91. Plenum Press New York.

Bailey, H. M., Barker, E. J. D., Libbenga, K. R., Van der Linde, P. C. G., Mennes, A. M. & Elliott, M. C. An auxin receptor in plant cells. *Biol. Plant.* (in press).

Ballio, A., Aducci, P. & Federico, R. (1982). Characterization of fusicoccin binding sites. *Abstracts 11th int. conf. on Plant Growth Substances*, Aberyswyth, p. 24.

Ballio, A., Bottalico, A., Framondino, M., Graniti, A. & Randazzo, G. (1971). Fusicoccin: structure-phytotoxicity relationships. *Phytopath. Medit.* **10**: 26–32.

Ballio, A., Chain, E. B., DeLeo, P., Erlanger, B. F., Mauri, M. & Tonolo, A. (1964). Fusicoccin: a new wilting toxin produced by *Fusicoccum amygdali* Del. *Nature* **203**: 297.

Ballio, A., D'Alessio, V., Randazzo, G., Bottalico, A., Graniti, A., Sparapano, L., Bosnar, B., Casinovi, G. C. & Gribanovski-Sassu, O. (1976). Occurrence of fusicoccin in plant tissues infected by *Fusicoccum amygdali* Del. *Physiol. Plant Pathol.* **8**: 163–9.

Ballio, A., De Michelis, M. I., Lado, P. & Randazzo, G. (1981). Fusicoccin structure–activity relationships: Stimulation of growth by cell enlargement and promotion of seed germination. *Physiol. Plant.* **52**: 471–5.

60 K. R. Libbenga et al.

Ballio, A., Federico, R., Pessi, A. & Scalorbi, D. (1980). Fusicoccin binding sites in subcellular preparations of spinach leaves. *Plant Sci. Lett.* **18:** 39–44.
Ballio, A., Federico, R. & Scalorbi, D. (1981). Fusicoccin structure-activity relationships: *In vitro* binding to microsomal preparations of maize coleoptiles. *Physiol. Plant.* **52:** 476–81.
Batt, S. & Venis, M. A. (1976) Separation and localization of two classes of auxin binding sites in corn coleoptile membranes. *Planta* **130:** 15–21.
Batt, S., Wilkins, M. B. & Venis, M. A. (1976). Auxin binding to corn coleoptile membranes: kinetics and specificity. *Planta* **130:** 7–13.
Baulcombe, D. C. & Key, J. L. (1980). Polyadenylated RNA sequences which are reduced in concentration following auxin treatment of soybean hypocotyls. *J. Biol. Chem.* **255:** 8907–13.
Beffagna, N., Coccuci, S. & Marrè, E. (1977). Stimulating effect of fusicoccin on K-activated ATPase in plasmalemma preparations from higher plant tissues. *Plant Sci. Lett.* **8:** 91–8.
Beffagna, N., Pesci, P. & Marrè, E. (1981). Influence of magnesium in the homogenization medium on the state of a divalent cation-stimulated, diethylstilbestrol-sensitive adenylnucleotidyl phosphatase activity from pea stem microsomal preparations. *Biochim. Biophys. Acta* **645:** 300–10.
Bevan, M. & Northcote, D. H. (1981). Some rapid effects of synthetic auxins on mRNA levels in cultured plant cells. *Planta* **152:** 32–5.
Bhattacharyya, K. & Biswas, B. B. (1978). Membrane-bound auxin receptors from *Avena* roots. *Ind. J. Biochem. Biophys.* **15:** 445–8.
Bhattacharyya, K. & Biswas, B. B. (1982). Induction of a high affinity binding site for auxin in *Avena* root membrane. *Phytochem.* **21:** 1207–11.
Blondeau, J. P., Rocher, P. & Robel, P. (1978). Competitive inhibition of specific steroid-protein binding: practical use of relative competition ratios for the derivation of equilibrium inhibition constants. *Steroids* **32:** 563–75.
Bogers, R. J., Kulescha, Z., Quint, A., Van Vliet, Th.B. & Libbenga, K. R. (1980). The presence of a soluble auxin receptor and the metabolism of 3-indoleacetic acid in tobacco-pith explants. *Plant Sci. Lett.* **19:** 311–17.
Bogers, R. J. & Libbenga, K. R. (1981). Molecular action mechanism of plant growth regulators. In *Aspects and Prospects of Plant Growth Regulators*, Monograph 6, ed. B. Jeffcoat, pp. 177–185. Wantage, Oxfordshire: Wessex Press.
Bouman, H., Mennes, A. M. & Libbenga, K. R. (1979). Transcription in nuclei isolated from tobacco tissues. *FEBS Letts.* **101:** 369–72.
Bouman, H., Van Paridon, H., Vogelaar, A., Mennes, A. M. & Libbenga, K. R. (1981). Size analysis of RNA synthesized in isolated tobacco-callus nuclei. *Plant Sci. Lett.* **22:** 361–7.
Buller, R. E., Schrader, W. T. & O'Malley, B. W. (1976). Steroids and the practical aspects of performing binding studies. *J. Steroid Biochem.* **7:** 321–6.
Chamness, G. C., Huff, K. & McGuire, W. L. (1975). Protamine-precipitated estrogen receptor: a solid-phase ligand exchange assay. *Steroids* **25:** 627–35.
Chamness, G. C. & McGuire, W. L. (1975). Scatchard plots: common errors in correction and interpretation. *Steroids* **26:** 538–42.
Cleland, R. E. (1972). The dosage-response curve of auxin-induced cell elongation: a re-evaluation. *Planta* **104:** 1–9.
Cleland, R. E. (1976). Kinetics of hormone-induced H⁺ excretion. *Plant Physiol.* **58:** 210–13.
Cleland, R. E. (1979). Auxin and H⁺ excretion: the state of our knowledge. In *Plant Growth Substances*, 10th int. conf. plant growth substances, Madison, ed. F. Skoog, pp. 71–8. Berlin Heidelberg New York: Springer-Verlag.
Cleland, R. E. (1982). The mechanism of auxin-induced proton efflux. In *Plant Growth*

Substances 1982, ed. P. F. Wareing, pp. 23–31. London, New York, Paris, San Diego, San Francisco, Sao Paulo, Sydney, Tokyo, Toronto: Academic Press.

Cleland, R. E. & Lomax, T. (1977). Hormonal control of H^+ excretion from oat cells. In *Regulation of Cell Membrane Activities in Plants*, ed. E. Marrè & O. Ciferri, pp. 161–72. Amsterdam: Elsevier.

Cross, J. W. & Briggs, W. R. (1976). An evaluation of markers for plasma membranes in membrane fractions from *Zea mays*. *Carnegie Inst. Wash. Yearb.* **75**: 379–83.

Cross, J. W. & Briggs, W. R. (1978). Properties of a solubilized microsomal auxin-binding protein from coleoptiles and primary leaves of *Zea mays*. *Plant Physiol.* **62**: 152–7.

Cross, J. W. & Briggs, W. R. (1979). Solubilized auxin-binding protein. *Planta* **146**: 263–70.

Cross, J. W., Briggs, W. R., Dohrmann, U. C. & Ray, P. M. (1978). Auxin receptors of maize coleoptile membranes do not have ATPase activity. *Plant Physiol.* **61**: 581–4.

Cuatrecasas, P., & Hollenberg, M. D. (1976). Membrane receptors and hormone action. *Adv. Protein Chem.* **30**: 251–451.

De Hertogh, R., Van der Heyden, I. & Ekka, E. (1975). 'Unbound' ligand adsorption on dextran-coated charcoal: practical considerations. *J. Steroid Biochem.* **6**: 1333–7.

De Lean, A. & Rodbard, D. (1979). Kinetics of cooperative binding. In *The Receptor*, ed. R. D. O'Brien, pp. 143–92. New York: Plenum Press.

De Michelis, M. I., Lado, P. & Ballio, A. (1975). Relationships between structural modifications of fusicoccin and their effects on cell enlargement and seed germination. *Physiol. Plant.* **52**: 471–5.

Dohrmann, U., Hertel, R. & Kowalik, H. (1978). Properties of auxin binding sites in different subcellular fractions from maize coleoptiles. *Planta* **140**: 97–106.

Dohrmann, U., Hertel, R., Pesci, P., Cocucci, S. M., Marrè, E., Randazzo, G. & Ballio, A. (1977). Localization of '*in vitro*' binding of the fungal toxin fusicoccin to plasma-membrane-rich fractions from corn coleoptiles. *Plant Sci. Lett.* **9**: 291–9.

Dohrmann, U. & Ray, P. M. (1976). Measurements of auxin binding to solubilized receptor sites of corn seedling tissue. *Carnegie Inst. Wash. Yearbook* **75**: 395–9.

Döllstädt, R., Hirschberg, K., Winkler, E. & Hübner, G. (1976). Bindung von Indolylessigsäure und Phenoxyessigsäure an Fraktionen aus Epikotylen und Wurzeln von *Pisum sativum* L. *Planta* **130**: 105–11.

Dupont, F. M., Bennett, A. B. & Spanswick, R. M. (1982). Proton transport in microsomal vesicles from corn roots. In *Plasmalemma and Tonoplast: their Functions in the Plant Cell*, ed. D. Marmé, E. Marrè & R. Hertel, pp. 409–16. Amsterdam: Elsevier.

Edwards, K. L. & Scott, T. K. (1974). Rapid growth responses of corn root segments: Effect of pH on elongation. *Planta* **119**: 27–37.

Eisenthal, R. & Cornish-Bowden, A. (1974). The direct linear plot: a new graphical procedure for estimating enzyme kinetic parameters. *Biochem. J.* **139**: 715–20.

Erdei, L., Toth, I. & Zsoldos, F. (1979). Hormonal regulation of Ca^{2+}-stimulated K^+ influx and Ca^{2+}, K^{2+}-ATPase in rice roots: *in vivo* and *in vitro* effects of auxins and reconstitution of the ATPase. *Physiol. Plant.* **45**: 448–52.

Evans, M. L. (1976). A new sensitive root auxanometer: preliminary studies of the interaction of auxin and acid pH in the regulation of intact root elongation. *Plant Physiol.* **58**: 599–601.

Evans, M. L. & Hokanson, R. (1969). Timing of the response of coleoptiles to the application and withdrawal of various auxins. *Planta* **85**: 85–95.

Evans, M. L., Mulkey, T. J. & Vesper, M. J. (1980). Auxin action on proton influx in corn roots and its correlation with growth. *Planta* **148**: 510–12.

Farrimond, J. A., Elliott, M. C. & Clack, D. W. (1978). Charge separation as a component of the structural requirements for hormone activity. *Nature* **274**: 401–2.

Feldman, H. A. (1972). Mathematical theory of complex ligand-binding systems at equilibrium: some methods for parameter fitting. *Analyt. Biochem.* **48:** 317–38.

Geuns, J. M. C. (1982). Plant steroid hormones – what are they and what do they do? *Trends in Biochem. Sci.* **7:** 7–9.

Hager, A., Menzel, H. & Krauss, A. (1971). Versuche und Hypothese zur Primärwirkung des Auxins beim Streckungswachstum. *Planta* **100:** 47–75.

Hanson, J. B. & Trewavas, A. J. (1982). Regulation of plant cell growth: the changing perspective. *New Phytol.* **90:** 1–18.

Hardin, J. W., Cherry, J. H., Morré, D. J. & Lembi, C. A. (1972). Enhancement of RNA polymerase activity by a factor released by auxin from plasma membrane. *Proc. Natl. Acad. Sci. USA* **69:** 3146–50.

Helmerhorst, E. & Stokes, G. B. (1980). Microcentrifuge desalting: a rapid quantitative method for desalting small amounts of protein. *Analyt. Biochem.* **104:** 130–5.

Hertel, R., Lomax, T. L. & Briggs, W. R. (1983). Auxin transport in membrane vesicles from *Cucurbita pepo* L. *Planta* **157:** 193–201.

Hertel, R., Thomson, K.-St. & Russo, V. E. A. (1972). *In-vitro* auxin binding to particulate cell fractions from corn coleoptiles. *Planta* **107:** 325–40.

Hodges, T. K. & Leonard, R. T. (1974). Purification of a plasma-membrane bound adenosine triphosphatase from plant roots. In *Methods in Enzymology*, vol. **32**, ed. S. P. Colowick & N. O. Kaplan, pp. 392–406. New York: Academic Press.

Hollenberg, M. D. & Cuatrecasas, P. (1979). Distinction of receptor from nonreceptor interactions in binding studies. In *The Receptor*, ed. R. D. O'Brien, pp. 193–214. New York: Plenum Press.

Housely, S. (1961). Kinetics of auxin-induced growth. In *Handbuch der Pflanzenphysiologie*, vol. **14**, *Wachstum und Wachsstoffe*, ed. H. Burström, pp. 1007–43. Berlin: Springer-Verlag.

Ihl, M. (1976). Indole-acetic acid binding proteins in soybean cotyledon. *Planta* **131:** 223–8.

Jablonović, M. & Noodén, L. D. (1974). Changes in compatible IAA binding in relation to bud development in pea seedlings. *Plant and Cell Physiol.* **15:** 687–92.

Jacobs, M. & Hertel, R. (1978). Auxin binding to subcellular fractions from *Cucurbita* hypocotyls: *in vitro* evidence for an auxin transport carrier. *Planta* **142:** 1–10.

Jacobs, S., Chang, K-J. & Cuatrecasas, P. (1975). Estimation of hormone receptor affinity by competitive displacement of labeled ligand: effect of concentration of receptor and of labeled ligand. *Biochem. Biophys. Res. Comm.* **62:** 31–41.

Jacobs, S. & Cuatrecasas, P. (1976). The mobile receptor hypothesis and 'cooperativity' of hormone binding. Application to insulin. *Biochim. Biophys. Acta* **433:** 482–95.

Jacobs, S. & Cuatrecasas, P. (1977). The mobile receptor hypothesis for cell membrane receptor action. *Trends in Biochem. Sci.* **2:** 280–2.

Jacobsen, H.-J. (1981). Soluble auxin-binding proteins in pea. *Cell Biol. Intern. Rep.* **5:** 768.

Jacobsen, H.-J. (1982). Soluble auxin-binding proteins in pea epicotyls. *Physiol. Plant.* **56:** 161–7.

Jacobsen, J. V. (1977). Regulation of ribonucleic acid metabolism by plant hormones. *Ann. Rev. Plant Physiol.* **28:** 537–64.

Kaethner, T. M. (1977). Conformational change theory for auxin structure-activity relationships. *Nature* **267:** 19–23.

Kasamo, K. & Yamaki, T. (1974). Auxin action on membrane bound Mg^{++}-activated ATPase. In *Plant Growth Substances 1973*, Proc. 8th int. conf. on plant growth substances, Tokyo, pp. 699–707. Tokyo, Japan: Hirokawa Publishing Co.

Kasamo, K. & Yamaki, T. (1976). *In vitro* binding of IAA to plasma membrane-rich fractions containing Mg^{++}-activated ATPase from mung bean hypocotyls. *Plant and Cell Physiol.* **17:** 149–64.

Katekar, G. F. (1979). Auxins: on the nature of the receptor and molecular requirements for auxin activity. *Phytochem.* **18**: 223–33.

Keightley, D. D. & Cressie, N. A. C. (1980). The Woolf plot is more reliable than the Scatchard plot in analysing data from hormone receptor assays. *J. Steroid Biochem.* **13**: 1317–23.

Kende, H. & Gardner, G. (1976). Hormone binding in plants. *Ann. Rev. Plant Physiol.* **27**: 267–90.

Klämbt, H. D. (1961). Wachstuminduktion und Wachsstoffmetabolismus im Weizenkoleoptilzylinder. *Planta* **56**: 618–31.

Kurkdjian, A. & Guern, J. (1978). Intracellular pH in higher plant cells I. Improvements in the use of the 5,5-dimethyloxazolidine-2-[^{14}C], 4-dione distribution technique. *Plant Sci. Lett.* **11**: 337–44

Lamb, C. J. (1978). Hormone binding in plants. *Nature* **274**: 312–14.

Larsen, P. (1961). Biological determination of natural auxins. In *Handbuch der Pflanzenphysiologie*, vol. **14**, *Wachstum und Wachsstoffe*, ed. H. Burström, pp. 521–82. Berlin: Springer-Verlag.

Leonard, R. T. & Hotchkiss, C. W. (1976). Cation-stimulated adenosine triphosphatase activity and cation transport in corn roots. *Plant Physiol.* **58**: 331–5.

Letham, D. S., Goodwin, P. B. & Higgins, T. J. V. (1978). *Phytohormones and Related Compounds. A Comprehensive Treatise*, vols. I, II. Amsterdam, Oxford, New York: Elsevier.

Libbenga, K. R. (1978). Hormone receptors in plants. In *Frontiers of Plant Tissue Culture 1978*, Proc. 4th int. congr. plant tissue and cell culture, ed. T. A. Thorpe, pp. 325–33. Manitoba: University of Calgary.

Libbenga, K. R., Maan, A. C. & Bogers, R. J. (1982). Auxin-binding sites from tobacco cell and tissue cultures. In *Plasmalemma and Tonoplast: their Functions in the Plant Cell*, ed. D. Marmé, E. Marrè & R. Hertel, pp. 285–92. Amsterdam: Elsevier.

Likholat, T. V., Pospelov, V. A., Morozova, T. M. & Salganik, R. I. (1974). Capacity of wheat coleoptile cells for specific binding of auxin and effect of the hormone on template activity of chromatin in seedlings of different ages. *Sov. Plant Physiol.* **21**: 779–84.

Löbler, M. (1984). Isolierung, Karakterisierung und Lokalisierung eines Auxin bindenden Proteins aus Mais Koleoptilen. Inaugural Dissertation Universität Bonn.

Maan, A. C., Vreugdenhil, D., Bogers, R. J. & Libbenga, K. R. (1983). The complex kinetics of auxin-binding to a particulate fraction from tobacco-pith callus. *Planta* **158**: 10–15.

Maan, A. C., Van der Linde, P. C. G., Harkes, P. A. A. & Libbenga, K. R. (1985). Correlation between the presence of membrane-bound auxin binding and root regeneration in cultured tobacco cells. *Planta*, in press.

Macri, V., Vianello, A. & Dell'Antone, P. (1982). Passive and ATP-dependent proton fluxes in pea stem microsomal vesicles. In *Plasmalemma and Tonoplast: their Functions in the Plant Cell*, ed. D. Marmé, E. Marrè & R. Hertel, pp. 417–22. Amsterdam: Elsevier.

Marrè, E. (1977). Effects of fusicoccin and hormones on plant cell membrane activities: observations and hypotheses. In *Regulation of Cell Membrane Activities in Plants*, ed. E. Marrè & O. Ciferri, pp. 185–202. Amsterdam: Elsevier.

Marrè, E. (1979). Fusicoccin: a tool in plant physiology. *Ann. Rev. Plant Physiol.* **30**: 273–88.

McRae, R., Foster, R. J. & Bonner, J. (1953). Kinetics of auxin interaction. *Plant Physiol.* **28**: 343–55.

Melanson, D. & Trewavas, A. J. (1982). Changes in tissue protein pattern in relation to auxin induction of DNA synthesis. *Plant Cell Environm.* **5**: 53–64.

Mennes, A. M., Bouman, H., Van der Burg, M. P. M. & Libbenga, K. R. (1978). RNA synthesis in isolated tobacco callus nuclei, and the influence of phytohormones. *Plant Sci. Lett.* **13:** 329–39.

Mennes, A. M., Voogt, E. & Libbenga, K. R. (1977). The isolation of nuclei from tobacco pith explants. *Plant Sci. Lett.* **8:** 171–7.

Mettler, I. J., Mandala, S. & Taiz, L. (1982). Proton gradients produced *in vitro* by microsomal vesicles from corn coleoptiles. Tonoplast origin? In *Plasmalemma and Tonoplast: their Functions in the Plant Cell,* ed. D. Marmé, E. Marrè & R. Hertel, pp. 395–400. Amsterdam: Elsevier.

Meyer, Y. & Chartier, Y. (1981). Hormonal control of mitotic development in tobacco protoplasts. *Plant Physiol.* **68;** 1273–8.

Moloney, M. M. & Pilet, P. E. (1981). Auxin binding in roots: a comparison between maize roots and coleoptiles. *Planta* **153:** 447–52.

Moloney, M. M. & Pilet, P. E. (1982). Auxin binding in maize roots. In *Plasmalemma and Tonoplast: their Functions in the Plant Cell,* ed. D. Marmé, E. Marrè & R. Hertel, pp. 293–302. Amsterdam: Elsevier.

Moloney, M. M., Henri, H. & Pilet, P. E. (1983). Regulation of *in vitro* H$^+$-transport in sealed pea root vesicles (Abstr.). *Plant Phyiol.* **72:** Suppl., 140.

Mondal, H., Mandal, R. K. & Biswas, B. B. (1972). RNA stimulated by indole acetic acid. *Nature New Biol.* **240:** 111–13.

Munson, P. J. & Rodbard, D. (1980). LIGAND: a versatile computerized approach for characterization of ligand-binding systems. *Analyt. Biochem.* **107:** 220–39.

Murphy, G. J. P. (1979). Plant hormone receptors: comparison of naphthaleneacetic acid binding by maize extracts and by a non-plant protein. *Plant Sci. Lett.* **15:** 183–91.

Murphy, G. J. P. (1980*a*). A reassessment of the binding of naphthaleneacetic acid by membrane preparations from maize. *Planta* **149:** 417–26.

Murphy, G. J. P. (1980*b*). Naphthaleneacetic acid binding by membrane-free preparations of cytosol from maize coleoptile. *Plant Sci. Lett.* **19:** 157–68.

Narayanan, K. R., Mudge, K. W. & Poovaiah, B. W. (1981*a*). *In vitro* auxin binding to cellular membranes of cucumber fruits. *Plant Physiol.* **67:** 836–40.

Narayanan, K. R., Mudge, K. W. & Poovaiah, B. W. (1981*b*). Demonstration of auxin binding to strawberry fruit membranes. *Plant Physiol.* **68:** 1289–93.

Nissl, D. & Zenk, M. H. (1969). Evidence against induction of protein synthesis during auxin-induced initial elongation of *Avena* coleoptiles. *Planta* **89:** 323–41.

Normand, G., Schuber, F., Benveniste, P. & Beauvais, D. (1977). Effect of red light on the binding of NAA on maize coleoptile microsomes. In *Regulation of Cell Membrane Activities in Plants,* ed. E. Marrè & O. Ciferri, Amsterdam: Elsevier.

O'Brien, R. D. (1979). *The Receptors. A comprehensive Treatise,* vol. **1,** *General Principles and Procedures.* New York: Plenum Press.

O'Malley, B. W. & Schrader, W. T. (1976). The receptors of steroid hormones. *Sci. American* **234:** 32–43.

Oostrom, H., Kulescha, Z., Van Vliet, Th.B. & Libbenga, K. R. (1980). Characterization of a cytoplasmic auxin receptor from tobacco-pith callus. *Planta* **149:** 44–7.

Oostrom, H., Van Loopik-Detmers, M. A. & Libbenga, K. R. (1975). A high affinity receptor for indoleacetic acid in cultured tobacco pith explants. *FEBS Lett.* **59:** 194–7.

Paton, W. D. M. (1961). A theory of drug action based on the rate of drug-receptor combination. *Proc. Royal Soc. Ser. B* **154:** 21–69.

Penny, P. & Penny, D. (1978). Rapid responses to phytohormones. In *Phytohormones and Related Compounds. A Comprehensive Treatise,* vol. II, ed. D. S. Letham, P. B. Goodwin & T. J. V. Higgins, pp. 537–97. Amsterdam: Elsevier.

Pesci, P. & Beffagna, N. (1982). ATPase and fusicoccin binding in pea stem microsomal preparations. In *Plasmalemma and Tonoplast: their Functions in the Plant Cell,* ed. D. Marmé, E. Marrè & R. Hertel, pp. 377–82. Amsterdam: Elsevier.

Pesci, P., Cocucci, S. M. & Randazzo, G. (1979). Characterization of fusicoccin binding to receptor sites on cell membranes of maize coleoptile tissues. *Plant Cell Environm.* **2:** 205–9

Pesci, P., Tognoli, L., Beffagna, N. & Marrè, E. (1979). Solubilisation and partial purification of a fusicoccin-receptor complex from maize microsomes *Plant Sci. Lett.* **15:** 313–22.

Pitman, M. G., Anderson, W. P. & Schaefer, N. (1977). H^+ ion transport in plant roots. In *Regulation of Cell Membrane Activities in Plants,* ed. E. Marrè & O. Ciferri, pp. 147–60. Amsterdam: Elsevier.

Poole, R. J. (1978). Energy coupling for membrane transport. *Ann. Rev. Plant. Physiol.* **29:** 437–60.

Posthumus, A. C. (1967). Crown-gall en Indolazijnzuur. Ph.D. thesis, University of Leiden.

Poulsen, H. S. (1981). Oestrogen receptor assay – limitation of the method. *Eur. J. Cancer* **17:** 495–501.

Ralet, P. G. F. A. H. & Brombacher, P. J. (1981). The use of coated charcoal in the determination of oestrogen receptor activity. *Eur. J. Nucl. Med.* **6:** 159–62.

Ray, P. M. (1977). Auxin-binding sites of maize coleoptiles are localized on membranes of the endoplasmic reticulum. *Plant Physiol.* **59:** 594–9.

Ray, P. M., Dohrmann, U. & Hertel, R. (1977a). Characterization of naphthaleneacetic acid binding to receptor sites on cellular membranes of maize coleoptile tissue. *Plant Physiol.* **59:** 357–64.

Ray, P. M., Dohrmann, U. & Hertel, R. (1977b). Specificity of auxin-binding sites on maize coleoptile membranes as possible receptor sites for auxin action. *Plant Physiol.* **60:** 585–91.

Rayle, D. L. & Cleland, R. E. (1970). Enhancement of wall loosening and elongation by acid solutions. *Plant Physiol.* **46:** 250–3.

Roy, P. & Biswas, B. B. (1977). A receptor protein for indoleacetic acid from plant chromatin and its role in transcription. *Biochem. Biophys. Res. Comm.* **74:** 1597–606.

Rubery, P. H. (1981). Auxin receptors. *Ann. Rev. Plant Physiol.* **32:** 569–96.

Rubinstein, B. & Cleland, R. E. (1981). Responses of *Avena* coleoptiles to suboptimal fusicoccin: kinetics and comparisons with indoleacetic acid. *Plant Physiol.* **68:** 543–7.

Sachs, T. & Thimann, K. V. (1964). Release of lateral buds from apical dominance. *Nature* **201:** 939–40.

Sakai, S. & Hanagata, T. (1983). Purification of an auxin-binding protein from etiolated mung bean seedlings by affinity chromatography. *Plant Cell Physiol.* **24:** 685–93.

Sakai, S. (1984). Characterization of 2,4-D binding to the auxin-binding protein purified from etiolated mung bean seedlings. *Agric. Biol. Chem.* **48:** 257–9.

Scatchard, G. (1949). The attraction of proteins for small molecules and ions. *Ann. N.Y. Acad. Sci.* **51:** 660–72.

Scherer, G. F. E. (1984a). H^+-ATPase and auxin-stimulated ATPase in membrane fractions from zucchini (*Cucurbita pepo* L.) and Pumpkin (*Cucurbita maxima* L.) hypocotyls. *Z. Pflanzenphysiol.* **114:** 233–7.

Scherer, G. F. E. (1984b). Stimulation of ATPase activity by auxin is dependent on ATP concentration. *Planta* **161:** 394–7.

Scherer, G. F. E. & Morré, D. J. (1978). *In vitro* stimulation by 2,4-dichlorophenoxy-acetic acid of an ATPase and inhibition of phosphatidate phosphatase of plant membranes. *Biochem. Biophys. Res. Comm.* **84:** 238–47.

Skoog, F. & Miller, C. O. (1957). Chemical regulation of growth and organ formation in plant tissues cultured *in vitro*. *Symp. Soc. Exp. Biol.* **11:** 118–31.

Slone, J. M. & Bilderback, D. E. (1983). Membrane-bound auxin-binding sites in Alaska pea stems. *Plant Physiol.* Suppl. **72:** 662.

Spanswick, R. M. (1981). Electrogenic ion pumps. *Ann. Rev. Plant Physiol.* **32:** 267–89.

Steggles, A. W. & King, R. J. B. (1970). The use of protamine to study [6,7-³H]oestradiol-17 β binding in rat uterus. *Biochem. J.* **118**: 695–701.

Stoddart, J. L. & Venis, M. A. (1980). Molecular and subcellular aspects of hormone action. In *Encyclopedia of Plant Physiology*, vol. 9, *Hormonal Regulation of Development I*, ed. J. MacMillan, pp. 445–510. Berlin: Springer-Verlag.

Stout, R. G. & Cleland, R. E. (1980). Partial characterization of fusicoccin binding to receptor sites on oat root membranes. *Plant Physiol.* **66**: 353–9.

Stout, R. G. & Cleland, R. E. (1982). MgATP-generated electrochemical proton gradient in oat root membrane vesicles. In *Plasmalemma and Tonoplast: their Functions in the Plant Cell*, ed. D. Marmé, E. Marrè & R. Hertel, pp. 401–7. Amsterdam: Elsevier.

Sze, H. (1981). Nigericin-stimulated ATPase activity in microsomal vesicles of tobacco callus. *Proc. Natl. Acad. Sci. USA* **77**: 5904–8.

Sze, H. & Hodges, T. K. (1977). Selectivity of alkali cation influx across the plasmamembrane of oat roots. *Plant Physiol.* **59**: 641–6.

Tappeser, B., Wellnitz, D. & Klämbt, D. (1980). Auxin affinity proteins prepared by affinity chromatography. *Z. Pflanzenphysiol.* **101**: 295–302.

Theologis, A. & Ray, P. M. (1982). Early auxin-regulated polyadenylylated mRNA sequences in pea stem tissue. *Proc. Natl. Acad. Sci. USA* **79**: 418–21.

Thompson, M., Krull, U. J. & Venis, M. A. (1983). A chemoreceptive bilayer lipid membrane based on an auxin-receptor ATPase electrogenic pump. *Biochem. Biophys. Res. Comm.* **110**: 300–4.

Tognoli, L., Beffagna, N., Pesci, P. & Marrè, E. (1979). On the relationship between ATPase and fusicoccin binding capacity of crude and partially-purified microsomal preparations from maize coleoptiles. *Plant Sci. Lett.* **16**: 1–14.

Tognoli, L. & Marrè, E. (1981). Purification and characterization of a divalent cation-activated ATP-ADPase from pea stem microsomes. *Biochim. Biophys. Acta* **642**: 1–14.

Trewavas, A. J. (1980). An auxin induces the appearance of auxin binding activity in artichoke tubers. *Phytochem.* **19**: 1303–8.

Trewavas, A. J. (1981). How do plant growth substances work? *Plant Cell Environm.* **4**: 203–28.

Trewavas, A. J. (1982). Growth substance sensitivity: the limiting factor in plant development. *Physiol. Plant.* **55**: 60–72.

Vanderhoef, L. N. (1979). Auxin-regulated elongation: a summary hypothesis. In *Plant Growth Substances*, Proc. 10th int. conf. plant growth substances, Madison, ed. F. Skoog, pp. 90–6. Berlin, Heidelberg, New York: Springer-Verlag.

Van der Linde, P. C. G., Bouman, H., Mennes, A. M. & Libbenga, K. R. (1984). A soluble auxin-binding protein from cultured tobacco tissue stimulates RNA synthesis *in vitro*. *Planta* **160**: 102–6.

Van der Linde, P. C. G., Maan, A. C., Mennes, A. M. & Libbenga, K. R (1985). Auxin receptors in tobacco. In *Proc. 16th FEBS meeting at Moskow*, ed. Y. A. Ovchinnikov, Part C, pp. 397–403. VNU Science Press.

Van Haastert, P. J. M. (1980). Distinction between the rate theory of signal transduction by receptor activation. *Neth. J. Zoöl.* **30**: 473–93.

Venis, M. A. (1977*a*). Receptors for plant hormones. *Adv. Bot. Res.* **5**: 53–88.

Venis, M. A. (1977*b*). Solubilization and partial purification of auxin-binding sites of corn membranes. *Nature* **266**: 268–9.

Venis, M. A. (1979). Purification and properties of membrane-bound auxin receptors in corn. In *Plant Growth Substances*, Proc. 10th int. conf. plant growth substances, Madison, ed. F. Skoog, pp. 61–70. Berlin, Heidelberg, New York: Springer-Verlag.

Venis, M. A. (1981). Cellular recognition of plant growth regulators. In *Aspects and Prospects of Plant Growth Regulators*, Monograph 6, ed. B. Jeffcoat, pp. 187–95. Wantage, Oxfordshire: Wessex Press.

Venis, M. A. & Watson, P. J. (1978). Naturally occurring modifiers of auxin-receptor interactions in corn: identification as benzoxazolinones. *Planta* **142:** 103–7.

Villalobo, A., Dufour, J. P. & Goffeau, A. (1982). Electrogenic proton transport in proteoliposomes of yeast plasmamembrane ATPase. In *Plasmalemma and Tonoplast: their Functions in the Plant Cell*, ed. D. Marmé, E. Marrè & R. Hertel, pp. 389–94. Amsterdam: Elsevier.

Vreugdenhil, D. (1981). Investigations on a membrane-bound auxin receptor. Ph.D. thesis, University of Leiden.

Vreugdenhil, D., Burgers, A., Harkes, P. A. A. & Libbenga, K. R. (1981). Modulation of the number of membrane-bound auxin-binding sites during the growth of batch-cultured tobacco cells. *Planta* **152:** 415–19.

Vreugdenhil, D., Burgers, A. & Libbenga, K. R. (1979). A particle-bound auxin receptor from tobacco pith callus. *Plant Sci. Lett.* **16:** 115–21.

Vreugdenhil, D., Harkes, P. A. A. & Libbenga, K. R. (1980). Auxin-binding by particulate fractions from tobacco leaf protoplasts. *Planta* **150:** 9–12.

Walton, J. D. & Ray, P. M. (1981). Evidence for receptor function of auxin-binding sites in maize. *Plant Physiol.* **68:** 1334–8.

Wardrop, A. J. & Polya, G. M. (1977). Properties of a soluble auxin-binding protein from dwarf bean seedlings. *Plant Sci. Lett.* **8:** 155–63.

Wardrop, A. J. & Polya, G. M. (1980a). Co-purification of pea and bean leaf soluble auxin-binding proteins with ribulose-1,5-biphosphate carboxylase. *Plant Physiol.* **66:** 105–11.

Wardrop, A. J. & Polya, G. M. (1980b). Ligand specificity of bean leaf soluble auxin-binding protein. *Plant Physiol.* **66:** 112–18.

Williamson, F. A., Morré, D. J. & Hess, K. (1977). Auxin binding activities of subcellular fractions from soybean hypocotyls. *Cytobiologie* **16:** 63–71.

Zeroni, M. & Hall, M. A. (1980). Molecular effects of hormone treatment on tissue. In *Encyclopedia of Plant Physiology*, vol. 9, *Hormonal Regulation of Development I*, ed. J. MacMillan, pp. 511–86. Berlin: Springer-Verlag.

Zurfluh, L. L. & Guilfoyle, T. J. (1980). Auxin-induced changes in the patterns of protein synthesis in soybean hypocotyl. *Proc. Natl. Acad. Sci. USA* **77:** 357–61.

Zurfluh, L. L. & Guilfoyle, T. J. (1982). Auxin-induced changes in the population of translatable messenger RNA in elongating sections of soybean hypocotyl. *Plant Physiol.* **69:** 332–7.

Appendix: Abbreviations

ASP	ammonium sulphate precipitation
BA	benzoic acid
BSA	bovine serum albumin
c-IRP	receptor protein for indoleacetic acid, located on the chromatin
CM	crude membrane
DCC	dextran-coated charcoal
DCCD	N-N'-dicyclohexylcarbodiimide
DES	diethylstilbestrol
ED	equilibrium dialysis
ER	endoplasmic reticulum
FC	fusicoccin
g f wt	gram fresh weight
IAA	3-indoleacetic acid
IBA	3-indolebutyric acid
IPA	3-indole-3'-propionic acid

k_{off}	dissociation rate constant
MCPA	2-methyl-4-chlorophenoxyacetic acid
n-IRP	receptor protein for indoleacetic acid, located in the nucleoplasm
NAA	naphthaleneacetic acid
NPA	*N*-1-naphthylphthalamic acid
PAA	phenylacetic acid
PCIB	*p*-chlorophenoxyisobutyric acid
PM	plasma membrane
PMSF	phenylmethylsulfonylfluoride
RuBPCase	ribulose-1,5-biphosphate carboxylase
SF	supernatant factor
TIBA	2,3,5-triiodobenzoic acid
TP	tonoplast
1-NAA	1-naphthylacetic acid (NAA)
2-OH-3,5-DIBA	2-hydroxy-3,5-diiodobenzoic acid
2,3-D	2,3-dichlorophenoxyacetic acid
2,4-D	2,4-dichlorophenoxyacetic acid
2,6-D	2,6-dichlorophenoxyacetic acid
2-Cl-PA	2-chlorophenoxyacetic acid
2-NAA	2-naphthylacetic acid
2-NOA	2-naphthoxyacetic acid
4-Cl-PA	4-chlorophenoxyacetic acid

2

Ethylene receptors

M. A. Hall

2.1   Introduction

The growth-regulating properties of ethylene were first documented over a century ago (Girardin, 1864) although ethylene effects
have been utilized, albeit unwittingly, since before the birth of Christ
(see e.g. Abeles, 1973). It is now clear that, in common with other
endogenous growth regulators, ethylene effects a wide range of developmental responses in plants e.g. abscission, fruit ripening, seed germination, extension growth and root initiation. Since the advent of gas
chromatography, which greatly simplified the measurement of ethylene,
the numbers of papers dealing with all aspects of ethylene physiology
have become legion. Nevertheless, whilst it can be said that such work
has succeeded in describing adequately the effects of ethylene at most
levels of organization, very little has emerged concerning the primary
site of action of the growth regulator. That some progress has been
made in the last few years owes rather more to serendipity than to any
other factor. The purpose of this article is to describe the work to date
and to attempt to place it in some perspective. The treatment is necessarily somewhat subjective since the properties of only two ethylene-
binding sites have been published in any detail.

2.2   Early studies on ethylene binding

It has always been assumed that the concentration of ethylene
in the intercellular spaces of plant tissues reflects the concentration of
the growth regulator within the cell and that the two are related via
the partition coefficient of ethylene between air and cell contents.
Further, since the cell contents are composed largely of water, it was
thought that the partition coefficient would approximate closely to that
between ethylene and water (about 0.1 at physiological temperatures).

Early studies had confirmed that this was indeed the case for animal tissues (see Jerie, Zeroni & Hall, 1978, for references).

In connection with studies on the ethylene status of plant tissues we devised a method for measuring the partition coefficent of ethylene from air into living material and applied this technique to a range of species (Jerie, Shaari, Zeroni & Hall, 1978). Much to our surprise the values obtained with plants in many cases greatly exceeded 0.1. Even allowing for the relatively crude nature of the technique, some of the figures obtained for the partition coefficient were much too large to be accounted for by, say, solution of ethylene in various lipophilic fractions of the cell.

We explored this matter in more detail with developing cotyledons of *Phaseolus vulgaris* (Jerie, Shaari & Hall, 1979). This tissue had a partition coefficient of around 2.5 which fell to around 0.05 if the cotyledons were steam-killed prior to assay. About 98% of the radioactive ethylene accumulated by the tissue was recoverable as such and thus the effect could not be attributed to metabolism. Once taken up by the tissue the ethylene was only slowly released into an air stream $(1-10\% \ h^{-1})$, although heating caused immediate release of virtually all the ethylene held in the tissue. Analysis of the uptake of ethylene indicated that the cotyledons had a high affinity for the gas and remarkably, the abilities of the ethylene analogues propylene and vinyl chloride to inhibit ethylene uptake competitively were closely related to their relative physiological effectiveness in developmental systems.

If tissue which had been prelabelled with [14C]ethylene was macerated and a cell-free extract prepared it was found that nearly all the radioactivity remained in the 1000 g supernatant after centrifugation; on the other hand more than 98% of the radioactivity pelleted at 65 000 g. The radioactivity from the 1000 g supernatant showed a discontinuous distribution when centrifuged in various sucrose gradients. Moreover, extracts prepared in a similar way but from cotyledons not treated with [14C]ethylene had the capacity of binding [14C]ethylene. A comparison of the results from these experiments (Table 2.1) indicated that the binding activity in the cell-free extracts represents the system responsible for the *in vivo* incorporation of ethylene, since the distribution of radioactivity in the gradient was essentially the same in both treatments (Bengochea *et al.*, 1980*b*).

It seemed, therefore, that here was an ethylene-binding preparation analogous to those already isolated for auxins (see e.g. Batt, Wilkins & Venis, 1976). At about the same time Sisler & Filka (1979) demonstrated the presence of similar systems in tobacco and more recently

Table 2.1 Comparison of the distribution of ethylene-binding activity on continuous and discontinuous sucrose gradients

Fraction number (1–ml fractions from top of gradient)	30% frozen/thawed sucrose gradient [^{14}C]C$_2$H$_4$ bound (Bq)			Discontinuous sucrose gradient [^{14}C]C$_2$H$_4$ bound (Bq)		
	Pretreated cotyledons	Untreated cotyledons	Protein[a] (mg)	Pretreated cotyledons	Untreated cotyledons	Protein[a] (mg)
1	12.2	10.3	18	6.4	6.3	3
2	44.2	25.3	11	7.3	4.8	9
3	63.2	46.4	12	8.3	4.5	12
4	61.3	68.1	15	7.8	6.4	8
5	53.7	47.6	16	7	11.8	3
6	33.7	43.9	8	10.1	15.9	2
7	21.7	30.1	8	12.2	16.8	2
8	13.3	22.1	2	13.5	19.8	1
9	10.2	14.9	2	15.2	25.5	0.5
10	8.3	11.4	1	13	28.1	0.5
11	9.5	11.3	2	25.7	30.2	2
12	13	8.4	1	23.6	39.6	0.5
13	5.3	5.3	1	13.6	29.4	2
14	5	5	4	5.6	14.1	4
15				4.9	11.7	0.5
16				4.4	10.7	0.5
Pellet	56	52.1	12			
Sample prior to 1st centrifugation (1 ml)		38.5	20			

(30% frozen/thawed: Pretreated fractions 62%[b]; Untreated fractions 65%[b]. Discontinuous: Pretreated fractions 58%[c]; Untreated fractions 63%[c].)

Samples were derived either from cotyledons pretreated for 2.5 h with 50 Bq [^{14}C]C$_2$H$_4$ ml^{-1} air or from cotyledons not so treated. Samples of the gradients derived from untreated cotyledons were exposed to 50 Bq [^{14}C]C$_2$H$_4$ ml^{-1} air for 5 h at 25°C and counted. [a] Untreated cotyledons only; [b] fraction subsequently layered on discontinuous gradient; [c] fraction used as binding preparation.
(From: Bengochea et al. (1980) Planta 148: 397–406.)

in *Phaseolus aureus* (Sisler, 1980). The findings were rather unexpected since most workers, including ourselves, had felt that ethylene-binding sites were unlikely to prove easy to detect for practical reasons.

Two factors permit the determination of ethylene binding in *Phaseolus vulgaris* and other species shown to exhibit the phenomenon. Thus, the rate constant of dissociation of the ethylene-binding site complex is very low so that not only can the unbound ligand be removed readily by flushing with air but also the complex can be manipulated over an extended period without significant loss of activity (see e.g. Table 2.1). Moreover, it is possible to keep the concentration of ligand in contact with the binding site constant. Assays were carried out in sealed vessels with a large gas volume and a small liquid volume containing the binding preparations. Thus, as ethylene in solution becomes bound, more enters the liquid by partitioning from the gaseous phase and because the partition coefficient of the binding medium is low, the concentration in solution changes little during equilibration. If the system behaved as auxin-binding sites do, namely if the ethylene-binding site had a high rate constant of dissociation, it would be impossible to remove the free ligand by flushing with air without also dissociating most of the bound ligand. Equally, if it were not possible to maintain the concentration of ligand in solution constant then the radioactivity bound would be barely detectable.

2.3   Characterization of the ethylene-binding sites in *Phaseolus vulgaris*

Using the cell-free system described above we were able to elucidate the properties of the ethylene-binding site. Incubation of the site with increasing concentrations of free ethylene in the presence of a constant concentration of radioactive ligand indicated that binding was saturable (Fig. 2.1) and Scatchard analysis of these and other data indicated the presence of a single binding site with a K_D in the region of 6×10^{-10} M (Fig. 2.2). Almost all the radioactivity bound in these experiments could be recovered as ethylene thus excluding the possibility that metabolism was responsible for the effect.

Chang, Jacobs & Cuatrecasas (1975) have pointed out that where the concentration of receptor sites is not at least ten times lower than the true dissociation constant then the apparent dissociation constant obtained from Scatchard plots in fact varies as a linear function of receptor concentration according to the equation:

$$[H_t]_\frac{1}{2} = K_D + [R_t]/2 \qquad (1)$$

where $[H_t]_{\frac{1}{2}}$ is the apparent dissociation constant and $[R_t]$ the concentration of binding sites.

In our system the concentration of binding sites (Fig. 2.2) would be of the order of 4.82×10^{-9} M and the apparent K_D 6.4×10^{-10} M. Accordingly, we measured the apparent dissociation constant at a range of binding site concentrations and obtained the plot shown in Fig. 2.3. Where $[R_t]/2 = 0$, i.e. the ordinate intercept, $[H_t]_{\frac{1}{2}} = K_D$. From the analysis we obtained a true k_D for the system of 0.88×10^{-10} M. This observation has some interesting consequences. If the binding site is indeed a functional receptor then the concentration of sites in the cell is itself going to determine what concentration of hormone will yield half-saturation of these sites. This might have a number of implications. For example, the situation could arise where the amount of hormone–receptor complex could vary independently of hormone concentration. Equally, the fact that there exist a number of different processes in plants with different concentration dependencies on ethylene might not imply different types of receptor protein but rather the same molecule present at different concentration.

Fig. 2.1. Effect of unlabelled ethylene on binding of $[^{14}C]C_2H_4$; 16.7 Bq $[^{14}C]C_2H_4$ ml^{-1} air were present in all incubations.
1 pmol $[^{14}C]C_2H_4 = 4.4$ Bq. Samples were incubated for 24 h at 25°C. Results are representative of three experiments. (From: Bengochea *et al.* (1980) *Planta* **148**: 397–406.)

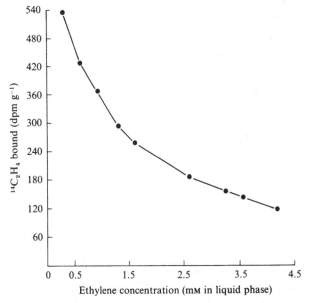

Using the same cell-free preparation we investigated the rate constants of association (k_1) and dissociation (k_{-1}) of ethylene with and from the binding site. These experiments yielded values for k_1 of $3.18 \pm 1.6 \times 10^4 \, \text{M}^{-1}\text{s}^{-1}$ and for k_{-1} of $1.38 \pm 0.14 \times 10^{-5} \text{s}^{-1}$ at 25°C. This in turn gave a value for $K_\text{D}((k_{-1})/(k_1))$ of 5.8×10^{-10} M which was in good agreement with the value of 6.4×10^{-10} M obtained by Scatchard analysis (Fig. 2.3). As mentioned above, these rate constants are very low in relation to findings with other plant hormones; although precise values have not been measured in such cases, examination of the data indicates that they are several orders of magnitude higher. Nevertheless, such values

Fig. 2.2. Scatchard plot of data obtained by incubating binding preparation with a range of concentrations of $[^{14}\text{C}]\text{C}_2\text{H}_4$ from 4×10^{-10} M to 5.4×10^{-9} M (in the liquid phase). Incubations were carried out for 24 h at 25°C. Results are representative of ten experiments. (From Bengochea *et al*. (1980) *Planta* **148**: 397–406.)

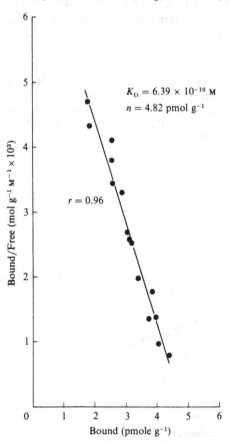

as those obtained with ethylene are not unprecedented and similar figures have been obtained with animal hormones (Cheng, 1975; Best-Belpomme, Mester, Weintraub & Baulieu, 1975).

The resolution of these values provided an explanation for the results described earlier which showed that ethylene bound *in vivo* remained attached to the binding site even during prolonged purification. In the same way, observations that prolonged flushing of binding-site preparations with nitrogen or air led to increased apparent numbers of binding sites can be explained by assuming that endogenous ethylene attached to the binding site would be removed in this fashion. Another interesting finding was that freshly prepared binding-site preparations were inactivated more slowly by heat than preparations which had been flushed with air, indicating that ethylene already bound to the site renders it more stable.

In other experiments it was demonstrated that the binding system was heat-labile and sensitive – at least in part – to proteolytic enzymes. The system has a broad pH optimum between 7.5 and 9.5. A range of cations and other growth regulators were without significant effect at physiological concentrations.

Dithiothreitol, dithioerythritol and mercaptoethanol all inhibit ethylene binding as does *p*-chloromercuribenzoate; the effect of the thiols

Fig. 2.3. Effect of binding-site concentration upon the apparent dissociation constant for ethylene. Each point is derived from a Scatchard analysis of ethylene binding at different dilutions of binding site. (From Bengochea *et al.* (1980) *Plant* **148**: 397–406.)

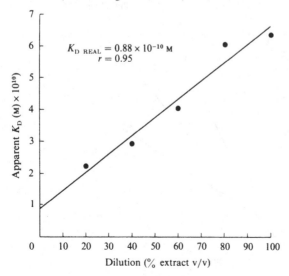

$K_{D \text{ REAL}} = 0.88 \times 10^{-10}$ M
$r = 0.95$

Apparent K_D (M) $\times 10^{10}$

Dilution (% extract v/v)

can be partially reversed by diamide. In these respects the binding site resembles that for auxin in maize (Ray, Dohrmann & Hertel, 1977; Venis, 1977). When the effect of a range of structural analogues of ethylene on binding was investigated (Bengochea *et al.*,1980*a*) it was found that alkanes were without effect. On the other hand a wide range of alkenes and alkynes inhibited binding. Without exception the inhibition was found to be competitive and results of such experiments are shown in Fig. 2.4. From these data it was possible to calculate a K_i for each of the analogues and the results are shown in Table 2.2. The effectiveness of the analogues in competing with ethylene for the binding site (K_i analogue)/(K_D) ethylene was compared with their relative effectiveness in yielding a half-maximal response in the pea stem extension test (Burg & Burg, 1967). It emerged that the two sets of figures resembled each other fairly closely. Equally, carbon dioxide which often acts as a competitive inhibitor of ethylene effects on plant development, also inhibited binding.

Fig. 2.4. Lineweaver-Burk plots for ethylene binding to cell-free preparations from *Phaseolus vulgaris*. Samples were treated with [^{14}C]C$_2$H$_4$ at a range of concentrations in the presence or absence of propylene or propyne at 25°C overnight. Binding was assayed as usual. (From: Bengochea *et al.* (1980) *Planta* **148**: 407–11.)

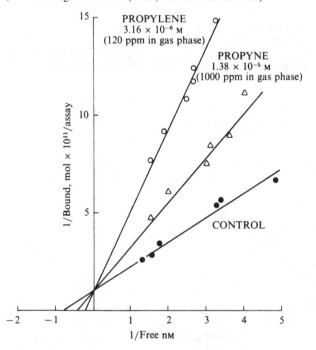

77

Table 2.2 *Comparison of inhibitor constants for structural analogues of ethylene in the* Phaseolus vulgaris *binding-site system*

		K_i(gas) (M)	K_i(liquid) (M)	k_i(gas) (μl l^{-1})	K_i(gas) (relative)	Concentration for half maximal activity on pea growth[a] (μl l^{-1}) (relative)	K_i(liquid) (relative)
Ethylene	$CH_2{=}CH_2$	$8.5\cdot10^{-9}$	$9.00\cdot10^{-10}$	0.18	1	1	1
Propylene	$CH_3CH{=}CH_2$	$1.04\cdot10^{-6}$	$5.60\cdot10^{-7}$	23.3	128	100	600
Vinyl chloride	$CH_2{=}CHCl$	$3.80\cdot10^{-6}$	$1.80\cdot10^{-6}$	85.1	466	1400	2022
Carbon monoxide	CO	$8.70\cdot10^{-6}$	$2.00\cdot10^{-7}$	195	1068	2700	222
Acetylene	$CH{\equiv}CH$	$8.25\cdot10^{-6}$	$1.03\cdot10^{-5}$	185	1013	2800	$11\,444$
Vinyl fluoride	$CH_2{=}CHF$	$9.30\cdot10^{-6}$	$4.37\cdot10^{-6}$	208	1139	4300	$4\,800$
Propyne	$CH_3C{\equiv}CH$	$2.16\cdot10^{-5}$	$6.70\cdot10^{-6}$	484	2651	8000	$7\,444$
Vinyl methyl ether	$CH_2{=}CH{-}O{-}CH_3$	$1.11\cdot10^{-3}$	$1.11\cdot10^{-4}$	$24\,864$	$136\,196$	$100\,000$	$123\,333$
1-Butyne	$CH_3CH_2C{\equiv}CH$	$1.18\cdot10^{-4}$	$6.97\cdot10^{-5}$	4054	$22\,206$	$110\,000$	$77\,427$
1-Butene	$CH_3CH_2CH{=}CH_2$	$4.90\cdot10^{-3}$	$4.90\cdot10^{-4}$	$109\,760$	$601\,227$	$270\,000$	$544\,444$
Vinyl ethyl ether	$CH_2{=}CH{-}O{-}CH_2CH_3$		$9.94\cdot10^{-3}$			$300\,000$	$11\,100\,000$
Carbon dioxide,	CO_2			$25\,760$	$141\,104$	$300\,000$	

Extracts were treated with [^{14}C]C$_2$H$_4$ at a range of concentrations in the presence or absence of the analogue in question at 25°C overnight. Binding was assayed as usual and the K_i for the analogue calculated from Lineweaver-Burk plots of the data. The K_i (liquid) was calculated from the K_i (gas) using the partition coefficient of the analogue in the assay medium.
[a] From Burg & Burg (1967). Reproduced with permission. Inactive at concentrations (μl l^{-1}) in parentheses: methane, ethane, propane, cyclopropane (10^5); cis-2-butene, trans-2-butene (10^4).
(From: Bengochea *et al.* (1980) *Planta* **148**; 407–11.)

Recent work by Sisler and his co-workers indicates that the ethylene-binding site in *Phaseolus aureus* is very similar in these respects to that in *Phaseolus vulgaris* (Sisler, 1982; Beggs & Sisler, 1983).

2.4   Subcellular location of the ethylene-binding site in *Phaseolus vulgaris* L.

The centrifugation studies outlined in Section 2.2 above indicated that the ethylene-binding site in *Phaseolus* is not free in the cytosol and consequently we investigated its subcellular distribution (Evans, Bengochea, Cairns, Dodds & Hall, 1982a). Rate zonal and isopycnic centrifugation coupled with analysis for marker enzymes indicated that the binding site was associated with elements of the endomembrane system and with protein body membranes. That at least part of the activity is associated with endoplasmic reticulum (ER) membranes was strongly indicated by the observation that a band of binding activity located at a median density of $1.175 \, \mathrm{g\,cm^{-3}}$ was significantly reduced by EDTA – a treatment which leads to dissociation of ribosomes from rough ER (Fig. 2.5). Moreover, two putative ER marker enzymes, namely antimycin-insensitive NADH- and NADPH-cytochrome c reductase both showed activity coincident with this band of ethylene-binding activity.

Kinetic analysis of the various bands of activity separated in these studies indicated that the K_D for ethylene in each case was identical and similar to that obtained from unfractionated preparations. Given the rather contentious situation in which the marker enzyme field finds itself we were rather diffident in definitely assigning the location of the binding site to the membranes referred to. We decided, therefore, to attempt localization by means of autoradiography. In this we were helped by a particular facet of the system, namely the low rate constant of dissociation of ethylene from the binding site. Thus, we were able to treat intact cotyledons with [^{14}C]ethylene and after an appropriate period of incubation to remove all unbound ethylene by flushing with air. Excised blocks of tissue were then fixed with osmium tetroxide which, coincidentally, reacts with carbon carbon double bonds thus immobilizing the bound radioactive ethylene in the tissue (Evans, Dodds, Lloyd, ap Gwynn & Hall, 1982b).

Sections from these tissue blocks were prepared for light- and electron-microscope autoradiography. Examples of the results obtained are shown in Figs. 2.6 to 2.8. The autoradiographs were subjected to rigorous statistical analysis and this confirmed the findings in the cell fractionation studies referred to above, namely that the binding activity is located on endoplasmic reticulum and protein body membranes. This is shown clearly by the specific activity values in Table 2.3.

The study was in itself significant since although such work has been performed with other plant hormones, in the main no effort has been made to ensure, as we have in this case, that the radioactivity is still associated with unchanged hormone. Indeed, it seems that given the high rate constants of dissociation of other plant hormones from their binding sites (with the probable exception of those for gibberellins described by Stoddart elsewhere in this volume), localization by the means described above is likely to prove impossible unless metabolism can

Fig. 2.5. Effect of EDTA or Mg^{2+} on binding fractions from rate-zonal centrifugation $2\,cm^3$ of sample (normally applied to isopycnic gradient without further treatment) treated with either $2\,cm^3$ 5 mM $MgCl_2$ (●), or $2\,cm^3$ 10 mM EDTA (△). Sucrose (□)*. (From Evans *et al.* (1982) *Plant, Cell & Environment* **5**; 101–7.)
* All values expressed per cm^3 gradient fraction.

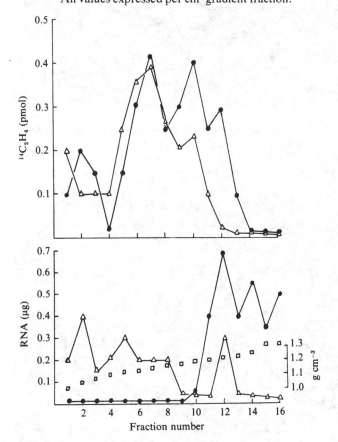

80   *M. A. Hall*

Fig. 2.6. Light-microscope autoradiograph: cotyledons (1 g fresh mass) incubated with 166 Bq cm^{-3} [^{14}C]C$_2$H2094 for 5 h, vented with air (16 dm^3 min^{-1} for 30 s) and incubated with non-radioactive C$_2$H$_4$ (10 nl cm^{-3}) for 15 min. Autoradiograph prepared using Kodak AR10 stripping film (× 2000). Arrows show position of examples of silver grains (S, amyloplast; PB, protein body; N, nucleus; CW, cell wall). (From Evans *et al.* (1982) *Planta* **154**: 48–52.)

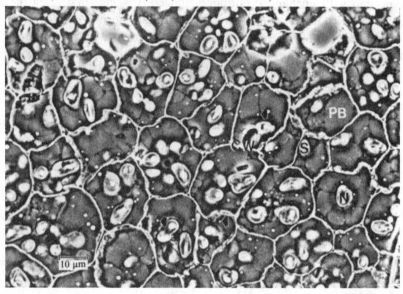

Fig. 2.7. Light-microscope autoradiograph: prepared as described for Fig. 2.6 but without incubation with [^{14}C]C$_2$H$_4$. (× 2000). Symbols as for Fig. 2.6. (From Evans *et al.* (1982) *Planta* **154;** 48–52.)

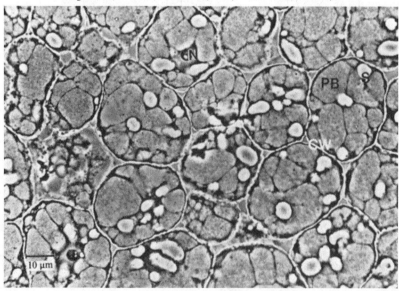

Fig. 2.8. Electron-microscope autoradiograph: cotyledons (1 g fresh mass) treated as described for Fig. 2.6. Specimen fixed in osmium tetroxide and glutaraldehyde and embedded in Spurr's resin. Sections stained with Reynolds' lead citrate (Reynolds, 1963) and 5% (w/v) aqueous uranyl acetate. (× 30000). Arrows show position of examples of developed silver grains (ER, endoplasmic reticulum; M, mitochondrion; PB, protein body). (From Evans *et al.*, (1982) *Planta* **154:** 48–52.)

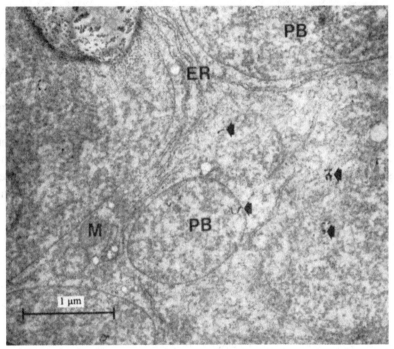

Table 2.3. *Electron-microscope autoradiography: specific activity values*

Organelle	Specific acitivity[a]
Endoplasmic reticulum	3.4
Protein body membrane	1.5
Protein body contents	0.6
Cytoplasm	0.9
Cell wall/plasma membrane	0.4
Mitochondria	0.7
Chloroplasts and nuclei	0.1
Bare plastic/other	0.5
Starch/lipid bodies	0.3

[a] Specific activity

$$= \frac{\text{frequency of silver grains over organelle}}{\text{area of organelle}}$$

(From: Evans, D. E., Dodds, J. H., Lloyd, P. C., ap Gwynn, I. & Hall, M. A. (1982) *Plants* **154;** 48–52.)

be prevented and some means found of 'fixing' the hormones at their binding sites. Very recently, Hornberg & Weiler (1984) have made an elegant demonstration of abscissic acid-binding sites in stomata from *Vicia faba* but as this relied on photoactivation of the ligand *in situ* it is unlikely that similar breakthroughs will be possible with the other natural growth regulators.

2.5   Studies on solubilized ethylene-binding proteins

Considerable progress has now been made on the solubilization and purification of the ethylene-binding proteins in *Phaseolus vulgaris* and *Phaseolus aureus* (Thomas, Smith & Hall, 1984; Sisler, 1984). As noted above both these proteins are membrane-bound and can be released only by treatment with detergents. The binding-site proteins precipitate if the detergent is removed indicating that they are hydrophobic, integral membrane proteins. We have routinely used Triton X-100, Triton X-114 and octyl glucoside and have succeeded in obtaining solubilized preparations capable of binding ethylene. The properties of the solubilized binding site are not markedly dissimilar to those of the

Fig. 2.9. Scatchard plot of ethylene binding by a preparation solubilized from membranes of *Phaseolus vulgaris* with Triton X-100. (From: Unpublished work of C. J. R. Thomas, & M. A. Hall.)

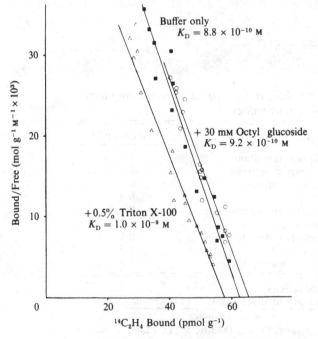

Buffer only
$K_D = 8.8 \times 10^{-10}$ M

+ 30 mM Octyl glucoside
$K_D = 9.2 \times 10^{-10}$ M

+0.5% Triton X-100
$K_D = 1.0 \times 10^{-9}$ M

Bound/Free (mol g^{-1} M^{-1} × 10^3)

$^{14}C_2H_4$ Bound (pmol g^{-1})

membrane-bound site, the only major differences being a downward shift in pH sensitivity and a decrease in the rate constants of association and dissociation. The affinity and specificity of the site for ethylene and its analogues are unchanged (Fig. 2.9). Determination of the Stoke's radius and sedimentation coefficient of the binding site–detergent complex indicates that the molecule is highly asymmetric (Stoke's radius $= 6.3 \times 10^{-9}$ nm; $(S_w^{20} = 2.01 \times 10^{-13})$ – and has a molecular weight of the order of 50000 daltons although that for the protein alone will be significantly less depending on the number of detergent molecules per molecule of protein (C. J. R. Thomas, A. R. Smith and M. A. Hall, unpublished). We have now achieved considerable purification of the ethylene-binding protein by a combination of gel-permeation chromatography on Sephacryl coupled with detergent partitioning, pH precipitation and fast protein liquid chromatography (Hall, Smith, Thomas & Howarth, 1984; Hall, 1984).

2.6   Is the ethylene-binding site in *Phaseolus vulgaris* a receptor?

A problem which arises with all the plant hormone-binding systems isolated hitherto is that only rarely has a specific mode of action for a plant hormone at the subcellular or biochemical level been elucidated. Thus, in contrast to the situation existing with animal hormones and their receptors, where it is often possible to relate the interaction of the hormone with its receptor and the initiation of a specific biochemical event, in plants it is necessary to relate the observed physical parameters of the binding system to the properties expected of the binding system constituting a functional receptor. In respect of the two principal criteria, namely affinity and specificity, the binding sites for ethylene in *Phaseolus vulgaris* and *Phaseolus aureus* emerge as likely candidates as functional receptors. Thus the K_D of 6.4×10^{-10} M in the aqueous phase corresponds to a concentration of about $0.14 \, \mu l \, l^{-1}$ in the gaseous phase. The concentrations of ethylene in the gaseous phase which usually yield threshold, half-maximal and saturation responses in developmental systems are 0.01, 0.1 and $1.0 \, \mu l \, l^{-1}$. Thus the value of $0.14 \, \mu l \, l^{-1}$ which yields half-saturation of the binding site corresponds well to the expected value of $0.1 \, \mu l \, l^{-1}$.

Equally the K_i values for structural analogues closely parallel the concentrations necessary to provide half-maximal effects in the pea-growth test (Burg & Burg, 1967) and other developmental systems (see Abeles, 1973). Moreover, the alkanes which are without effect on developmental phenomena are equally ineffective in inhibiting binding of ethylene to the site.

The rate constants of association and dissociation are not consistent with those expected for an ethylene receptor in some systems such as root growth where the effects of ethylene are observable shortly after application of the growth regulator and are reversed equally rapidly when it is removed (Hall, Kapuya, Sivakumaran & John, 1977). On the other hand, in many other systems the effect of ethylene takes some time to develop and is not rapidly reversible.

The subcellular location of the binding site is also consistent with many of the known effects of ethylene at the biochemical level, where the secretion and/or synthesis of proteins is affected – for example in leaf abscission (e.g. Morré, 1968).

Nevertheless, it must be admitted that until and unless the ethylene-binding activity in *Phaseolus vulgaris* can be shown to be associated with the initiation of a specific biochemical response then the binding site cannot be accurately described as a functional receptor in the true sense. The situation is exacerbated in the system described since there is no known function for ethylene in cotyledon development. On the other hand we are investigating two systems where at least some of the primary events initiated by ethylene *are* well documented, namely abscission zones and ripening fruit. Unfortunately, the concentration of binding sites in such tissues appears to be very low, rendering rigorous characterization impossible using currently available methods. Only by much increasing the sensitivity of our methods of detection can such sites be investigated and to this end we are preparing antibodies to puri-fied ethylene-binding protein from *Phaseolus vulgaris* with the intention of developing immunoassay techniques for use both with abscission zones of *Phaseolus vulgaris* and, if there is sufficient interspecific similarity between the active sites of ethylene-binding proteins, to fruit tissues in other species.

2.7   The relationship between ethylene-binding and ethylene-metabolizing systems

Until quite recently there was no rigorous evidence that higher plants can metabolize ethylene. This was in contrast to the other endo-genous plant hormones – for all of which metabolizing systems are known to exist – but was not thought incongruous since ethylene diffuses rapidly into and out of plant tissue (see e.g. Hall, 1977) and thus there appeared to be an alternative means of controlling endogenous ethylene concentra-tions. Many of the early reports which indicated that plants could meta-bolize ethylene (e.g. Jansen, 1969) were later found to be inaccurate and the incorporation of radioactivity shown to be due to the presence of radioactive impurities in the [^{14}C]ethylene used.

Nevertheless, Beyer (1975a) using rigorously purified substrate was able to demonstrate incorporation of radioactivity from ethylene into *Pisum sativum* tissue, albeit at very low rates when the ethylene was applied at physiological concentrations. The metabolizing system leads to the incorporation of ethylene into water-soluble tissue metabolites (tissue incorporation, TI) and the oxidation of ethylene to CO_2 (OX). Later work (Blomstrom & Beyer, 1980) demonstrated that the first identifiable product was ethylene glycol and its glucose conjugate. Withal, the fact remained that the rates of ethylene metabolism observed were orders of magnitude too low to propose that the system is involved in controlling endogenous ethylene concentrations. However, Beyer and his co-workers were able to demonstrate not only that this type of metabolizing activity was present in other tissues (such as Morning Glory flowers (Beyer & Sundin, 1978), Carnations (Beyer, 1977), and cotton and bean abscission explants (Beyer, 1975b)) but also that changes in rates of ethylene metabolism in tissues were reflected in changes in their sensitivity to the growth regulator. this led him to propose (Beyer, 1975a; Beyer & Blomstrom, 1980) that ethylene metabolism is an integral part of the mode of action of the hormone. This proposal was further supported by the observation (Beyer, 1979) that Ag^+ and CO_2 which both have anti-ethylene properties modify ethylene metabolism in *Pisum* in parallel with their inhibitory effects on ethylene-induced growth retardation. This work also indicated that TI and OX are separate systems since they are affected differentially by Ag^+ and CO_2.

Jerie & Hall (1978) were able to demonstrate ethylene metabolism in developing cotyledons of *Vicia faba*. The first product was ethylene oxide which suggested that the system was similar to that in *Pisum* since ethylene glycol is readily formed non-enzymatically from ethylene oxide. However, in other respects the two systems were very different. Thus, the rate of oxidation of ethylene at physiological concentrations was some three to five orders of magnitude higher in *Vicia* than in *Pisum*. Equally, all of the ethylene metabolized in *Vicia* is converted initially to ethylene oxide and metabolism to CO_2 is insignificant (S. K. Musa & M. A. Hall, unpublished).

It has proved possible to isolate cell-free preparations from *Vicia* capable of metabolizing ethylene (Dodds, Musa, Jerie & Hall, 1979; Dodds, Heslop-Harrison & Hall, 1980). The preparations have a high affinity for ethylene ($K_m = 4.17 \times 10^{-10}$ M) and physiologically active structural analogues of ethylene competitively inhibit oxidation of the hormone in much the same way as binding is inhibited in *Phaseolus vulgaris*. In more recent work (Smith, Venis & Hall, 1985) we have shown that the enzyme is an NADPH-linked monooxygenase and we have obtained rates of oxidation *in vitro* as great as those seen *in vivo*. Further investigations of the system in *Pisum* both by Beyer (1978) and ourselves (Evans, Smith, Taylor & Hall, 1984) indicate that the low rates of ethylene metabolism at physiological concentrations (i.e. $0.01–1.0\,\mu l\,l^{-1}$) in this system are a consequence of its low affinity for the hormone. Thus, using intact *Pisum* seedlings we have measured a K_m for ethylene of 1.64×10^{-6} M ($3640\,\mu l\,l^{-1}$ gas phase) in the TI component. On the other hand the V_{max} for TI in both *Pisum* and *Vicia* are very similar. Moreover, propylene which is metabolized by both *Pisum* and *Vicia* and which competes with ethylene in both systems, gives a K_i of 3.76×10^{-7} M in pea, and 1.8×10^{-6} M in broad bean.

Thus, in all respects other than affinity of ethylene, namely V_{max}, affinity for ethylene analogues and primary product, the TI system in *Pisum* is very similar to that in *Vicia* and by analogy to the binding-site system in *Phaseolus*. Hence in *Pisum* and the other species investigated by Beyer a correlation exists between ethylene metabolism and ethylene-induced responses whereas this is not proven for *Vicia* and *Phaseolus*. On the other hand in terms of affinity for ethylene the systems in *Vicia* and *Phaseolus* are much better candidates for the role of ethylene receptors.

In the light of the evidence it seems to stretch coincidence too far to suppose that these three systems are totally unrelated, especially when their close phylogenetic relationship is taken into consideration, but if they do indeed represent different facets of a common mechanism how can their differences be explained?

The answer to this question is as yet totally unknown and overmuch speculation here would probably be overtaken by events in such a rapidly progressing field long before this book appears. There are, nevertheless, several interesting recent observations which may provide a key to the problem. Beyer & Blomstrom (1980) demonstrated that ethylene oxide synergizes with ethylene in retarding growth in *Pisum* and this type of observation has now been extended to other tissues (Beyer, 1984). Likewise, we have observed a similar phenomenon in relation to the acceleration of leaf abscission in *Vicia*. now, if the ethylene-metabolizing systems are indeed involved in the response to the hormone then clearly if the product of metabolism is also involved – perhaps at the same site, perhaps not – then the reservations raised by the low affinity of the *Pisum* system for ethylene may not be so serious since in such a case the response would not follow simple Michaelis–Menten kinetics.

Equally, while CO_2 at high concentrations inhibits OX in *Pisum* and TI in *Vicia* it can promote TI in *Pisum*. It appears to do so by increasing the affinity of the TI system for ethylene. This might suggest that the CO_2 produced by the OX system acts as some sort of effector for the TI system. These observations have an interesting analogy in one developmental system affected by ethylene, namely the breaking of dormancy in *Spergula* seed. Here, low concentrations of CO_2 (atmospheric or somewhat below) are necessary in order for the ethylene to exert its effect (Jones & Hall, 1979). Such proposals still leave many questions unanswered, in particular why, if metabolism is related to mode of action, does high-affinity binding but no metabolism occur in *Phaseolus vulgaris*?

The author's view is that on the principle that the simplest solution is the most likely, the similarities between the various systems are more compelling than their differences and that they do indeed, represent different aspects of a common mechanism. A reconciliation of all the facts is, however, likely to be some way off.

2.8   References

Abeles, F. B. (1973). *Ethylene in Plant Biology.* New York: Academic Press.

Batt, S., Wilkins, M. B. & Venis, M. A. (1976). Auxin binding to corn coleoptile membranes: kinetics and specificity. *Planta* **130:** 7–13.

Beggs, M. J. & Sisler, E. C. (1983). Ethylene binding properties of mung bean extracts. *Plant Physiol.* **72:** S226.

Bengochea, T., Acaster, M. A., Dodds, J. H., Evans, D. E., Jerie, P. H. & Hall, M. A. (1980*a*). Studies on ethylene binding by cell free preparations from cotyledons of *Phaseolus vulgaris* L.: II. Effects of structural analogues of ethylene and of inhibitors. *Planta* **148:** 407–11.

Bengochea, T., Dodds, J. H., Evans, D. E., Jerie, P. H., Niepel, B., Shaari, A. R. & Hall, M. A. (1980*b*). Studies on ethylene binding by cell free preparations from cotyledons of *Phaseolus vulgaris* L.: I. Separation and characterisation. *Planta* **148:** 397–406.

Best-Belpomme, M., Mester, J., Weintraub, H. & Baulieu, E. E. (1975). Oestrogen receptors in chick oviduct. Characterisation and subcellular distribution. *Eur. J. Biochem.* **57:** 537–49.

Beyer, E. M. Jr. (1975*a*). [^{14}C]C_2H_4: its incorporation and metabolism by pea seedlings under aseptic conditions. *Plant Physiol.* **56:** 273–8.

Beyer, E. M. Jr. (1975*b*). Ethylene [^{14}C] metabolism during abscission. *Proc. 29th Beltwide Cotton Production Research Conference*, pp. 51–2.

Beyer, E. M. Jr. (1977). [^{14}C]C_2H_4 its incorporation and oxidation to [^{14}C]CO_2 by cut carnations. *Plant Physiol.* **60:** 203–6.

Beyer, E. M. Jr. (1978). Rapid metabolism of propylene by pea seedlings. *Plant Physiol.* **61:** 893–5.

Beyer, E. M. Jr. (1979). Effect of silver ion, carbon dioxide and oxygen on ethylene action and metabolism. *Plant Physiol.* **63:** 169–73.

Beyer, E. M. Jr. (1984). Why do plants metabolise ethylene? In *Ethylene: Biochemical, Physiological and Applied Aspects,* ed. Y. Fuchs & E. Chalutz, pp. 65–74. The Hague: Martinus Nijhoff/Dr W. Junk Publishers.

Beyer, E. M. Jr. & Blomstrom, D. C. (1980). Ethylene metabolism and its possible physiological role in plants. In *Plant Growth Substances 1979,* ed. F. Skoog. Berlin, Heidelberg, New York: Springer-Verlag.

Beyer, E. M. Jr. & Sundin, (1978). [^{14}C]C_2H_4 metabolism in morning glory flowers. *Plant Physiol.* **61:** 896–9.

Bloomstrom, D. C. & Beyer, E. M. Jr. (1980). Plants metabolise ethylene to ethylene glycol. *Nature* **288:** 66–8.

Burg, S. P. & Burg, E. A. (1967). Molecular requirements for the biological activity of ethylene. *Plant Physiol.* **42:** 144–52.

Chang, K-J., Jacobs, S. & Cuatrecasas, P. (1975). Quantitative aspects of hormone receptor interactions of high affinity. Effect of receptor concentration and measurement of dissociation constants of labelled and unlabelled hormones. *Biochim. Biophys. Acta* **406:** 294–303.

Cheng, K-W. (1975). Properties of follicle-stimulating-hormone receptor in cell membranes of bovine testis. *Biochem. J.* **149:** 123–32.

Dodds, J. H., Heslop-Harrison, J. S. & Hall, M. A. (1980). Metabolism of ethylene to ethylene oxide by cell-free preparations from *Vicia faba* L. cotyledons: effects of structural analogues and of inhibitors. *Plant Sci. Lett.* **19:** 175–80.

Dodds, J. H., Musa, S. K., Jerie, P. H. & Hall, M. A. (1979). Metabolism of ethylene to ethylene oxide by cell-free preparations from *Vicia faba* L. *Plant Sci. Lett.* **17:** 109–14.

Evans, D. E., Bengochea, T., Cairns, A. J., Dodds, J. H. & Hall, M. A. (1982a). Studies on ethylene binding by cell free preparations from cotyledons of *Phaseolus vulgaris* L.: subcellular localisation. *Plant, Cell & Env.* **5:** 101–7.

Evans, D. E., Dodds, J. H., Lloyd, P. C., ap Gwynn, I. & Hall, M. A. (1982b). A study of the subcellular localisation of an ethylene binding site in developing cotyledons of *Phaseolus vulgaris* L.: by high resolution autoradiography. *Plants* **54:** 48–52.

Evans, D. E., Smith, A. R., Taylor, J. E. & Hall, M. A. (1984). Ethylene metabolism in *Pisum sativum* L.: kinetic parameters, the effects of propylene, silver and carbon dioxide and comparison with other systems. *Plant Growth Regulation* **2:** 187–96.

Girardin, J. P. L. (1864). Einfluss des Leuchtgases auf die Promenaden- und Strassenbaüme. *Jahresberichter Agrikulturen-Chemie Versuchsstation, Berlin* **7:** 199–201.

Hall, M. A. (1977). Ethylene involvement in senescence processes. *Ann. Appl. Biol.* **85:** 424–8.

Hall, M. A., Kapuya, J. A., Sivakumaran, S. & John, A. (1977). The role of ethylene in the response of plants to stress. *Pest. Sci.* **18:** 217–3.

Hall, M. A. (1985). Studies on the mechanism of action of ethylene. Proceedings of the 16th FEBS Meeting, Moscow 5, Part C, pp. 383–9. VNU Science Press BV.

Hall, M. A., Smith, A. R., Thomas, C. J. R. & Howarth, C. J. (1984). Binding sites for ethylene. In *Ethylene: Biochemical, Physiological and Applied Aspects,* ed. Y. Fuchs & E. Chalutz, pp. 55–64. The Hague: Martinus Nijhoff/Dr W. Junk Publishers.

Hornberg, C. & Weiler, E. W. (1984). High-affinity binding sites for abscisic acid on the plasmalemma of *Vicia faba* guard cells. *Nature* **310:** 321–4.

Jansen, E. F. (1969). Metabolism of ethylene in the avocado. In *Food Science Technology Vol. 3, Proc. 1st Int. Congr. Food Sci. Tech.,* ed. J. M. Leitch, pp. 475–81, New York: Gordon and Breash, Science Publishers Inc.

Jerie, P. H. & Hall, M. A. (1978). The identification of ethylene oxide as a major metabolite of ethylene in *Vicia faba* L. *Proc. Roy. Soc. Lond. Ser. B* **200**: 87–94.

Jerie, P. H., Sharri, A. R. & Hall, M. A. (1979). The compartmentation of ethylene in developing cotyledons of *Phaseolus vulgaris* L. *Planta* **144**: 503–7.

Jerie, P. H., Shaari, A. R., Zeroni, M. & Hall, M. A. (1978). The partition coefficient of $[^{14}C]C_2H_4$ in plant tissue as a screening test for metabolism or compartmentation of ethylene. *New Phytol.* **81**: 499–504.

Jerie, P. H., Zeroni, M. & Hall, M. A. (1978). Movement and distribution of ethylene in plants in relation to the regulation of growth and development. *Pest. Sci.* **9**: 162–8.

Jones, J. F. & Hall, M. A. (1979). Studies on the requirement for CO_2 and ethylene for germination of *Spergula arvensis* seeds. *Plant Sci. Lett.* **16**: 87–93.

Morré, D. J. (1968) Cell wall dissolution and enzyme secretion during leaf abscission. *Plant Physiol.* **43**: 1545–9.

Ray, P. M., Dohrmann, U. & Hertel, R. (1977). Charasterisation of naphthaleneacetic acid binding to receptor sites on cellular membranes of maize coleoptile tissue. *Plant Physiol.* **59**: 357–64.

Reynolds, E. S. (1963). The use of lead citrate at a high pH as an electron-opaque stain in electron microscopy. *J. Cell Biol.* **17**: 208–12.

Sisler, E. C. (1980). Partial purification of an ethylene-binding component from plant tissue. *Plant Physiol.* **66**: 404–6.

Sisler, E. C. (1982). Ethylene binding properties of a Triton X-100 extract of mung bean sprouts. *J. Plant Growth Reg.* **1**, 211–18.

Sisler, E. C. (1984). Distribution and properties of ethylene-binding component from plant tissue. In *Ethylene: Biochemical, Physiological and Applied Aspects*, ed. Y. Fuchs & E. Chalutz, pp. 45–54. The Hague: martinus Nijoff/Dr W. Junk Publishers.

Sisler, E. C. & Filka, M. A. (1979). Partial purification of ethylene binding component from plant tissue. *Plant Physiol.* **63**: S131.

Smith. P. G., Venis, M. A. & Hall, M. A. (1985). Oxidation of ethylene by cotyledon extracts from *Vicia faba* L.: co-factor requirements and kinetics. *Planta* **163**: 97–104.

Thomas, C. J. R., Smith, A. R. & Hall, M. A. (1984). The effect of solubilisation on the character of an ethylene binding site from *Phaseolus vulgaris* L. *Planta* **160**: 474–9.

Venis, M. A. (1977). Receptors for plant hormones. In *Adv. Bot. Res.* **5**: 53–87. Ed. H. W. Woolhouse, London: Academic Press.

3

Gibberellin receptors

John L. Stoddart

3.1   Introduction

In the widest sense, the term 'receptor' can be applied to any cellular site which enters into an association, transient or permanent, with a specific molecule, regardless of the consequences of this event. In practice the definition is restricted to interactions between the relevant compound and sites which can be related to a biological response of the containing cell or tissue. Furthermore, the site is required to differentiate between active and inactive molecular configurations.

Thus, in considering gibberellins (GAs), we will be concerned with various categories of subcellular interaction, either demonstrated or postulated, which can be related to the known growth responses in various plant tissues.

Throughout this chapter the terms 'hormone' and 'growth regulator' are used interchangeably. Strictly, in accordance with animal usage, hormones are compounds which act remotely from their sites of synthesis, and do so in catalytically small amounts. It is arguable whether GAs conform to this definition but designation as 'hormones' is so widespread in the literature that no useful purpose would be served by adherence to semantic distinctions. Suffice it to say that the implied animal analogy can be counterproductive in some situations, and the reasons behind this remark will emerge from the arguments contained in the body of the chapter.

3.2   Characteristics of the gibberellin response

3.2.1   Structural and metabolic considerations

At the time of writing the number of described, naturally occurring gibberellin structures is 66. All are variants of the *ent*-gibberellane skeleton (Fig. 3.1 [1]), with widely varying growth-regulating properties.

The biological implications of the various structural forms have been
examined by Brian, Grove & Mulholland (1967) and by Crozier, Kuo,
Durley & Pharis (1970) and it is evident that the effects of a given substituent can vary widely, depending upon the bioassay system in which
the compound is evaluated. An ionizable carboxyl group at the 7 position
seems to be universally required for high biological potency and this
feature determines the partitioning behaviour at various pH values.
Methylation of the carboxyl group reduces biological activity (Hiraga,
Yamani & Takahashi, 1974).

All highly active GAs possess a 19,10 lactone bridge projected below
the plane of the A-ring which is approximately counterbalanced, in
stereochemical terms, by the reduced D-ring at the opposite end of the
molecule, the latter being in the alpha configuration. Thus, the two
ends of the molecule have a similar topology, which is not materially
altered by rotation through 180°C about C-6. When the D-ring has the
opposite stereochemical configuration, the resultant molecule is devoid
of biological activity (Brian *et al.* 1967) and the same is true of non-
lactone GAs which generally have limited potency in bioassays and may
need to be endogenously converted before acting.

Hydroxylation is an important determinant of biological efficacy. The
highest activities are associated with β-hydroxylation at the 3-position
and the effectiveness is enhanced when the 13-position is also substituted

Fig. 3.1. Structural formulae of gibberellins and glycosyl conjugates:
[*1*] *ent*-gibberellane skeleton; [*2*] gibberellin A₁; [*3*] gibberellin A₁
glucosyl ether; [*4*] gibberellin A₁ glucosyl ester.

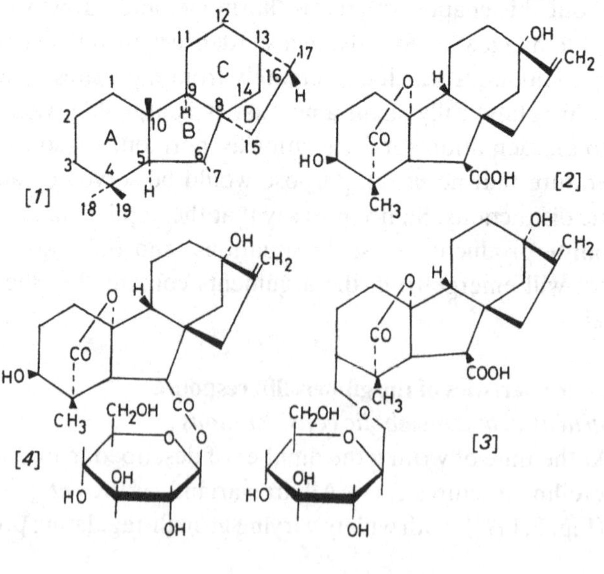

(e.g. structure [2] in Fig. 3.1). Only moderate activity is obtained from a single 13 hydroxylation. Addition of a hydroxyl group at the 2β position has a strong negative effect, exemplified by the complete loss of biological potency when GA_1 ([2] in Fig. 3.1) is 2-hydroxylated to yield GA_8. Sponsel, Hoad & Beeley (1977) have provided an extensive discussion of 2-hydroxylation and note that insertion in the α orientation has a much less drastic effect. A role for the 2-hydroxylating enzyme system in regulating endogenous biological activity has been postulated by Nadeau & Rappaport (1972), Davies & Rappaport (1975) and Frydman & MacMillan (1975). Insertion of a 3-hydroxyl group in the α configuration, yielding a biologically inactive molecule, also inhibits further hydroxylation at the 2 position. Clearly, the stereochemistry of substitution at these two A-ring sites is of vital importance in determining biological effectiveness. If, for example, GA action requires a reversible association of the regulator with its site of action, then a mechanism for continually removing free biologically active species from the action-site compartment would be advantageous. This 'pool-flushing' system would permit rapid responses to environmentally induced changes in the rate of GA synthesis, and the 2-hydroxylating system provides a suitable means for achieving this objective.

Conjugation with other molecules, most frequently sugars, is a common consequence of feeding GAs to plants and the occurrence of GA glycosides has been demonstrated in a range of tissues. Conjugation can occur *via* ether (Fig. 3.1 [3]) or ester (Fig. 3.1 [4]) linkage and the glycosyl portion of endogenous conjugates is, most commonly, glucose (e.g. Sembdner, Gross, Liebisch & Schneider, 1980). It is uncertain whether conjugates with other sugars occur naturally but the question has some interest in connection with later sections of this chapter (section 3.4). The biological function of conjugates is unresolved. There is good evidence to show that many tissues are capable of hydrolysing these compounds to release the GA aglycone (see Sembdner *et al.* 1980) but it seems probable that this process occurs in the vacuole (Ohlrogge, Garcia-Martinez, Adams & Rappaport, 1980) and reabsorption of GAs into the cytoplasm becomes conjectural. In seeds most of the conjugated GAs release biologically inactive species such as GA_8 on hydrolysis but a small proportion of GA_3 glucoside is found in some species, such as *Pharbitis*, and this may be hydrolysed during germination (Barendse & Gilissen, 1977). It is not known whether the released GA_3 contributes materially to the processes of germination.

Notwithstanding the arguments about possible physiological function, it should be noted that the addition of a sugar moeity to the GA skeleton has a fundamental effect upon molecular behaviour, producing a marked tendency to align with the aqueous component at polarity interfaces within the cell.

3.2.2   Compartmentation

Expressions of the cellular or tissue content of a substance, such as a GA, are generally meaningless in the absence of information on subcellular distribution. The total pool may be subdivided into a series of compartments, either physicochemical or ultrastructural, and when relating levels of growth-regulator activity to cellular response it must be established that the mass of compound determined and the primary action site are situated in the same compartment, with an appropriate quantitative relationship.

(1) Ultrastructural compartmentation
Almost all of the available data on this aspect are derived from studies on the fate of exogenously applied labelled GAs and must be interpreted with caution, due to the possibility that such a means of supply may give rise to unusual distribution patterns.

As an example of a typical subcellular distribution resulting from the exogenous supply of a growth regulator we may consider the fractionation of lettuce (*Lactuca sativa* L.) hypocotyl sections floated for 16 h on a solution of [^{3}H]-GA$_1$ (Stoddart, 1979*a*). After homogenization and differential centrifugation, over 95% of the total radioactivity present in the tissue was recovered in the 100 000 g supernatant. Approximately 4% was located in the low-speed (2000 g) pellet and less than 1% of the total was recovered in the 20 000 g and 100 000 g pellets. When an aliquot of the high-speed supernatant was further fractionated by gel-filtration on Sephadex G-200 only 0.2% of the loaded radioactivity was eluted with high molecular weight components. It is possible, of course, that the conditions of this extraction were not conducive to the persistence of non-covalent particulate interactions but the procedures employed were identical to those frequently used to produce metabolically competent organelles and the results are similar to those obtained by most workers (e.g. Ohlrogge *et al.* 1980). These latter workers have also shown that during the course of such feeding experiments an increasing proportion of the 'soluble' GA can be localized in the vacuole. They claim that, after incubation of barley or cowpea protoplasts in [^{3}H]-GA$_1$ solution for 16 h, as much as 100% of the total uptake was present in this compartment.

In terms of the primary action of GA, this is probably a misleading statistic because growth regulator supplied exogenously must traverse the rest of the cell in order to take up residence in the vacuole and biological action, depending upon transient associations, could occur during this process.

The plastid has also received considerable attention as a possible control site in the regulation of GA action. Early observations that significant amounts of GA-like activity could be associated with this organelle (Stoddart, 1968; Frydman & Wareing, 1973; Cooke & Saunders, 1975) were followed by a series of publications suggesting that interchange with the cytoplasm was regulated *via* phytochrome located in the chloroplast envelope (e.g. Evans & Smith, 1976). A fuller review of this aspect of GA compartmentation is given by Graebe & Ropers (1978) but a general overview of the evidence suggests that, whilst it is clear that some GA-like activity is associated with the plastid, its identification as a primary synthetic site is questionable; as is the support for the residence of a phytochrome-mediated efflux mechanism in the inner envelope.

There is an increasing body of evidence to suggest that a significant proportion of the total cellular content of exogenously fed GA can be recovered with the cell-wall fraction (Stoddart, 1979*a,b*; Stoddart & Williams, 1979) and, under circumstances where uptake of GA and the growth response are temporally separated by using inhibitors or low-temperatures, then the proportion of total growth regulator recovered in the wall fraction can be as high as 20%. However, a value of around 5% is more normal. The physiological implications of this will be discussed in section 3.4.1.

(2) Physicochemical compartmentation

The residence of a GA in the soluble phase of the cell, which accounts for the majority of the endogenous pool, does not preclude the existence of compartments with differing availabilities to the growth-controlling sites. In considering such a possibility it is necessary to be aware of the physical and chemical properties of the GA molecule.

Gibberellins are weak carboxylic acids with pKa values around 4.0. Consequently they exist in an ionized (dissociated) form throughout the cellular pH range (with the possible exception of the vacuole). This characteristic is utilized by most of the solvent-extraction schemes employed in the isolation of GAs, where the preferential solubility of the molecule in organic solvents at pH values below the pKa (i.e. in the undissociated state) is exploited. At cellular pH values, therefore, there is a predisposition for GAs to remain in the polar phase of the cell, normally represented by the cytosol, plastid stroma, nuclear matrix and vacuolar sap. The ionized state militates against transport across membranes by diffusion and the relatively low lipid solubility of dissociated GAs tends to reinforce this effect. The preponderant distribution of GA within the soluble phase of the cell is, therefore, entirely consistent with its molecular properties at the governing pH. Superimposed upon this basic influence there are, however, a number of other more subtle determinants of molecular behaviour.

The positioning and number of the hydroxyl groups and the ionization state of the carboxyl group are major determinants of biological potency, and are discussed at length by Brian *et al.* (1967). However, for the purposes of this section, it is sufficient to emphasize that the addition of a substituent at a critical molecular site may have small effects upon polarity which effectively remove the GA from the chemical compartment containing active molecular species.

On a superficial level, it can be suggested that the biological effectiveness of a given GA structure may be a function of its ability to partition into the lipid environment of the cell. This hypothesis is only partially supported by the available data, for a range of GAs, on partition coefficients between aqueous and organic solvents within the cellular pH range (Durley & Pharis, 1972). In this context it should be appreciated that ion-transport effects can result in localized low pH conditions in the vicinity of membrane surfaces, which do not reflect those prevailing in the cytosol as a whole (Pauls, Chambers, Dumbroff & Thompson, 1982).

The process of conjugation, already discussed in section 3.2.1, imposes restraints on the behaviour of the resultant complex. Presence of a highly polar sugar places stringent limitations on possible orientations in the proximity of a lipid interface, especially in ether linkages where the glycosidic moeity is in the same region as critical A-ring features. Conversely, ester-linked conjugates may be transported more readily across growth-limiting membrane barriers because of the masking of the ionizable carboxyl function. It is additionally possible that attachment to a sugar molecule may provide access to a range of membrane-based carbohydrate transporter systems and thus facilitate movement between cells or to critical sites within the cell. Thus, conjugates represent a further chemical compartment within the soluble phase of the cell but it is uncertain whether the GA skeleton can return from this compartment to the active pool.

3.2.3   Dose–response relationships

The number of action sites for any hormone is usually regarded as being finite and less than the total population of potential molecular occupants (ligands) in the action compartment. Response is considered to be proportional to the number of sites occupied and saturated when none remain available. The rate of occupation, and the intensity of the biological response, will relate to the concentration of free hormone up to the point of saturation, after which the correlation will cease to be valid. These dose–response relationships take a characteristic form for the individual classes of animal hormones and the various groups of plant growth regulators. The data are, however, usually expressed in terms one step removed from the above by relating action to external or applied hormone levels. This assumes uniformity of uptake and transport over the considered concentration range, as well as the absence of intervening losses due to metabolism.

In Fig. 3.2 dose–response relationships are depicted for estradiol action in the rat uterus, an intensively studied animal system, and two morphologically distinct plant tissues which respond to GAs.

Fig. 3.2. Dose–response curves for estradiol action in the rat uterus compared with gibberellin action in the barley aleurone and lettuce hypocotyl systems. UP = specific uterine protein; BA = β-amylase secretion by aleurone layers; LH = elongation of lettuce hypocotyl sections. Unjoined open circles indicate relative levels of radioactivity in uterine cell nuclei. Horizontal bars indicate concentration range for response.

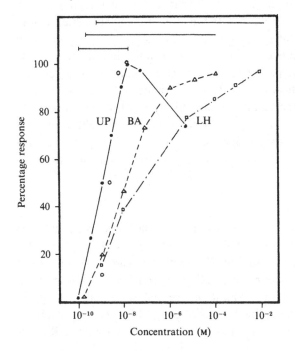

Production of specific uterine protein in response to estradiol shows typically rapid saturation characteristics, reaching a maximum at 10^{-8} M and spanning only two orders of magnitude. At higher concentrations there is a clear depression of response. With other animal systems the progression from zero to maximum effects can be even more abrupt, exemplified by insect morphological responses to ecdysteroids where the entire relationship is encompassed within one order of magnitude of concentration. A useful overall review of this area has been provided by Roy & Clark (1980).

In contrast, the two GA responses depicted in Fig. 3.2 extend over at least a 100-fold wider concentration range and show no obvious saturation effects. Examination of the open circle symbols associated with the estradiol curve shows that the production of uterine protein has a simple magnitude relationship to the hormone concentration. On the other hand, it is generally the case in plants, and especially where response is measured in terms of cell extension, that growth is related logarithmically to growth-regulator concentration (e.g. Reeve & Crozier, 1975).

Superficially, the dose–response behaviour of the GAs does not suggest that they operate via a low capacity, high-affinity site similar to that occupied by animal steroid hormones. Indeed, the curves are more descriptive of a receptor mechanism which is capable of accepting growth regulator in very large amounts, with the ultimate biological effect conditioned more by the intrinsic capacity of the system to respond than by the acceptance limits of the primary subcellular interaction. Kende & Gardner (1976) have postulated that the response differences between GAs and steroids indicate a different underlying cellular action mechanism. Whilst this conclusion may be entirely valid, it is also possible that response curves of the GA type could result from a multicomponent system. For example, they might be explained by the presence of a series of specific receptors with differing saturation characteristics and, possibly, different biological consequences.

3.2.4   Tissue sensitivity

The ability of a tissue to respond to a hormone is conditioned by its intrinsic capacity for growth (or secretion) and the rapidity with which maximum rate is achieved. These factors can be quantified in a compound manner under the heading of 'sensitivity'. Tissues from closely related sources can exhibit differential sensitivity to a given growth regulator, as exemplified by the hook (highly sensitive) and basal (insensitive) regions of the etiolated pea epicotyl (Musgrave, Kays &

Kende, 1969). Additionally, the absolute sensitivity of plant tissues can change with age; an effect well illustrated by lettuce hypocotyl sections. When freshly excised material is incubated at 30°C, with an optimal concentration of GA, growth equivalent to 250% of the initial section length can be observed over a 52 h period. If, on the other hand, fresh sections are maintained in distilled water for a similar period and then exposed to GA, the growth increment is restricted to about 20% (Silk & Jones, 1975). Proportional responses can be obtained with delay periods of intermediate lengths and comparable effects were seen in sections maintained in 10 mM potassium chloride (author's unpublished observations). The implication of ageing, rather than progressive deterioration in excised tissue, is supported by the observation that when d5 corn (Phinney & West, 1961) seedlings, at the 6th leaf stage are given an overall spray of GA, the subsequent elongation is restricted to the 5th and 6th leaf sheaths and the emerging leaf lamina, with the subtending tissues showing no response whatsoever (Stoddart, unpublished observations). The extensibility of rice coleoptiles also diminishes with age (Furuya, Masuda & Yamamoto, 1972). Clearly, the inference is that the ability to increase growth rate in the presence of GA depends on the developmental stage or age-related changes in cell structure.

There is good evidence that environment may also exert a strong conditioning effect. Stoddart, Tapster & Jones (1978) constructed a series of GA dose–response curves for the lettuce hypocotyl section system over the range between 5°C and 30°C. Plots of growth rate against temperature, constructed for a range of GA concentrations, revealed discontinuous relationships with inflexions between 12°C and 13°C. The slope of the second linear segment of each plot increased with growth regulator concentration. The data from these experiments are summarized in Fig. 3.3 which illustrates the change in the Q_{10} for growth produced by increasing GA concentration. At the optimal GA level the growth response at a given temperature was almost trebled in comparison with the controls. Mutant genotypes, such as 'slender' in barley (Stoddart & Foster, 1978), suggest that there is a positive restraint mechanism operative on growth at low temperatures and the sensitivity data described for lettuce are consistent with GA acting as a modifier of such a system.

Light also modifies sensitivity towards GA. Irradiation of dark-grown pea seedlings with red light greatly reduces their sensitivity to GA Musgrave *et al.* 1969) and also diminishes accumulation of the growth regulator in the most responsive hook region. The relationship between light quality and GA response has been studied in greater detail in the lettuce hypocotyl. Growth in untreated sections was strongly inhibited at the red end of the spectrum with peaks at 420 and 480 nm whilst less intense effects were also noted at 720 nm. When treated with GA_3 at 10^{-6} M the sections grew to the same final length at all wavelengths, indicating that the growth regulator was able to override the light-mediated inhibition. Thus, for the lettuce system, light did not modify 'sensitivity' per se.

3.3   The soluble-receptor hypothesis

3.3.1   *Animal steroid receptor mechanisms*

After intense activity over the past 10 years, consequent upon the availability of radioactive compounds at extremely high specific activities (of the order of 100 Ci/mM), we now have a very clear concept of the subcellular events involved in the responses to a range of animal steroid hormones. For a recent review of this area the reader is referred to Roy & Clark (1980) and, in order to facilitate subsequent discussion of the concept in a plant context a brief review of the salient features of such a mechanism is appropriate at this point.

Response involves a structure-specific primary interaction with a cytosol receptor protein, typically an 8s dimer, which is thought to migrate to, and dissociate at, the nuclear membrane. On entry into the nucleus

Fig. 3.3. Effect of gibberellin concentration on the Q_{10} for growth in the lettuce hypocotyl section system. Q_{10} = growth at temperature (X) compared with that 10°C lower.

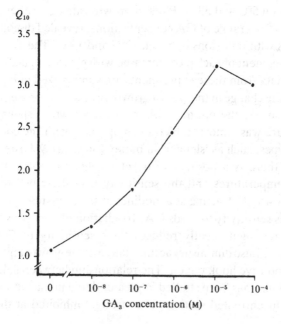

the 4s subunit/steroid hormone complex makes a two-point attachment of chromatin and non-histone protein. As a consequence there are changes in intranuclear RNA synthesis, resulting in an increased flow of transcripts for currently expressed DNA regions, or the opening up of new tracts of genetic information. In the chick oviduct, for example, the ultimate effect of steroid action is increased production of mRNA for ovalbumin (Comstock, Rosenfeld, O'Malley & Means, 1972).

The key specific element of the system, the receptor protein, is thought to be present in all somatic cells of the organism but in much higher concentrations in target tissues. The hormone binds to the receptor as a high-affinity, low-capacity event with a dissociation constant around 1 nM and a steep saturation curve (Fig. 3.2). This event is viewed against a background of low-affinity, high-capacity binding to other cellular components and is normally distinguished by being freely exchangeable. A further critical feature is the ability of the binding site to distinguish between subtle variations in molecular architecture which confer or destroy biological effectiveness. To distinguish between high- and low-affinity sites a linear transformation of the Edsall & Wyman (1958) binding equation is employed, usually in the manner described by Scatchard (1949). Here B/F is plotted against B, where B = bound hormone (mol) and F = the concentration of free hormone. This procedure usually yields a curved distribution which can be partitioned into two linear segments, whose x and y intercepts provide measures of the capacities of the high- and low-affinity sites in the preparation and whose slopes represent the reciprocals of the dissociation constants.

3.3.2 The search for 'steroid-type' mechanisms in plants

Why should a mode of action for GAs in plants be expected to resemble that for steroid hormones in animals? This line of thought has been conditioned by the knowledge that the biosynthetic sequences for the two groups of compound are common up to the C15 level in the isoprenoid pathway, and by the evident structural and physicochemical similarities. Proponents of such homologies have, however, tended to minimize contrary evidence, exemplified by the divergent dose–response characteristics mentioned previously.

(1) Soluble receptors for gibberellins

A number of workers have studied the possibility that soluble receptors, analogous to those described for steroids, may exist in the plant cytosol. These investigations were greatly facilitated by the development of methods for the synthesis of labelled GAs of high purity and specific activity (*c.* 50 Ci/mM) by Pitel & Vining (1970) and Nadeau & Rappaport (1974), thus allowing a search for the existence of low-capacity sites.

Using [1,2-³H]-GA₁ synthesized by these techniques, Stoddart, Breidenbach, Nadeau & Rappaport (1974) examined subcellular interactions of the growth regulator in etiolated pea epicotyls. Excised tissue was infiltrated with [³H]-GA₁ for 12 to 18 h at 20°C, washed to remove surface radioactivity, homogenized and centrifuged at 20 000 g. The resultant supernatant was fractionated by gel-permeation chromatography on Sephadex G-200 in a manner analogous to that employed in animal studies (e.g. Gorell, Gilbert & Sidall, 1972). A typical distribution of protein and radioactivity in the resultant profile is depicted in Fig. 3.4, where binding is indicated by labelling in the peak of excluded material and also in a second zone of intermediate molecular weight. These fractions accounted for around 1% of the total GA₁ uptake and the complexes were recovered from both the highly responsive hook region and the GA-insensitive basal tissues. Light is also known to reduce responsiveness of the hook region towards GA (Musgrave *et al.* 1969) and feeding of epicotyl sections with labelled GA₁ in the light resulted in reduced recovery of radioactivity in both of the G200 zones. Similar data have been published by Konjevic, Grubisic, Markovic & Petrovic (1976) and

Fig. 3.4. Elution profile for a 20 000 g supernatant from dark-grown pea epicotyl hooks infiltrated with [³H]gibberellin A₁ fo 12 h at 20°C. V_0 = void volume of column. (Redrawn from Stoddart *et al.* 1974.)

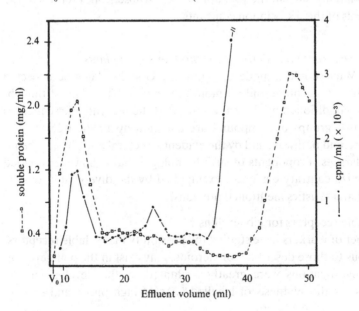

by Keith & Srivastava (1980). *Prima facie*, the association of radio-activity, fed as GA_1, with components having molecular weights higher than GA_1, or its primary metabolites, represents evidence for binding to macromolecular components, amongst which may be specific receptors. Confirmation of the biological relevance of this finding rests upon kinetic examination of the properties of the complex associations revealed by gel filtration.

Firstly, are the binding sites specific for biologically active molecular structures? Here the evidence is fairly firm. When pea tissues are supplied with radioactive GA_1 they metabolize it rapidly to the 2-hydroxyl form, GA_8, which is biologically inactive. Stoddart *et al.* (1974) fed labelled precursor by standing the excised epicotyls in the GA_1 solution, which was then translocated to the hook region. Thus, due to metabolism *en route*, the free GA pool in the hook zone contained approximately equal proportions of labelled GA_1 and GA_8 at the end of the uptake period and both molecular species were, presumably, available to the binding sites. However, recovery of the radioactivity from both of the G-200 binding zones showed, on TLC separation, that only the biologically active GA_1 was present. Further examination of this apparent specificity was conducted using compounded mixtures of:

(a) $[^3H]$-GA_1 plus $[^3H]$-epi-GA_1 (a chemical rearrangement in which the 3-hydroxyl group is in the opposite stereochemical configuration and which is biologically inactive), and

(b) $[^3H]$-GA_1 plus $[^3H]$-16-keto-GA_1 (a C/D ring rearrangement which shows weak (*c.* 10%) biological activity in the pea system).

In both cases the mixtures were adjusted to contain equal proportions of radioactivity in the two components. In the first case only $[^3H]$-GA_1 was recovered in the bound fractions and in the second both compounds were present in each zone. Keith & Srivastava (1980), using epicotyl slices incubated at 5°C, showed a similarly close relationship between biological activity and the ability to compete with $[^3H]$-GA_1 for the available binding sites over a 3-day exchange period. Thus, there was clear evidence of biological specificity for some of the binding activity.

However, these macromolecular associations, once formed, tend to be non-exchangeable in equilibrium dialysis systems. The initial data of Stoddart *et al.* (1974) suggested that this was not the case but the results were later found to be confounded by pore blockage in the dialysis membrane, possibly due to coating with polysaccharides present in the pea extracts. Subsequent extensive studies by the same authors (see Stoddart, 1975) emphasized the refractory nature of the complexes and this finding was confirmed by Keith & Srivastava (1980). Furthermore, these authors showed that the GA_1 associations were stable over an extended period and that they had half-lives of 38.5 h at 0°C and 9 h at 18°C. Treatment with DNase, RNase and phospholipase did not reduce binding levels but both protease and heat treatment were able to effect substantial release of radioactivity. Binding levels were reduced by applying competition during the uptake process using mixtures [^{3}H]-GA_1 with 100-fold excesses of unlabelled GA_1, GA_5 or GA_8. In these experiments the binding profiles on G-200 were depressed by cold GA_1 and GA_5 but not the biologically inactive GA_8. Unfortunately, because of complications introduced by the intervening uptake and translocation mechanisms, these data are not necessarily descriptive of the properties of the binding site. They cannot be used to imply *in vivo* exchangeability but do support the apparent specificity for biologically active molecular structures. A considerable step forward in the study of soluble GA-binding phenomena has been taken by Keith, Foster, Bonettemaker & Srivastava (1981), using a 100 000 g supernatant from *Cucumis sativus* hypocotyls. When concentrated by 60% ammonium sulphate precipitation, this fraction contained binding sites which were excluded on Sephadex G-50 columns and which were susceptible to protease and heat treatment in a similar fashion to the complexes from pea epicotyls. More significantly, this preparation could be used to demonstrate [^{3}H]-GA_4 binding by *in vitro* equilibrium dialysis and the 'receptor' activity was subject to competition from unlabelled GA species in the external medium. The competitive effects of unlabelled additions to the dialysis system was in strict accordance with their known biological activities although, surprisingly, IAA was able to compete-out about 40% of the [^{3}H]-GA_4 binding when added at 6×10^{-8} M. Scatchard (1949) plot analysis indicated the presence of saturable and unsaturable components in the enriched supernatant and a dissociation constant of 10^{-7} M was determined for the former category. Values of 0.37×10^{-12} M/mg soluble protein or 0.78×10^{-12} M/g fresh weight were obtained as estimates of binding-site concentration. The half-life of the associations was much shorter than that described for the pea system, being about 10 minutes at 0°C. Specific binding was maximal around pH 7.5. The cucumber system satisfies many of the requirements attaching to a putative cytosol receptor. It is not sedimentable at high 'g' forces, it binds labelled GA with high affinity, has a binding-site concentration which is approximately one thousandth of that found for some animal steroid hormone receptors, recognises biologically active molecular configurations and is freely exchangeable. On the other hand, there is still some doubt whether the saturation kinetics of the high affinity site match the known GA_4 dose–response characteristics of intact cucumber hypocotyl tissue.

The ready exchangeability of the binding sites in cucumber extracts is puzzling when compared with the intractability of the associations formed in the pea epicotyl and with the apparent absence of similar 'receptor' interactions in highly GA-responsive tissues such as the lettuce (*Lactuca sativa* L.) hypocotyl (Stoddart, unpublished observations). Finally, it is important to differentiate the protein interactions under study from those involved in the normal enzymatic processing of the ligand molecule. Stoddart (1982) has shown that binding preparations from d5 maize are capable of catalysing a GA_1 to GA_8 conversion when supplied with appropriate cofactors and this possibility needs to be examined across the range of reported receptor preparations. The case for the existence of a cytosol receptor for GAs is becoming stronger but still remains suggestive rather than decisive. It can be argued equally, however, that the available data do not yet rule out the existence of a steroid analogue receptor system.

3.4   Interactions of gibberellins with insoluble components
     The idea that GAs may be acting as quasi steroid hormones predisposes experimental approaches towards searching for soluble high-affinity, freely exchangeable associations with subcellular components. Such investigations will, therefore, tend only to reveal interactions conforming to these criteria and the detection of alternative action mechanisms may require a more flexible approach.

3.4.1   Cell wall events
     Adams *et al.* (1975) with *Avena* and, more latterly, Stuart & Jones (1977) using the lettuce hypocotyl have provided clear evidence that the ultimate effect of GA in stimulating cell elongation resides in an increase in cell wall plasticity, and that the time scale over which this event occurs matches closely the onset and progress of extension growth. Lockhart (1960) has shown a similar effect of GA in pea stem internodes and has further noted that the enhancement of plasticity could be prevented in the presence of the 80s ribosomal protein biosynthesis inhibitor, cycloheximide: an effect confirmed in the *Avena* studies. Thus, there are strong suggestions that GA mediates elongation *via* an effect on cell wall properties and that this intervention requires concurrent cytoplasmic protein synthesis.

     The possibility that GA may interact directly with the cell wall has been examined in lettuce hypocotyl tissue by Stoddart (1979*a,b*) and

Stoddart & Williams (1979). Sections maintained in the light in solutions of [³H]-GA$_1$ were homogenized and fractionated by differential centrifugation. Most of the contained radioactivity was localized in the soluble fraction but a significant percentage could be recovered consistently with the wall debris. Repeated washing and density-gradient centrifugation of this material indicated that the label was firmly associated with cell wall material in the 2000 g pellet (2KP) and not simply physically entrapped. Wall labelling accounted for 80% of the total sedimentable radioactivity in the original extracts.

Incorporation into 2KP material occurred only in intact cells and incubation of [³H]-GA$_1$ with wall preparations, or wall plus supernatant, was almost totally ineffective. Thus, although exchange phenomena are not ruled out, there are firm indications that GAs can enter into a form of stable association with a component(s) of the cell wall. Uptake and incorporation of [³H]-GA$_1$ into the 2KP fraction were linear up to 24 h and a similar relationship was established between the logarithm of 2KP-GA content and growth regulator concentration in the original medium (over the range from 10^{-7} to 10^{-3} M). Most importantly, it was also possible to demonstrate a high correlation between pelletable radioactivity and the growth rate of source tissue.

Hypocotyl sections become virtually unresponsive to GAs at temperatures below 10°C (Stoddart *et al.* 1978). In sections exposed to [³H]-GA$_1$ at 5°C or 30°C for 16 h, uptake was seen to be proportional to temperature with sections at 30°C containing four times the amount of growth regulator found in the sections at 5°C. Dilution of the [³H]-GA$_1$ with varying amounts of unlabelled growth regulator allowed sections to be loaded with a varying amount of GA under circumstances where growth was permitted or restrained. At the end of this pulse period the sections were washed and transferred to a GA-free medium, all at 30°C, for a further 30 h. Both growth and incorporation into the 2KP fraction were ascertained at the end of the pulse and chase periods. The decreased uptake of GA at 5°C, and the subsequent transfer to a GA-free medium at a growth-favouring temperature, combined to locate a much higher percentage of [³H]-GA$_1$ uptake in the 2KP fraction. At the end of the chase period, 30°C pulsed sections had 9% of uptake in the wall pellet compared with over 25% in samples initially held at 5°C. It was also evident that, although total radioactivity fell slowly during the chase period, due to efflux to the external medium, radioactivity continued to be transferred to the 2KP fraction after removal from the GA solution. It can be established, therefore, that 2KP labelling is not a consequence of growth but must be either causal or a metabolic process unrelated to the growth-stimulating activity of the GA. The chase data also show that association with the 2KP material did not occur during ingress of the GA but was a consequence of outward movement from the cytoplasm.

This transfer of label from the soluble pool to the 2KP fraction has been studied further using the protein biosynthesis inhibitor MDMP (D-2-(4-methyl 2,6-dinitroanilino)-N-methylpropionamide) in the external medium. This resulted in parallel inhibition of growth and 2KP incorporation and at 10^{-4} M MDMP both processes were running at about 15% of control. Addition of the inhibitor at various points in the course of the growth response resulted in immediate parallel reductions in growth rate and pellet labelling. These data, therefore, mirror the properties of the physical measurements made on wall plasticity in the presence of GA (Adams *et al.* 1975; Stuart & Jones, 1977). The 2KP labelling in lettuce is stable to extraction with salt and to a range of hydrolytic enzymes; it is also unaffected by treatment with ethanol/acetone mixture. On the other hand, over 80% of the contained radioactivity could be released with 1 M KOH, 0.1 M Na_2CO_3 or quarternary ammonium hydroxide tissue solubilizers (Stoddart, 1979*b*). About 10% was resistant to all extraction procedures.

Autoradiography at the light- and electron-microscope levels (Stoddart, 1983) indicates that label supplied as [^{3}H]-GA$_1$ can result in preferential distribution of silver grains over wall areas in hypocotyl sections, thus supporting the inference of the centrifugation studies.

When evaluating such phenomena for possible relationship to the biological action of GAs we need precise data on the linkage between GA and the wall matrix. These are not yet available but it is evident that such a relatively large interaction, with close correlations to the growth process, must receive serious consideration. In particular, it may prompt further studies on the control and substrate specificity of the GA-glycosylation reactions which are so widespread in plants.

The interfibrillar matrix of the cell wall constitutes a favourable site within which to look for mediating effects on wall plasticity and the localization of GA in the pectic fraction places this incorporation in an appropriate place. To suggest how GA might intervene at this site to modify the turnover of wall linkages, or otherwise modify the rheology of the matrix, would, at present, be an entirely speculative exercise but it is an area of active study and the critical data should become available.

Moll & Jones (1981) using a magnetic transducer and a sophisticated flow-through system have shown that the effect of GA in stimulating growth in the lettuce hypocotyl is intimately related to the movement of Ca^{2+} out of the cell wall. High concentrations of $CaCl_2$ inhibit elongation and this effect can be overcome by chelators, such as EDTA, or GA_3. This suggests that growth is a function of Ca^{2+} abundance in the cell wall (presumptively, in the matrix) and that GAs act by stimulating uptake of Ca^{2+} from the wall into the cell. The consequent decrease of Ca^{2+} activity in the wall would then stimulate growth. As Ca^{2+} is known to associate with pectic components, the localization of GA in this fraction may relate to the Ca^{2+} displacement.

An alternative wall-based mechanism for GA action has been postulated by Fry (1979) working with cultured *Spinacia* cells. He noted that wall polymers from these cells contained esterified ferulic acid and that, under the influence of peroxidase, these compounds can be induced to form diferuloyl bridges between polymers, thus catalysing gelation. This process is envisaged as being unfavourable to growth and it is postulated, on the basis of evidence presented in the paper, that GA stimulates cell growth by inhibiting the action of peroxidase on the ferulic acids. In consequence cross-links are not formed, gelation is prevented and growth permitted. This hypothesis provides an attractive explanation for age-dependent loss of GA sensitivity but Jones (1982) has shown that in sections of lettuce hypocotyl, the probable rate-limiting enzyme for ferulic acid synthesis (phenylalanine-ammonia lyase, PAL) did not change in a manner which could be related to GA response or light-induced cell wall hardening.

3.4.2   *Effects of gibberellins on cell membranes*

Although the overall properties of the GA molecule at cellular pH values predispose it towards residence in the soluble phase there are, nevertheless, parts of the structure which retain apolar characteristics and this has led to suggestions that GAs may act by virtue of an association with cellular membrane. No *in vivo* evidence has been advanced to support this hypothesis and it is a general observation that, when tissue homogenates are fractionated, little GA is recovered in the membrane components. For example, Ginzburg & Kende (1968) found that only 0.4% of the total contained radioactivity was recovered in the various membrane classes isolated from homogenates of pea seedlings fed with $[^3H]$-GA_1. Of course, the low levels of recovery in these fractions might be a consequence of leakage during extraction and no equilibrium dialysis studies based upon purified membrane fractions have been reported, to date.

Initial suggestions of a direct role for GAs in mediating membrane properties came from Wood & Paleg (1972, 1974), using liposome systems based upon soybean phospholipid with various additions, such as dicetyl phosphate and various sterols. Glucose was incorporated as a tracer and the liposomes were cleaned of free sugar species by gel filtration prior to use. Leakage of glucose from these structures was enhanced, up to a maximum of 100%, by increasing concentrations of GA_3 but this process was considerably influenced by the minor sterol inclusions in the liposome mix. The plant-derived β-sitosterol was the most effective. Temperature had a large effect on leakage, with an abrupt transition to a constant efflux rate of about 5% at temperatures below 25°C. A progressive downward shift of the transition temperature was achieved with increasing concentrations of GA_3, up to a maximum depression of 11°C with 2.5×10^3 M. The GA was shown to be held in the lipid matrix by electrostatic attraction between the GA_3, carboxyl group and the cationic trimethylamine group of the lecithin. This association was measured in a deuterochloroform environment and it is questionable whether it would also occur in aqueous medium.

Although, superficially, these studies point to a role for GAs in modulating membrane permeability *in vivo*, there are dangers in extrapolating from artificial structures and also some inconsistencies in the observed properties of the liposome / GA associations. Most notably, it was found that GA_1 and the biologically inert GA_8 were equally effective in modifying the leakage rate of liposomes, and high activity was also exhibited by indoleacetic acid and a range of sterols. These findings cast doubt upon the specificity of the process and the high effective concentration range (5×10^{-4} M to 2.5×10^{-3} M) coupled with a steep saturation curve (1.5 to 2 orders of magnitude) reinforce the impression that this may not be a good model for GA action in plant cells. A new light has been cast recently on these effects by Pauls *et al.* (1982) using spin-labelling techniques and differential scanning calorimetry on defined phospholipid liposomes. A GA_4/GA_7 mixture was found progressively to lower both phase-transition temperature and cooperativity, with GA_8 being ineffective in this respect. Ambient pH had a marked influence and the GA effects were maximal around pH 3.5. The concentration range for response was again high, ranging between 1 and 30 mM. However, the authors point out that ion-pumping activities in the vicinity of cell membrane surfaces would produce low pH conditions and thus lower the effective concentration range. A further important concept was also advanced, namely that effects on membranes are likely to be equilibrium events, with the lipid phase content matching the concentration of GA in the surrounding cytosol. In such circumstances, it would be impossible to demonstrate selective localization in isolated membranes, thus offering an explanation for the failure of previous attempts to achieve this.

On a less-defined level, the temperature/GA sensitivity data discussed in section 3.4 clearly point to a mediating effect of cell membranes at some point in the overall GA response mechanism. Whether this occurs at the level of the primary receptor, and reflects phase-change characteristics of the membrane bearing the site, or whether it merely describes the general conditioning effect of cell membrane properties on the capacity for growth, however stimulated, remains to be established. That such interactions are important determinants of growth is confirmed by studies with mutant barley (Stoddart & Foster, 1978). In wild-type material growth is virtually arrested at temperatures below 10°C and the plants adopt a small-leaved, dark-green habit. However, in the 'slender' mutants, growth continues at a high rate even at 5°C, indicating that growth restraint at low temperatures is a genetically determined trait. Treatment of normal seedlings with GA at low temperatures produces plants which resemble those possessing the 'slender' allele. Presumably, such effects are expressed *via* changes in membrane architecture and these lines should be valuable as probes of the relationships between GAs and cell membranes.

3.5   Conclusions

As will be clear from the foregoing sections, there is no unanimity of opinion on the likely subcellular mode of action of the GAs. The evident structural affinities with animal steroid hormones have biased efforts heavily towards a search for this type of mechanism with, as Kende & Gardner (1976) have emphasized, little positive progress. The recent findings of Keith *et al.* (1981) have provided some impetus in this direction but the crucial question, 'How does the binding phenomenon under study relate to the biological action of the ligand?', must be answered before their data can be placed in its proper perspective. This, of course, is equally true of putative receptor systems associated with other plant growth regulators, such as auxins (Venis, 1977), cytokinins (Fox & Erion, 1975) and ethylene (Bengochea *et al.* 1980). It also applies to alternative mechanistic proposals for GAs and, in many respects, is the most difficult aspect of studies in hormone primary action. When proteinaceous receptor sites are under investigation, it is imperative that they are distinguished from interactions with enzymes metabolizing the compound in question. This is particularly important when very high specific radioactivities are being employed, as is usually the case with GA receptor studies. Many of the tissues in current use will rapidly modify [^{3}H]-GA$_1$ and this aspect must, self-evidently, receive early attention.

In addition, there has to be some resolution of the divergence between the gross dose–response relationships for GA-stimulated growth and the saturation kinetics determined for the candidate binding sites.

Changes in cell wall plasticity following GA treatment may be either an early or a late consequence of the entry of the growth regulator into the cell and, in order to clarify this aspect, more information is required on the way in which linkage turnover in the wall matrix may determine the rate and duration of the extension phase. This would provide a framework within which direct or indirect effects of GA could be assessed. In the same context, we need linkage information for the 2KP–GA associations in the lettuce system before this effect can be properly evaluated. Although not specifically considered in this chapter, mainly because of the almost total lack of investigations on receptor aspects, there also seems to be a need to consider cell wall effects in the barley aleurone layer. This uniquely responsive and intensively studied system may, by virtue of its intrinsic atypicality, provide us with the critical insights required to solve the enigma of GA action.

Similarly, some perception of the level at which cell membranes are implicated in the response chain, and the way in which they permit environmental mediation of the GA effect, could throw an important analytical sidelight on the problem.

Traditionally, single-gene mutants have been indispensable tools for biochemical/physiological studies. They have been used to telling effect in the elucidation of the biosynthetic pathways for GAs and a number of lines relating to the chain of biological action are now available. These constitute the most potentially fruitful avenue of approach to the continuing enigma of the subcellular action of these most potent plant hormones.

3.6   References

Adams, P. A., Montague, M. J., Tepfer, M., Rayle, D. L., Ikuma, H. & Kaufman, P. B. (1975). Effect of gibberellic acid on the plasticity and elasticity of *Avena* stem segments. *Plant Physiol.* **56**: 757–60.

Barendse, G. W. M. & Gilissen, H. A. M. (1977). The diffusion of gibberellins into agar and water during early germination of *Pharbitis nil* Choisy. *Planta* **137**: 169–75.

Bengochea, T., Dodds, J. H., Evans, D. E., Jerie, P. H., Niepel, B., Shari, A. R. & Hall, M. A. (1980). Studies on ethylene binding by cell-free preparations from cotyledons of *Phaseolus vulgaris* L. Separation and characterisation. *Planta* **148**: 397–406.

Brian, P. W., Grove, J. F. & Mulholland, T. P. C. (1967). Relationships between structure and growth promoting activity of the gibberellins and some allied compounds in four test systems. *Phytochemistry* **6**: 1475–99.

Comstock, J. P., Rosenfeld, G. C., O'Malley, B. W. & Means, A. R. (1972). Estrogen-induced changes in translation and specific messenger RNA levels during oviduct differentiation. *Proc. Natl. Acad. Sci. USA* **69:** 2377–80.

Cooke, R. J. & Saunders, P. F. (1975). Phytochrome mediated changes in extractable gibberellin activity in a cell-free system from etiolated wheat leaves. *Planta* **123:** 299–302.

Crozier, A., Kuo, C. C., Durley, R. C. & Pharis, R. P. (1970). The biological activities of 26 gibberellins in nine plant bioassays. *Can. J. Bot.* **48:** 867–77.

Davies, L. J. & Rappaport, L. (1975). Metabolism of tritiated gibberellins in d-5 dwarf maize. *Plant Physiol.* **55:** 620–5.

Durley, R. C. & Pharis, R. P. (1972). Partition coefficients of 27 gibberellins. *Phytochemistry* **11:** 317–26.

Edsall, J. T. & Wyman, J. (1958). *Biophysical Chemistry*, vol. 1, pp. 43–56. New York, London: Academic Press.

Evans, A. & Smith, H. (1976). Localisation of phytochrome in etioplasts and its regulation *in vitro* of gibberellin levels. *Proc. Natl. Acad. Sci. USA* **73:** 138–42.

Fox, J. E. & Erion, J. L. (1975). A cytokinin binding protein from higher plant ribosomes. *Biochem. Biophys. Res. Commun.* **64:** 694–700.

Fry, S. C. (1979). Phenolic components of the primary cell wall and their possible role in the hormonal regulation of growth. *Planta* **146:** 343–51.

Frydman, V. M. & MacMillan, J. (1975). The metabolism of gibberellins A_9, A_{20} and A_{29} in immature seeds of *Pisum sativum* cv. Progress No. 9. *Planta* **125:** 181–95.

Frydman, V. M. & Wareing, P. F. (1973). Phase change in *Hedera helix* L. Gibberellin-like substances in the growth phase. *J. Exp. Bot.* **24:** 1131–8.

Furuya, M., Masuda, Y. & Yamamoto, R. (1972). Effects of environmental factors on mechanical properties of the cell wall in rice coleoptiles. *Dev. Growth Differ.* **14:** 95–105.

Ginzburg, C. & Kende, H. (1968). Studies on the intracellular localisation of radioactive gibberellin. In *Biochemistry and Physiology of Plant Growth Substance*, ed. F. Wightman & G. Setterfield, pp. 333–40. Ottawa: Runge Press.

Gorell, T. A., Gilbert, L. I. & Sidall, J. B. (1972). Binding proteins for an ecdysone metabolite in the crustacean hepatopancreas. *Proc. Natl. Acad. Sci. USA* **69:** 812–15.

Graebe, J. E. & Ropers, H. J. (1978). Gibberellins. In *Phytohormones and Related Compounds – a Comprehensive Treatize. The Biochemistry of Phytohormones and Related Compounds*, vol. 1, ed. D. S. Letham, P. B. Goodwin & T. V. J. Higgins, pp. 107–204. Amsterdam: Elsevier.

Hiraga, K., Yamane, H. & Takahashi, N. (1974). Biological activity of some synthetic gibberellin glycosyl esters. *Phytochemistry* **13:** 2371–6.

Jones, R. L. (1982). Gibberellin control of cell elongation. In *Plant Growth Substances 1982*, ed. P. F. Wareing, pp. 121–30. London: Academic Press.

Keith, B., Foster, N. A., Bonettemaker, M. & Srivastava, L. M. (1981). *In vitro* gibberellin A_4 binding to extracts of cucumber hypocotyls. *Plant Physiol.* **68:** 344–8.

Keith, B. & Srivastava, L. M. (1980). *In vitro* binding of gibberellin A_1 in dwarf pea epicotyls. *Plant Physiol.* **66:** 962–7.

Kende, H. & Gardner, G. (1976). Hormone binding in plants. *Annu. Rev. Plant Physiol.* **27:** 267–90.

Konjevic, R., Grubisic, D., Markovic, R. & Petrovic, J. (1976). Gibberellic acid binding proteins from pea stems. *Planta* **131:** 125–8.

Lockhart, J. A. (1960). Intracellular mechanism of growth inhibition by radiant energy. *Plant Physiol.* **35:** 129–35.

Moll, C. & Jones, R. L. (1981). Calcium and gibberellin-induced elongation of lettuce hypocotyl sections. *Plant Physiol.* **152:** 450–6.

Musgrave, A., Kays, S. E. & Kende, H. (1969). *In vivo* binding of radioactive gibberellins in dwarf pea shoots. *Planta* **89**: 165–77.

Nadeau, R. & Rappaport, L. (1972). Metabolism of gibberellin A_1 in germinating bean seeds. *Phytochemistry* **11**:1611–16.

Nadeau, R. & Rappaport, L. (1974). The synthesis of ^{3}H-gibberellin A_3 and ^{3}H-gibberellin A_1 by the palladium catalysed actions of carrier-free tritium on gibberellin A_3. *Phytochemistry* **13**:1537–45.

Ohlrogge, J. B., Garcia-Martinez, J. L., Adams, D. & Rappaport, L. (1980). Uptake and subcellular compartmentation of gibberellin A_1 applied to leaves of barley and Cowpea. *Plant Physiol.* **66**: 422–7.

Pauls, K. P., Chambers, J. A., Dumbroff, E. B. & Thompson, J. E. (1982). Perturbation of phospholipid membranes by gibberellins. *New Phytol.* **91**: 1–17.

Phinney, B. O. & West, C. A. (1961). Gibberellins and plant growth. In *Encyclopaedia of Plant Physiology, Vol. XIV*, ed. W. Ruhland, pp. 1189–91. Berlin, Heidelberg, Göttingen: Springer-Verlag.

Pitel, D. W. & Vining, L. C. (1970). Preparation of gibberellin A_1-3,4-^{3}H. *Can. J. Biochem.* **48**: 259–63.

Reeve, D. R. & Crozier, A. (1975). Gibberellin bioassays. In *Gibberellins and Plant Growth*, ed. H. N. Krishnamoorthy, pp. 35–64. New Delhi: Wiley Eastern.

Roy, A. K. & Clark, J. H. (1980). *Gene Regulation by Steroid Hormones*, pp. 1–316. New York, Heidelberg, Berlin: Springer-Verlag.

Scatchard, G. (1949). The attractions of proteins for small molecules and ions. *Ann. N.Y. Acad. Sci.* **51**: 660–72.

Sembdner, G., Gross, D., Liebisch, H.-W. & Schneider, G. (1980). Biosynthesis and metabolism of plant hormones. In *Hormonal Regulation of Development. 1. Encyclopaedia of Plant Physiology. New series*, ed. J. MacMillan, pp. 281–444. Berlin, Heidelberg, New York: Springer-Verlag.

Sembdner, G., Weiland, J., Aurich, O. & Schreiber, K. (1968). Isolation, structure and metabolism of a gibberellin glucoside. In *Plant Growth Regulators, S.C.I. Monograph No. 31*, pp. 70–86.

Silk, W. K. & Jones, R. L. (1975). Gibberellin response in lettuce hypocotyl sections. *Plant Physiol.* **56**: 267–72.

Sponsel, V. M., Hoad, G. V. & Beeley, L. J. (1977). The biological activities of some new gibberellins (GAs) in six plant bioassays. *Planta* **135**: 143–7.

Stoddart, J. L. (1968). The association of gibberellin-like activity with the chloroplast fraction of leaf homogenates. *Planta* **81**: 106–12.

Stoddart, J. L. (1969). Incorporation of kaurenoic acid into gibberellins by chloroplast preparations of *Brassica oleracea*. *Phytochemistry* **8**: 831–7.

Stoddart, J. L. (1975). Characterisation of high molecular weight ^{3}H gibberellin A_1 complexes. *Rep. Welsh Pl. Breed. Stn for 1975*, pp. 80–1.

Stoddart, J. L. (1979a). Interaction of ^{3}H gibberellin A_1 with a subcellular fraction from lettuce (*Lactuca sativa* L.) hypocotyls. I. Kinetics of labelling. *Planta* **146**: 353–61.

Stoddart, J. L. (1979b). Interaction of ^{3}H gibberellin A_1 with a subcellular fraction from lettuce (*Lactuca sativa* L.) hypocotyls. II. Stability and properties of the association. *Planta* **146**: 363–8.

Stoddart, J. L. (1982). Gibberellin perception and its primary consequences: the current status. In *Plant Growth Substances 1982*, ed. P. F. Wareing, pp. 131–40. London: Academic Press.

Stoddart, J. L. (1983). Sites of gibberellin biosynthesis and action. In *The Biochemistry and Physiology of Gibberellins*, vol. 2, ed. A. Crozier, pp. 1–38. New York: Praeger Scientific.

Stoddart, J. L., Briedenbach, W., Nadeau, R. & Rappaport, L. (1974). Selective binding of ^{3}H gibberellin A$_1$ by protein fractions from dwarf pea epicotyls. *Proc. Natl. Acad. Sci. U.S.A.* **71**: 3255–9.

Stoddart, J. L. & Foster, C. A. (1978). A barley mutant with an accelerated growth rate at low temperatures. Abstract of Demonstration, *9th International Conference on Plant Growth Regulators*, Lausanne, Switzerland.

Stoddart, J. L., Tapster, S. M. & Jones, T. W. A. (1978). Temperature dependence of the gibberellin response in lettuce hypocotyls. *Planta* **141**: 283–8.

Stoddart, J. L. & Venis, M. A. (1980). Molecular and subcellular aspects of hormone action. In *Hormonal Regulation of Development. 1. Molecular Aspects of Plant Hormones. Encyclopaedia of Plant Physiology, New Series*, vol. 9, ed. J. MacMillan, pp. 445–510. Berlin, Heidelberg, New York: Springer-Verlag.

Stoddart, J. L. & Williams, P. D. (1979). Interaction of ^{3}H gibberellin A$_1$ with a subcellular fraction from lettuce (*Lactuca sativa* L.) hypocotyls. III. Requirement for protein synthesis. *Planta* **147**: 264-8.

Stuart, D. A. & Jones, R. L. (1977). The roles of extensibility and turgor in gibberellin- and dark-stimulated growth. *Plant Physiol.* **59**: 61–8.

Venis, M. A. (1977). Solubilisation and partial purification of auxin binding sites of corn membranes. *Nature* **66**: 268–9.

Warner, T. J. & Ross, J. D. (1981). Phytochrome control of maize coleoptile section elongation: the role of cell wall extensibility. *Plant Physiol.* **68**: 1024–6.

Wood, A. & Paleg, L. G. (1972). The influence of gibberellic acid on the permeability of model membrane systems. *Plant Physiol.* **50**: 103–8.

Wood, A. & Paleg, L. G. (1974). Alteration of liposomal membrane fluidity of gibberellic acid. *Aust. J. Plant Physiol.* **1**: 31–40.

4

The biochemistry and physiology of cyclic AMP in higher plants

Russell P. Newton and Eric G. Brown

4.1 Introduction

In order to appreciate the significance of the existence of a cyclic nucleotide in higher plants, its function in other types of living organism must first be comprehended. Adenosine 3':5' cyclic monophosphate (cAMP) was discovered in 1957 (Rall, Sutherland & Berthet, 1957). Sutherland's group found that the activity of liver glycogen phosphorylase was increased in the presence of a heat-stable compound synthesized in response to adrenalin. Chemical studies showed the structure of the factor to be (I). This simple molecule has since been shown in mammals

I

Adenosine 3':5'-cyclic monophosphate (cAMP)

to mediate the activity of a considerable number of short term, rapidly acting, 'non-maintenance' hormones (Jost & Rickenberg, 1971). According to modern concepts of how this functions (Fig. 4.1), an appropriate stimulus evokes a response from an endocrine gland in the form of secretion of a hormone, the so-called 'primary messenger'; the hormone then circulates in the blood stream until it reaches the cell membrane of its target tissue. Here the hormone binds at a specific receptor on the outer membrane surface, and this binding produces a conformational change in the receptor. In turn, a change is induced in a transducer unit, which consists of an integral lipoprotein with a binding site for GTP. This is followed by the activation of a catalytic unit, namely the enzyme adenylate cyclase, which is located on the inner surface of the membrane. Adenylate cyclase catalyses the conversion of ATP to cyclic AMP (cAMP), the 'secondary messenger', which is released into the cell where it is capable of binding to the regulatory units of cAMP-dependent protein kinase. This results in the release of free, active catalytic units of protein kinase, which phosphorylate key pacemaker enzymes such as glycogen phosphorylase, thereby modifying their individual activities and evoking a physiological response (Nimmo & Cohen, 1977).

An integral part of this control system is the enzyme phosphodiesterase, which is capable of decreasing rapidly the concentration of cAMP by hydrolysing it to AMP, thereby quenching the intracellular signal. Cyclic nucleotide phosphodiesterase exists in multiple forms, many of which are allosterically regulated by a variety of factors including activation by an endogenous calcium-dependent modulator protein, calmodulin, and by proteinaceous inhibitors (Wells & Harman, 1977).

Fig. 4.1. The 'secondary messenger concept' in which cAMP mediates the effect of the hormonal primary messenger in mammalian tissue. The hormone molecule attaches to a receptor unit on the target cell membrane and activates the adenylate cyclase catalytic unit *via* a transducer unit. The newly synthesized cAMP is released into the target cell where it unmasks the activity of a protein kinase by binding with the regulatory unit of this enzyme. In turn, the activated protein kinase catalyses the phosphorylation of other inactive enzyme proteins, so activating them. It is the net effect of these activated enzymes which produces the characteristic physiological response.

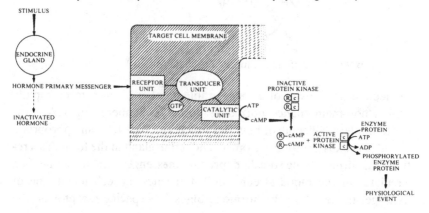

Some years after its discovery in mammals, cAMP was shown to occur in the bacterium *Escherichia coli* (Makman & Sutherland, 1965). Its role in bacteria is also regulatory, bringing about changes in protein synthesis at both the transcriptional and translational levels. An example of the former involves the *lac* operon of *E. coli* in which cAMP complexes with a specific receptor protein (CRP) which consequently undergoes an allosteric change (Fig. 4.2). The complex binds with a region of the DNA close to the *lac* promoter, enabling RNA polymerase to reach the DNA thus initiating *lac* transcription. In the *E. coli* system cAMP levels are responsive to changes in glucose concentration in the medium, and thus cAMP is effectively behaving as the first messenger (Pastan & Perlman, 1972).

Thus, although cAMP was discovered in its role as a mediator of hormone action in mammals, i.e. a secondary messenger, it is now known that in bacteria an equally significant function of this cyclic nucleotide is as a primary messenger. Similarly, cAMP plays a primary messenger role in the slime mould *Dictyostelium discoideum* (see next chapter).

Fig. 4.2. Regulatory role of cAMP during transcription in bacterial protein synthesis. This example shows the transcription of the *lac* operon in *E. coli*. The cyclic nucleotide complexes with the receptor protein (CRP) which, in turn, binds with the region of DNA close to the *lac* promoter. This facilitates access of RNA polymerase and so initiates transcription.

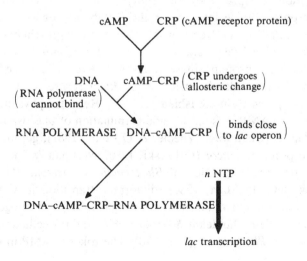

Cyclic AMP in bacteria: effect on transcription

β-galactosidase synthesis in *E. coli*

Occurrence of these regulatory roles of cAMP in animals and micro-organisms led to suggestions that this nucleotide may play a regulatory role in plants.

cAMP has been shown to be present in a variety of lower plants. There are also a number of reports of the occurrence of cyclic GMP in these organisms. Whereas most of these reports are qualitative some give quantitative estimates of the concentration of the cyclic nucleotide studied. For example, intracellular concentrations of 0.27–2.7 pmol of cAMP/mg protein have been determined for the cyanobacterium *Anabaena variabilis* (Hood, Armour, Ownby, Handa & Bressan, 1979). Interestingly, in rapidly growing cultures of this organism up to 90% of the total cAMP is not in the cells but in the culture medium (Hood, Armour, Ownby, Handa & Bressan, 1979). This release of cAMP into the medium appears to be a common phenomenon with cultures of green and blue-green algae (see, e.g. Francko & Wetzel, 1980; Bressan, Handa, Cherniak & Filner, 1980*a*; Bressan, Handa, Quader & Filner, 1980; Handa, Bressan, Quader & Filner, 1981). Similar observations have been made with several aquatic macrophytes (Francko & Wetzel, 1981). At present, however, the most detailed study of the cAMP metabolism of a lower plant is undoubtedly that concerning the slime mould *Dictyostelium discoideum*. During the amoeboid stage of development of this organism, cAMP functions as an aggregation factor ('acrasin'). The relationship is discussed, in detail, in the ensuing chapter.

There have been a number of published reports of the occurrence and physiological involvement of cAMP in mosses. For example, evidence has been presented that in *Bryum argenteum*, cAMP is involved in sex induction (Chopra & Bhatla, 1981). This cyclic nucleotide has also been reported to be involved in the metabolism of a variety of fungi (see, e.g. Pall, 1981). Recently, a cAMP-binding protein has been obtained from bakers' yeast and identified as a regulatory subunit of a cAMP-dependent protein kinase (Hixson & Krebs, 1980). cAMP-binding proteins have also been found in *Mucor* (Forte & Orlowski, 1980) and in *Coprinus* (Uno & Ishikawa, 1981). Reported physiological involvements of cAMP in fungi include germination of blastospores of *Candida albicans* (Chataway, Wheeler & O'Reilly, 1981), germination of sporangiospores of *Mucor* (Orlowski, 1980; Wertman & Paznokas, 1981), induction of germination of *Blastocladiella emersonii* (Gomes, Mennucci & deCosta Maia, 1980), dimorphic transition in *Candida albicans* (Niimi, Niimi, Tokunaga & Nakayama, 1980), carotenogenesis in *Blaskeslee trispora* (Dandekar & Modi, 1980), and the cellular distribution of Ca^{2+} in *Phycomyces* (Tu, 1978). The role of cAMP in lower photosynthetic organisms has been discussed within a recent review (Francko, 1983).

4.2 Occurrence of cAMP in higher plants

The first attempts to demonstrate the existence of cAMP in higher plants (Pollard, 1970; Narayanan, Vermeersch & Pradet, 1970; Salomon & Mascarenhas, 1971, 1972) were not generally accepted as adequate proof. Most of the reports were either presumptive evidence derived from observed physiological effects of endogenously supplied cAMP or cAMP analogues, or conclusions based solely on chromatographic identification commented on by others as insufficiently rigorous. In the former category, experiments which for example showed that cAMP delayed petiole abscission in *Coleus* (Salomon & Mascarenhas, 1971) merely demonstrate that the nucleotide is capable of mimicking hormonal effects. They do not show that it is a secondary messenger of auxin. In the second category [8-^{14}C]adenine was supplied to seeds (Pollard, 1970) or coleoptiles (Salomon & Mascarenhas, 1972) and a radioactive product, indistinguishable in a number of chromatographic systems from cAMP, was isolated. Several derivatives of this radioactive product were indistinguishable from the corresponding derivatives of authentic cAMP. These reports were criticized on the grounds that most of the procedures would not distinguish between adenosine 3':5'-cyclic monophosphate, the secondary messenger in mammals, and its 2':3' isomer. Thus the putative [^{14}C]cAMP originating from [^{14}C]adenine could have been the 2':3'-isomer, arising from enzymatic degradation of RNA after the labelled purine had been incorporated into RNA.

Other methods of identification included procedures involving enzymatic hydrolysis of the analyte by cAMP phosphodiesterase to AMP, which latter was then determined enzymatically (Narayanan *et al.*, 1970). Such methods were open to the criticism that they are valid only if the enzyme preparation has been shown to have absolute specificity for the substrate.

A sequential chromatographic and electrophoretic procedure for the extraction and purification of cAMP from plant tissues has been developed in our laboratories (Brown & Newton, 1973). The procedure involved extraction of freeze-dried leaf material into an ice-cold monophasic mixture of methanol, chloroform and formic acid. After rendering the homogenate biphasic the aqueous phase was separated and subjected to a charcoal adsorption and elution step. This was followed, sequentially, by anion and cation exchange chromatography, thin-layer chromatography (TLC), high-voltage electrophoresis and a further TLC step.

The identity of the purified sample of cAMP was confirmed by cochroma-
tography with an authentic sample in five paper chromatographic systems, three thin-layer systems and high-voltage electrophoresis in three
different buffers. Collectively these steps would separate cAMP from
all known naturally occurring adenine nucleotides, including the 2':3'-
cyclic isomer. By this method, purified material was obtained which
was inseparable from a reference sample of cAMP, and which would
bind to a cAMP-specific binding protein (described in next section).
Nevertheless, some authors claimed that such identification data were
equivocal (Keates, 1973a, b; Lin, 1974; Amrhein, 1974a) and suggested
that hitherto unidentified adenine nucleotides, paralleling cAMP in
behaviour in these systems, existed in higher plants. While no suggestions
of the identity of such compounds were forthcoming, such claims seemed
to retard progress in this field considerably.

Identification of the compound as adenosine 3':5'-cyclic monophosphate (I) was further supported by later studies (Ashton & Polya, 1977;
Brown, Al-Najafi & Newton, 1979) in which the kinetics of hydrolysis
were compared with those of a reference sample. More recently, proof
of identity was obtained. Using the same sequential procedure described
above, a large-scale extraction of *Phaseolus vulgaris* leaf tissue was carried out. The purified product was shown to have the characteristic UV
absorption spectrum of an adenosine derivative and to be capable of

Fig. 4.3. Mass spectrometric identification of adenosine 3':5'-cyclic
monophosphate isolated from a higher plant tissue (Newton *et al.*
1980). The cyclic nucleotide, extracted from seedlings of *Phaseolus
vulgaris*, was purified by the procedure of Brown & Newton (1973),
and trimethylsilylated by heating with acetonitrile and *N,O*-
bis(trimethylsilyl)trifluoroacetamide. The peak at m/z 545 is the
molecular ion. The ion products at m/z 310 and 378 are of crucial
significance in that they correspond to structures III and IV. These
are unique to the 3':5'-cyclic structure. This compares closely with the
spectrum obtained from a synthetic sample except that there are
additional ions in the above spectrum. These correspond to the
molecular ion (m/z 473) of cAMP-(TMS)$_2$ and its fragments at m/z
251 and m/z 250. This is attributable to incomplete silylation of some
of the sample molecules.

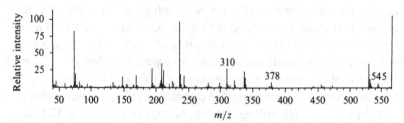

binding to the cAMP-binding protein (Newton, Gibbs, Moyse, Wiebers & Brown, 1980). After derivatizing the compound with trimethylsilane, it was subjected to mass spectrometry (MS). The spectrum (Fig. 4.3) showed a molecular ion (m/z 545) which corresponded to the addition of three trimethyl silyl groups (II); this spectrum was identical to that

II

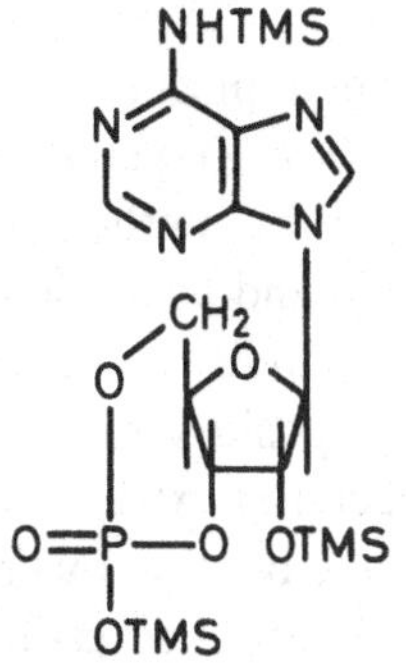

produced in the same manner from commercially obtained cAMP, and only differed from the published spectrum (Lawson, Stillwell, Tacker, Tsuboyama & McCloskey, 1971) of cAMP-(TMS)₃ in showing the presence of the molecular ion (m/z 473) and additional ion products (m/z 251 and 250) corresponding to cAMP-(TMS)₂. Presence of this was due to incomplete silylation of some of the sample molecules. The crucial feature of the mass spectrum is the presence of ion products of m/z 310 and m/z 378, which correspond to the structures shown in III and IV. These are unique to the 3′:5′-cyclic structure and cannot derive

m/z 310

III

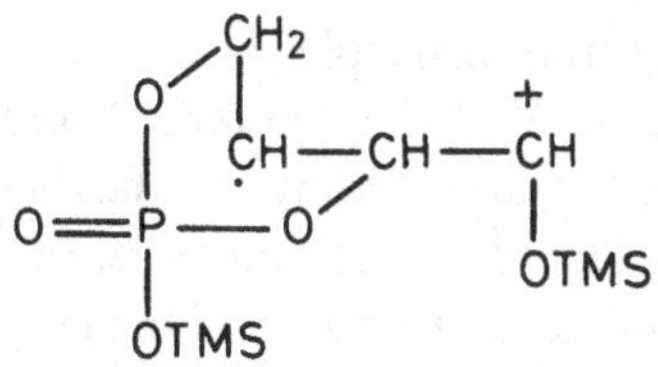

m/z 378

IV

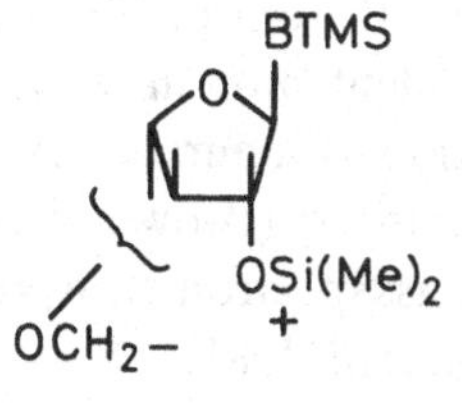

from cyclic deoxyAMP because their structures include the 2'-hydroxy-trimethylsilylated group which is absent in cyclic deoxyAMP and 2':3'-cAMP. The MS data thus confirmed unequivocally the previous identification of the compound as cAMP. Further confirmation of identity has subsequently been presented by others using mass spectrometry (Janistyn, 1981; Johnson, McLeod, Parker & Letham, 1981) and by infrared (IR) spectroscopy (Janistyn, 1981).

Just as identification of cAMP in plants has been, until recently, controversial, similar arguments have centred on the *in vivo* concentration of the cyclic nucleotides. This is mainly because the methods of assay utilized were developed for use with mammalian and bacterial extracts and were applied to plant extracts without their validity being checked. While such assays have been monitored for false positive results caused by contaminants present in mammalian and bacterial extracts, no such precautions were taken with plant extracts. While bearing this reservation in mind, it is true that the majority of the published estimates of the cAMP content of higher plant tissues show a very similar order of magnitude (Table 4.1). This is despite the variety of authors, techniques, tissues, species and age of plants.

One of the most commonly employed and convenient methods of estimating cAMP content is by a saturation binding assay. This technique utilizes a commercially available, specific cAMP-dependent protein kinase. The method (Brown, Albano, Elkins & Sgherzi, 1971) involves competition, for a limited number of binding sites on the cAMP-binding protein, between a standard quantity of radioactively labelled cAMP and varied quantities of unlabelled cAMP. A standard curve is constructed and by measuring the binding of radioactive cAMP in the presence of a tissue extract, the cAMP content of the extract can be calculated. An analogous competition principle is involved in the radioimmunoassay for cAMP (Steiner, Wehman, Parker & Kipnis, 1972) where the binding protein is replaced by an antibody specific for cAMP.

Other assay procedures utilized include a bioluminescence assay in which the cAMP is converted enzymatically to ATP and the latter determined by a luciferin-luciferase assay (Johnson, Hardman, Broadus & Sutherland, 1970); a protein kinase assay has also been used (Kuo & Greengard, 1972) in which the transfer of ^{32}P from $[\gamma\text{-}^{32}P]$-ATP to a protein, in the presence of a cAMP-dependent protein kinase and a tissue extract, is monitored. Direct estimation of a purified extract, by UV spectrophotometry, has also been employed (Brown & Newton, 1973), and recently a gas chromatography/mass spectrometric procedure has been developed (Janistyn, 1981; Johnson *et al.* 1981).

Table 4.1. *Reported concentrations of cAMP in plant tissues*

Species (tissue)	Concentration			Assay method	Reference
	pmol/g wet wt	pmol/g dry wt	Other units		
Acer	123–140			PK + Biolum	Raymond, Narayanan & Pradet (1973)
Agave	1			BP + PK	Ashton & Polya (1977)
Avena (coleoptile)	20–170			BP	Brewin & Northcote (1973a)
Avena (etioplast)			67 pmol/mg protein	BP	Wellburn, Ashby & Wellburn (1973)
Cicer (seed)	36–123			BP	Srivastava, Azhar & Krishna Murti (1972)
Citrus limeltrodes	230–245			BP	Kessler & Levenstein (1974)
Cucumis sativus	216–252			BP	Kessler & Levenstein (1974)
Cucumis sativus	53			BP	Wilson, Moustafa & Renwick (1978)
Datura	132			BP	Kessler & Levenstein (1974)
Daucus	125–161			BP	Raymond, Narayanan & Pradet (1973)
Glycine (callus culture)	2–40			BP	Brewin & Northcote (1973a)
Helianthus annuus			131 nM	RIA	Drlica, Gardner, Kadoc, Vijay & Troy (1974)
Helianthus tuberosus	25–62			RIA	Giannattasio, Mandato & Macchia (1974)
Hordeum	209–217			RIA	Miller & Galsky (1974)
Kalanchoe	2–6			BP + PK	Ashton & Polya (1977)
Lactuca	222–280			PK + Biolum	Raymond, Narayanan & Pradet (1973)
Lactuca	265			BP	Kessler & Levenstein (1974)
Lemna gibba	1040			BP	Kessler & Levenstein (1974)
Lolium	2–12			BP + PK	Ashton & Polya (1977)
Nicotiana tabaccum	122			BP	Kessler & Levenstein (1974)
Nicotiana tabaccum			142 nM	RIA	Drlica et al. (1974)

Table 4.1. contd.

Species (tissue)	Concentration			Assay method	Reference
	pmol/g wet wt	pmol/g dry wt	Other units		
Nicotiana tabaccum	70–102			BP + Biolum	Raymond et al. (1973)
Nicotiana tabaccum		400		RIA	Lundeen, Wood & Braun (1973)
Nicotiana tabaccum	70			BP + RIA	Pollard & Singh (1979)
Nicotiana tabaccum	60			BP	Saloman & Mascarenhas (1972)
Nicotiana tabaccum (callus culture)	84–149			GC/MS	Sen & Stevenson (1978)
Phaseolus vulgaris		2.2		Spec. + BP	Brown, Al-Najafi & Newton (1979); Brown & Newton (1973)
Phaseolus vulgaris	8–31			BP	Sen & Stevenson (1978)
Pinus radiata	9.6–38.4			BP	Wilson, Moustafa & Renwick (1978)
Pisum sativum	370			BP	Kessler & Levenstein (1974)
Solanum tuberosum	3–30			BP	Isherwood & Ring (1977)
Robinia (sap)			9.0 nM	BP	Becker & Ziegler (1973)
Sinapsis alba	114			BP	Janistyn & Drumm (1975)
Triticum vulgare	170–180			BP	Kessler & Levenstein (1974)
Vinca rosea			125–126 nM	RIA	Drlica et al. (1974)
Zea	2.1–3.5			PK	Tarantowicz-Marek & Kleczowski (1978)
Zea		0.5–4.0		BP	Bachofen (1973)
Zea	35			BP	Edlich & Gräser (1978)
Zea	267–360			BP	Sen & Pilet (1981)
Zea	382–710	4.8–8.9		GC/MS	Janistyn (1981)
Zizyphus vulgaris		5×10^5		BP	Hanabusa, Cyong & Takahashi (1981)

PK = protein kinase method; BP = binding-protein saturation assay; Biolum = bioluminescence method; RIA = radioimmunoassay; Spec = spectrophotometric determination; GC/MS = gas chromatography/mass spectrometry.

The validity of applying to plant extracts the cAMP-specific binding-protein saturation assay for determination of cAMP concentration has now been examined (Brown *et al.* 1979). The standard curve obtained in these studies (Fig. 4.4) intercepts the abscissa at a value of -3 pmol, and this value corresponds to the total amount of [8-^{3}H]-cAMP present. Addition of a dialysed aqueous extract from *Phaseolus* seedlings did not alter this intercept and a linear relationship remained. However, the resulting decrease in the slope of the curve indicates an increase in binding capacity, possibly due to occlusion or non-specific binding of cAMP by plant protein, or to the presence of an endogenous, specific binding protein in the extract. In the presence of a partially purified extract, previously incubated with cAMP phosphodiesterase to remove endogenous cAMP, the intercept remained the same but the binding

Fig. 4.4. Examination of the validity of applying to plant extracts the binding-protein assay for cAMP. The curves show the effect of *Phaseolus* extracts on the binding capacity of cAMP-binding protein. A standard binding-protein curve was determined in the presence and absence of *Phaseolus* extracts at various stages of the cAMP purification procedure of Brown & Newton (1973). Fraction I is the least highly purified fraction, and fraction VI the most highly purified. Standard curve, ▲—▲. Curve in the presence of: fraction I, ■—■; fraction III, □—□; fraction VI, ▲—▲ (i.e. identical to standard curve); fraction III + 4 pmol of cAMP, ○—○; fraction VI + 4 pmol of cAMP, ●—●. Each fraction had been preincubated with cyclic nucleotide phosphodiesterase to remove endogenous cAMP.

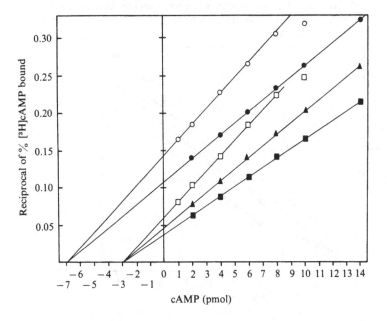

capacity was decreased (Fig. 4.4). At concentrations above 8 pmol of cAMP per assay, linearity was lost. These observations indicate the presence in the plant extract of a factor interfering with the binding of cAMP. After treatment with cAMP phosphodiesterase to remove endogenous cAMP, the final purified fraction from *Phaseolus* caused no detectable alteration to the standard plot. From this it can be deduced that the interfering substance(s) had been eliminated during the purification procedure. This conclusion is supported further by the observation that standard curves prepared in the presence of the partially purified fraction and the totally purified fraction, both previously incubated with phosphodiesterase, and with internal standards of 4 pmol of cAMP per assay, all gave intercepts at −7 pmol. This intercept corresponds to the 3 pmol of tritiated cAMP per assay plus the 4 pmol standard. The presence of the partially purified extract caused a loss of linearity and a decrease in binding capacity, while addition of the totally purified fraction caused no alteration to the standard plot.

Binding plots constructed from data determined in the presence of fractions from *Phaseolus* at various stages of the purification and contain-

Fig. 4.5. Standard binding plots determined in the presence of *Phaseolus* extracts containing endogenous cAMP. The slope of the curve obtained in the presence of purified fraction VI is identical to that of the standard curve and that obtained in the presence of fraction IV. Plot determined in the presence of: fraction II, ■—■; fraction III, □—□; fraction IV, ○—○; fraction VI, ●—●. The intercepts at −5.2 and at −3.0 pmol (standard curve) indicate the presence of 2.2 pmol of cAMP in the fraction VI sample (i.e. 2.2 nmol/g dry wt of tissue extracted).

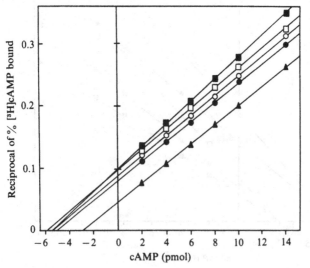

ing endogenous cAMP are shown in Fig. 4.5. The slopes of the curves obtained in the presence of the two most highly purified fractions are identical to the slope of the standard curve. It can be concluded, therefore, that neither of these fractions contain contaminants which interfere with the assay. In contrast, the presence of the less-purified extract caused a change in the slope of the standard curve indicating interference in the assay. These observations show that plants do contain substances which interfere with the binding-protein assay, but that the use of internal standards and purified extracts obviates this problem. The validity of applying the binding-protein procedure to purified plant extracts was further confirmed by comparing the results (2.2 nmol/g dry wt) with those obtained by direct spectrophotometry of similar extracts (2.4 nmol/g dry wt).

A final criticism in the literature concerning cAMP levels in plants is the suggestion that presence of the compound is due to bacterial contamination (Bonnafous, Olive, Borgna & Mousseron-Cadet, 1975). This is refuted by the work of Ashton & Polya (1977) who examined the bacterial content of the plant tissues under consideration and calculated that less than 0.1% of the cAMP was contributed by bacteria. Subsequent work (Ashton & Polya, 1978) with axenic cell cultures of rye grass confirmed this view, which is also supported by studies with axenic cultures of soybean callus tissue (Brewin & Northcote, 1973a) and of tobacco cells (Lundeen, Wood & Braun, 1973).

4.3   Phosphodiesterase

Of the two classes of enzymes concerned with cyclic nucleotide metabolism, the synthetic cyclases and the degradative phosphodiesterases, it is the latter category which has received the more extensive investigation in higher plants. Indeed, the first report of phosphodiesterase activity, in pea seedlings, was published (Lieberman & Kunishi, 1969) before the earliest claims that cAMP was present in plants. This was followed rapidly by demonstrations of phosphodiesterase activity in algae (Amrhein, 1974b), tobacco pith (Wood, Lin & Braun, 1972), potato (Ashton & Polya, 1975; Shimoyama, Sakamoto, Nasu, Shigehisa & Ueda, 1972), barley (Vandepeute, Huffaker & Alvarez, 1973), carrot leaves (Venere, 1972), soybean callus (Brewin & Northcote, 1973b), Jerusalem artichoke (Giannattasio, Sica & Macchia, 1974), and *Phaseolus* seedlings (Brown, Al-Najafi & Newton, 1975, 1977).

Protagonists of a plant regulatory system involving cAMP cited the presence of phosphodiesterase in plants as evidence for the existence of such a system. Others, however, claimed that phosphodiesterases

Table 4.2. *Subcellular distribution of cyclic nucleotide phosphodiesterase in spinach*

Fraction	Specific activity (nmol/min/mg protein)	Total activity in fraction (nmol/min)
Broken chloroplasts	0.27	4.2
Intact chloroplasts	3.30	20.7
Peroxisomes	11.0	17.3
Golgi bodies	4.69	14.7
Nuclei	0.04	0.9
Microsomes	0.98	31.4

of plants showed significant differences in their properties from those of their bacterial and mammalian counterparts and that these differences were indicative of a role in the hydrolysis of 2':3'-cAMP, formed during RNA catabolism, rather than in regulating the concentration of 3':5'-cAMP in tissues. The enzyme from pea seedling (Lin & Varner, 1972), for example, was reported to have greater activity towards 2':3'-cAMP than towards the 3':5'-isomer and produced 3'-AMP as the main hydrolysis product. This enzyme exhibited maximal activity at a much lower pH than cyclic nucleotide phosphodiesterases from bacterial or mammalian sources. Following a survey of the pH optima and substrate specificities of phosphodiesterases from a range of plant tissues and species, Amrhein (1974b) suggested that 3':5'-cAMP was not the natural substrate of higher plant phosphodiesterase. Recent work does not support this conclusion. *Phaseolus vulgaris* has been shown to contain a phosphodiesterase which is active towards a number of 3':5'-cyclic nucleotides but it is inactive towards 2':3'-cAMP (Brown et al., 1975, 1977). Other properties of this enzyme, including its K_m, pH optimum, and sensitivity to methylxanthines, were found to be more like those of mammalian 3':5'-cAMP phosphodiesterase than those of the enzyme described by Lin & Varner (1972). In addition, the *Phaseolus* enzyme produces 5'-AMP as the major product of hydrolysis. A further similarity to the mammalian enzyme system is the existence of an endogenous proteinaceous activator of the phosphodiesterase. The protein extracted from *Phaseolus* not only activates *Phaseolus* phosphodiesterase, but also stimulates bovine brain calmodulin-dependent phosphodiesterase. Thus, it can be concluded that in *Phaseolus* at least, there exists an enzyme comparable to mammalian cyclic nucleotide phosphodiesterase.

Because of the relative ease with which spinach tissue can be fractionated, studies of the subcellular distribution of phosphodiesterase were carried out with *Spinacea* seedlings rather than those of *Phaseolus*

Table 4.3. *Substrate specificity of spinach cyclic nucleotide phosphodiesterase*

Substrate (1 mM)	Phosphodiesterase 1c (nmol/min/mg protein)	Phosphodiesterase 1m (nmol/min/mg protein)
3':5'-cAMP	0.80	0.5
3':5'-cGMP	2.0	0.02
2':3'-cAMP	0.02	1.35
2':3'-cGMP	0.02	1.05

(Brown, Edwards, Newton & Smith, 1979, 1980). Phosphodiesterase activity was found in the intact and broken chloroplast fractions, the peroxisome, Golgi and microsome fractions, but negligible activity was recorded in the nuclear and mitochondrial fractions (Table 4.2). The largest total activity was found in the microsomes, but the highest specific activity was found in the peroxisomes. Gel chromatography of the phosphodiesterase of each fraction resulted in two peaks of activity, and further purification by DEAE cellulose showed that one peak (1c), found in the chloroplast fraction, was different from the corresponding peak (1m), found in the rest of the subcellular fractions. Investigation of these phosphodiesterases yielded information indicating that they differed in their pH optima, sensitivity to inhibition by theophylline, and activation by Ca^{2+}. They also differed in their response to incubation with trypsin; this protease initially inhibited 1m but initially activated 1c. Probably of greatest significance is the fact that they differ in substrate specificity (Table 4.3), with the chloroplast phosphodiesterase showing more activity towards 3':5'-cyclic nucleotides than towards the 2':3'-isomers. The chloroplast enzyme also exhibited greater activity toward guanine nucleotides than towards adenine nucleotides, whereas the microsomal enzyme has greater activity towards 2':3'-cyclic nucleotides than towards the 3':5'-isomers and has a greater activity towards adenine nucleotides. A further difference between 1c and 1m is the response to endogenous proteinaceous effectors of phosphodiesterase. Whereas 1m was unaffected by any of the proteins tested, 1c (Fig. 4.6) was sensitive to both an endogenous activator and an endogenous inhibitor.

The conclusion that the phosphodiesterases of the chloroplast and the rest of the cell are different enzymes was confirmed by a more detailed study. Both enzymes were shown to be part of a multienzyme complex (Brown *et al.* 1980). However, the chloroplast phosphodiesterase was present with ATPase, ribonuclease and acid phosphatase, whereas the microsomal phosphodiesterase was in a complex with ATPase, ribonuclease and nucleotidase. The two phosphodiesterases also differed in molecular weight, isoelectric point, chromatographic behaviour, and sensitivity to chaotropic agents. Study of the microsomal complex suggested that the 2':3'-cyclic nucleotide products of the ribonuclease were the natural substrates of the phosphodiesterase. Thus, in microsomes it appears that at least one phosphodiesterase does function in the sequential breakdown of messenger RNA, i.e. it has a role the same as that suggested for plant phosphodiesterases by Lin & Varner (1972) and Amrhein (1974b). The chloroplast phosphodiesterase, however, in view of its greater activity towards the 3':5'-isomers and lack of activity with 2':3'-isomers, would appear not to be involved in the same role. In addition to substrate specificity, other properties comparable with those of the mammalian 3':5'-phosphodiesterase include the pH optimum, inhibition by theophylline, activation by Ca^{2+} and by trypsin, and sensitivity to endogenous proteinaceous inhibitors and activators, and suggest that the chloroplast enzyme is primarily involved in the hydrolysis of 3':5'-cAMP. The subcellular location of the two types of spinach phosphodiesterase may explain the findings of the earlier work with

Fig. 4.6. Inhibition and activation of cyclic nucleotide phosphodiesterase by protein fractions from spinach chloroplasts. Peak 1c cyclic nucleotide phosphodiesterase from *Spinacea oleracea* was freed from effectors and assayed in the presence and absence of chloroplast protein fractions separated by gel-filtration on Sephadex G-100. The effect of any intrinsic phosphodiesterase activity in these fractions was taken into account in calculating the percentage activation or inhibition. Protein content, ■—■; cyclic nucleotide phosphodiesterase activity, ○—○; % inhibition or activation, ●—●

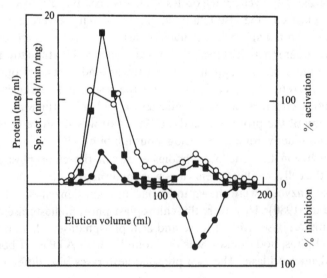

pea seedlings (Lin & Varner, 1972), in which the methods used could have caused selective extraction and purification of the microsomal enzyme.

Reports of 3':5'-cyclic nucleotide phosphodiesterase in plants are not confined to the work with *Phaseolus* and *Spinacea*. Three phosphodiesterases for example have been demonstrated in potato and while one has greatest activity as a NAD pyrophosphatase and a second most activity with 2':3'-cAMP as substrate, the third displays its major activity when 3':5'-cAMP is the substrate (Ashton & Polya, 1975). In *Portulaca* there are also at least three phosphodiesterases (Endress, 1978). Again one has greater activity with 2':3'-cyclic nucleotides as substrate, while a second, non-allosteric enzyme, shows more activity towards 3':5'-cyclic nucleotide substrates. The third phosphodiesterase displayed positive cooperativity, and although in the presence of papaverine it had higher activity towards 2':3'-cyclic nucleotide substrates, in the presence of nucleotides such as cyclic GMP a higher activity towards 3':5'-cAMP and 3':5'-cyclic GMP became evident. Thus, in at least some plant tissues, phosphodiesterases exist that have as their predominant function hydrolysis of 3':5'-cAMP and these enzymes are susceptible to intracellular regulation.

4.4 Adenylate cyclase

Adenylate cyclase catalyses formation of cAMP, from ATP, according to the equation:

$$[ATP \rightarrow 3':5'\text{-cAMP} + PP_i]$$

Because it is only present as a very small proportion of the total cell protein and in most cases is membrane-bound rather than 'soluble', this enzyme is often difficult to detect in biological extracts. The difficulty is exacerbated by the concomitant presence in crude biological extracts of active phosphodiesterases and ATPases which compete for the same substrate, and by cyclic nucleotide phosphodiesterases which hydrolyse the product of the cyclase reaction, that is cAMP.

The first reported occurrence of adenylate cyclase in the tissues of higher plants was in 1970 (Pollard, 1970). It was based on the observation that [8-^{14}C]adenine is incorporated, by preparations of barley aleurone layer, into a compound chromatographically indistinguishable from cAMP. Several similar demonstrations with a variety of plant tissues followed (Salomon & Mascarenhas, 1971; Brewin & Northcote, 1973a; Srivastava & Krishna Murti, 1972). However, on the grounds that the chromatographic procedures used were not sufficiently rigorous, some critics questioned identification of the radioactive product as cAMP.

Others, on the basis of their own negative results, went further and concluded that adenylate cyclase does not occur in plants (Hintermann & Parish, 1979; Yunghams & Marré, 1977). Of the earlier studies in which positive results were obtained, only two involved direct conversion of ATP to cAMP by a cell-free system (Janistyn, 1972; Giannattasio & Macchia, 1973). More recent work in our laboratory (Brown *et al.* 1979; Smith, Brown, Newton, Al-Najafi & Edwards, 1977) succeeded in demonstrating the presence of adenylate cyclase in cell-free enzymatic preparations obtained from *Hordeum* and *Phaseolus* seedlings. In both cases, the identity of the product of the reaction as cAMP was confirmed by a dual-labelling isotope technique in conjunction with the rigorous chromatographic and electrophoretic method used to obtain pure samples for the mass spectrometric identification of cAMP in plant extracts (Newton *et al.* 1980).

Although not obviated from a quantitative point of view, the difficulty arising from contaminating phosphates and ATPases was considerably lessened by having an internal ATP-regenerating system such as phosphocreatine/creatine phosphate present in the assay. In contrast to animal tissues, in plant tissues methylated xanthines such as theophylline have little or no effect in suppressing the residual activity of cyclic nucleotide phosphodiesterases.

Application of these techniques to chloroplasts, isolated in an intact state, showed adenylate cyclase to be present in these organelles where it is associated with the membrane fraction. This demonstration of adenylate cyclase activity within intact chloroplasts did, however, necessitate first disrupting the organelles in order to overcome problems of substrate permeability. A more recent modification of the application of the technique to chloroplasts has been to include in the adenylate cyclase incubation mixture, a cAMP binding protein to sequester the nascent cAMP before it can be attacked by residual phosphodiesterase activity. Demonstration of adenylate cyclase in chloroplast membranes confirms the earlier reported observation that *Avena* etioplasts convert [^{14}C]adenine into [^{14}C]-cAMP (Wellburn, Ashby & Wellburn, 1973). The association of the enzyme with membranes is also compatible with the work of Al-Azzawi & Hall (1976) who demonstrated, cytochemically, the presence of adenylate cyclase in the plasma membranes, endoplasmic reticulum, and nuclear membranes of maize tissue. It is pertinent to note, in this context, that the adenylate cyclase of animal tissues is also essentially located in the membrane fractions of the cell.

There are reports that some plant growth substances affect the activity of adenylate cyclase in plants. Janistyn (1972), working with homogenates of maize coleoptiles, reported that the rate of cAMP synthesis was more than doubled in the presence of 0.1 M IAA. Studying adenylate cyclase activity in homogenates of Jerusalem artichoke tubers, Giannattasio & Macchia (1973), observed that mM concentrations of gibberellic acid elicited a 300% increase.

4.5   cAMP-binding proteins

In those microorganisms in which cAMP functions as a primary messenger, a central role is played by a receptor which binds this nucleotide specifically and, in so doing, initiates a chain of metabolic events (Pastan & Perlman, 1972). In mammalian tissues, where cAMP serves as a secondary messenger to a number of hormones, a specific binding protein is again involved as the initial receptor. In this latter case, the protein involved is a regulatory component unit of a cAMP-dependent protein kinase (Langan, 1973). Presence of a cAMP-binding protein in a higher plant was first demonstrated as long ago as 1973 (Anderson & Pastan, 1973). Until recently, however, no cAMP-dependent protein kinase activity had been found in plant tissues and this was taken to be a fundamental difference between plant and animal tissues. We had previously suggested (Brown & Newton, 1981) that the failure to detect cAMP-dependent protein kinases in plants may be a reflection of the experimental approaches adopted. It is noteworthy in this context that an active protein kinase extracted from wheat embryos and shown to be insensitive to cAMP was found, after extensive purification, to cochromatograph with the cAMP-binding fraction (Carrutu, Manzocchi, Lanzani & Giannattasio, 1974). A similar observation was made with extracts obtained from tubers of Jerusalem artichoke (Giannattasio, Carrutu & Tucci, 1974). Furthermore, in seeking for cAMP-dependent protein kinases in plants, relatively few substrates have been used and most of these were of animal rather than of plant origin. Notwithstanding these technical problems Kato, Uno, Ishikawa & Fujii (1983) have purified three protein kinases from *Lemna*, one of which is activated by cAMP, cyclic GMP, and cyclic IMP. Another is partially inhibited by these nucleotides and the third is independent of cAMP. The protein fraction exhibiting the cAMP-activated kinase activity shows presence of a single cAMP-binding protein. It should be borne in mind that of all the protein kinases extracted from animal tissues, only a small proportion are demonstrably cyclic nucleotide dependent (Langan, 1973). Other factors contributing to the difficulty in finding cyclic nucleotide dependent protein kinases in plants may be (i) possible requirement by the kinase for absent modulator proteins, (ii) that the protein kinase has already been activated, *in vivo*, before extraction, and (iii) that the activating cyclic nucleotide is not cAMP but another purine or pyrimidine cyclic nucleotide, e.g. arthropod protein kinases require activation by cyclic GMP (Kuo, Wyatt & Greengard, 1971). A stable protein which can modulate the activity of cyclic nucleotide-dependent protein kinases from animal tissues has been isolated from a variety of animal and microbial sources. This same protein modulator has also been obtained from a higher plant source, namely sweet potato (Donnelly, Kuo, Reyes, Liu & Greengard, 1973*b*). It acts to alter the substrate specificity of both cAMP- and cyclic GMP-dependent protein kinases, increasing the activity towards some substances and decreasing it towards others. Several factors were found to be involved in determining whether the effect was stimulatory or inhibitory. These include type and concentration of substrate and type and concentration of cyclic nucleotide (Donnelly, Kuo, Miyamoto & Greengard, 1973*a*).

It is possible that the cAMP-binding proteins found in higher plants have properties and functions analogous to their counterparts from microorganisms rather than those of the corresponding proteins in animals. A pointer in this direction is the finding, in extracts of wheat embryos and potato, of highly specific 5′-nucleotidases which are competitively inhibited by cAMP and other cyclic nucleotides (Polya, 1975; Polya & Sia, 1976; Polya, Ashton & Sia, 1976). The K_i values for these enzymes with cAMP and cyclic GMP are of a micromolar order. With respect to molecular weight, subunit composition and K_d values for cAMP and cyclic GMP, the enzyme from potato is similar to the cAMP receptor protein of *E. coli*. The plant enzyme and the microbial enzyme also behave in very similar ways during purification (Polya, 1975).

Experiments with extracts from *Phaseolus* and *Hordeum* seedlings have also demonstrated the presence of cAMP-binding activity amongst the proteins of higher plant tissues (Brown *et al.* 1979; Smith, Brown, Newton & Edwards, 1979). Further work with the protein fraction of *Hordeum vulgare* resulted in the partial purification of a specific cAMP-binding protein (Brown, Newton & Smith, 1980). The apparent molecular weight of this protein is 1.7×10^5. Its pH optimum for binding is at pH 6.5 with a sharp fall in activity below pH 6 and above pH 7.5. Under optimal conditions, binding is maximal at 60 min incubation and half-maximal at 15 min incubation.

The cAMP-binding protein from *Hordeum* shows high specificity (Brown *et al.* 1980). Even at concentrations in excess of 60 times that of cAMP, there is no binding of a range of related nucleotides. Amongst those examined in this context were 2′:3′-cAMP, 2′-AMP, N^6-2′-*O*-dibutyryl cAMP and the 3′:5′-cyclic nucleotides of xanthosine, thymidine, and 2′-deoxythymidine. Some slight affinity was, however, exhibited for adenosine, 3′-AMP, 5′-AMP, ADP, 3′:5′-cyclic dAMP (2′-deoxyadeno-sine-3′:5′-cyclic monophosphate) and 3′:5′-cGMP. With 3′:5′-cyclic IMP there was negligible binding at low concentrations but significant binding at higher concentrations. Examination of equilibrium binding data by means of a Scatchard plot, yielded a dissociation constant of 8 nM for the cAMP-binding protein complex. This is of the same order as that (2–3 nM) of a typical mammalian cAMP-binding protein, obtained from bovine skeletal muscle (Gilman & Murad, 1974). On the evidence of the K_d values, the binding protein from barley seedlings binds cAMP much more tightly than does the protein obtained from Jerusalem arti-choke tubers (Giannattasio *et al.* 1974) or that from wheat embryo (Polya & Sia, 1976; Polya & Bowman, 1981). Recently, a commercial prepar-ation of wheat germ has been used as a source of a cAMP-binding protein (Polya & Bowman, 1981). The protein so obtained has an appar-ent molecular weight of 5.2×10^5 in media of low ionic strength and 2.8×10^5 in solutions of high ionic strength. These values are of the same order as that determined for the *Hordeum* protein (see above). Further resolution of the wheat germ protein yielded two active binding fractions, have K_d values for cAMP of 0.1–1.0 μM. The two fractions exhibited a marked adenine-analogue binding specificity and cAMP was the best ligand of the naturally occurring nucleotides tested. Whereas one of these proteins was phosphorylated in reactions catalysed by endo-genous protein kinases, there was no evidence of modulation of this activity by cAMP. Recently, Endress (1983) has extracted cAMP-binding proteins from callus cultures of *Portulaca grandiflora*. The binding ac-tivity of these is increased by cAMP and other nucleotides, and by adeno-sine and kinetin. Endress suggests that there is an allosteric-regulated cooperativity of the binding protein (perhaps hormone mediated) and that this may be related to activation of tyrosinase or DOPA decarboxy-lase in the betalain-synthesizing callus with which he has worked.

4.6   cAMP and the physiological processes of higher plants
      The earliest investigations of the physiological role of cAMP in plants were influenced by the secondary messenger concept of the cyclic nucleotide effects in mammals; speculation followed that a similar mediatory relationship with phytohormones existed. Various approaches have been adopted to evaluate experimentally such a hypothesis. These include examination of the effects of exogenously applied cAMP, or of its more stable, membrane-penetrable derivative N^6-2′-O-dibutyryl

cAMP, and of the effect of phytohormones on the synthesis of cAMP. It is generally accepted that with mammalian systems, the criteria for ascribing a secondary messenger role to cAMP, i.e. linking a hormone and physiological response, are that (a) administration of the hormone to a whole organism causes an elevation of the cAMP level in the target tissue, (b) administration of cAMP or a cAMP derivative to the target tissue produces the same physiological response of the hormone, (c) addition of the hormone to a cell-free homogenate of the target tissue produces an increase in the rate of conversion of ATP to cAMP, (d) simultaneous addition of cAMP and the hormone potentiates the effect of the hormone in a cell-free system, and (e) addition of phosphodiesterase inhibitors to either a cell-free system or a whole organism elicits the same physiological response as the hormone. The physiological response so monitored is often the activity of a single enzyme. In the majority of cases, the involvement can also be demonstrated of a cAMP-dependent protein kinase which increases or decreases the target enzyme activity.

In the plant kingdom, even omitting the requirement for a cyclic nucleotide dependent protein kinase, there is no single example known in which criteria (a) to (e) are fulfilled. Most of the investigations into a possible phytohormone-cAMP relationship have centred on gibberellin (Table 4.4). Gibberellin is reported to stimulate the incorporation of [8-^{14}C]adenine into cAMP and to elevate the concentration of cAMP in intact tissues (Table 4.4). The activity of a number of enzymes, particularly those hydrolytic enzymes released at the onset of germination, is induced or stimulated by GA_3, and, in the absence of the phytohormone, the same effect is elicited by exogenously supplied cAMP. In addition to the indications that cAMP can play a role in seed germination, there is also evidence to suggest that cAMP produces the same effect as gibberellin in the growth response of lettuce, barley, soybean, *Tradescantia* and *Avena* (Table 4.4). However, while it is clear that the cyclic nucleotide is capable of simulating the effects of gibberellin, it is not established that cAMP mediates the effects of the hormone. Even if it does, there is a significant divergence from the secondary messenger system functioning in animals.

Auxins have also been reported to elevate the level of cAMP in a number of plant tissues (Table 4.5), and various effects induced by IAA can be similarly elicited by cAMP. Again, this evidence serves only to indicate that the cyclic nucleotide is capable of simulating the effects of the hormone, and is by no means conclusive evidence of a mediatory role *in vivo*. A number of the enzymatic activities of diverse species

Table 4.4. *Physiological effects involving cAMP and gibberellins*

Effect	Tissue	Reference
GA₃ stimulates [¹⁴C]-Ad → '[¹⁴C]-cAMP'	Barley	Pollard (1970)
GA₃ elevates cAMP concentration	Maize etiolated	Tarantowicz-Marek & Kleczkowski (1978)
	Jerusalem artichoke	Giannattasio & Macchia (1973)
GA₃ and cAMP stimulate — ATPase	{ Barley seed	Earle & Galsky (1971)
	{ Phaseolus	Maslowski, Maslowski & Urbanski (1974)
α-amylase	Barley seed	Galsky & Lippincott (1969); Pollard (1972); Duffus & Duffus (1969); Nickels & Galsky (1970); Kessler & Kaplan (1972)
Protease	Barley seed	Nickels, Schaeffer & Galsky (1971)
Acid phosphase	Barley seed	Nickels *et al.* (1971)
Isocitrate lyase	{ Castor beans	Marriott & Northcote (1977)
	{ Hazel seeds	Potempa & Galsky (1973)
Ribonuclease	Cowpea seedlings	Kapoor & Sachar (1976)
Polyribosome formation	{ Maize	Tarantowicz-Marek & Kleczkowski (1975)
GA₃- and cAMP-induced synthesis of α-amylase blocked by abscisic acid	{ Barley	Rao & Khan (1975)
	{ Barley seed	Barton, Veerbeek, Ellis & Khan (1973)
	Rice endosperm	Roy, Ghose & Sircar (1973)
GA₃ and cAMP synergistically enhance dark germination	Lettuce seeds	Kamisaka & Masuda (1971)
		Hall & Galsky (1974)
GA₃ and cAMP enhance light germination	*Phacelia* seeds	Safran & Galsky (1974)
	{ *Avena*	Hartung (1973)
GA₃ and cAMP enhance IAA growth stimulation	{ Lettuce	Kamisaka, Sano & Katsumi (1972)
		Bianco & Bulard (1974)
	{ Barley	Gilbert & Galsky (1972)
		Holm & Miller (1972)
GA₃ and cAMP stimulate elongation of lypocotyls	{ Lettuce	Kamisaka, Sakurai & Masuda (1973)
	{ Soybean (derooted)	Holm (1972)
GA₃ and cAMP promote pollen tube elongation	*Tradescantia*	Malik, Chabra & Vermani (1976)

138

Table 4.5. *Physiological effects involving cAMP and auxins*

Effect	Tissue	Report
IAA elevates cAMP concentration	*Avena*	Saloman & Mascarenhas (1972)
	Soybean callus	Brewin & Northcote (1973a)
	Cicer	Azhar & Krishna Murti (1971)
	Maize	Janistyn (1972)
Delay petiole abscission	*Coleus*	Saloman & Mascarenhas (1971)
Enhance elongation of etiolated coleoptiles	Corn	Janistyn (1978)
	Wheat	Weintraub & Lawson (1972)
IAA and cAMP — Promote growth	Sunflower callus	Truelsen, Langhe & Verbeek-Wyndaele (1974)
	Oats	Ockerse & Mumford (1972)
	Artichoke	Kamisaka, Sakurai & Masuda (1973)
Synergistic effect on differentiation	Lettuce	Basile, Wood & Braun (1973)

and tissues have been reported to be activated by cAMP (Table 4.6), and there are at least two reports of plant enzymes being inhibited by cAMP. In no case however is there a valid dose/response curve or any firm evidence of a direct allosteric effect upon the enzymes. Thus, although modulation has been demonstrated, there is no reason to assume that this is part of an *in vivo* process.

Several other effects of exogenous cAMP have been reported (Table 4.7). While these authors have confirmed satisfactorily the effects cited, there is no supporting evidence that the nucleotide mediates the activity of any agent known to control these responses *in vivo*. One possible exception is the induction of betacyanin synthesis in *Amaranthus*, in which the requirement for light is lost in the presence of dibutyryl cAMP. With intact *Phaseolus* and *Zea* plants and isolated plastids of *Avena*, there appears to be a close relationship between the cAMP concentration and illumination (Table 4.8). It has also been suggested that there is a light-dependent gradient of cAMP in the root and shoot of maize. While some investigations (DeNicola, Amico & Piatteli, 1975; Elliott & Murray, 1975) suggest that cAMP is not a secondary messenger in light-induced betacyanin synthesis, it has been reported that the cAMP level in mustard seedlings responds to a 5-min pulse of red light which operates exclusively through phytochrome (Janistyn & Drumm, 1975). In studies in our own laboratory (unpublished), we have observed that the K_m of chloroplast cAMP phosphodiesterase decreases by an order of magnitude when *Spinacea oleracea* is kept in the dark for 20 h. This may indicate that it is the phosphodiesterase rather than the cyclase which has the major regulatory influence on cAMP concentrations.

The overall picture of the physiological effect(s) of cAMP in plants is of a range of effects, some of which parallel effects induced by hormones, while others do not. In the majority of these reports, the observations recorded appear to be valid, but even in observations verified by several different authors, there is no case in which a satisfactory molecular mechanism has been demonstrated. Thus, while these reports form a basis upon which to build further investigations, they can be regarded only as an indication of possible functions of cAMP and not as substantive evidence of an established role.

4.7 Discussion

As has been described in the foregoing sections of this chapter, plant tissues contain all the essential components of cyclic nucleotide-mediated regulatory systems. In addition to cAMP, a specific cAMP-binding protein, adenylate cyclase, and cyclic nucleotide phosphodiester-

Table 4.6. *Effect of exogenous cAMP on plant enzyme activity*

Enzyme	Tissue	Report
Plant enzymes induced or activated		
Acid phosphatase	Barley seed	Nickels, Schaeffer & Galsky (1971)
Protease	Barley seed	Marriott & Northcote (1977)
Iscitrate lyase	Castor beans	Potempa & Galsky (1973)
	Hazel seeds	Higgins, Goodwin & Carr (1974)
Nitrate reductase	Mung bean roots	Habaguchi (1977)
Polyphenol oxidase	Carrot callus	Kapoor & Sachar (1976)
Ribonuclease	Cowpea seedling	Duffus & Duffus (1969); Kessler & Kaplan (1972)
α-amylase	Barley seed	Earle & Galsky (1971)
ATPase	Barley seed	
Plant enzymes inhibited		
5'(3')-ribonucleotide phosphohydrolase	Wheat seedling	Polya & Ashton (1974)
5'-nucleotidase	Potato	Polya (1975)

Table 4.7. *Some other observed physiological effects of cAMP*

Effect	Tissue	Report
Reversal of flowering inhibition	Lemna	Posner (1973)
		Oota & Kondo (1974)
Enhancement of RNA synthesis	Maize	Tarantowicz-Marek & Kleczkowski (1975)
Induction of betacyanin synthesis	Amaranthus	DeNicola, Amico & Piatteli (1975
		Elliott & Murray (1975)
Synthesis of terpenoid phytoalexins	Sweet potato root	Ogumi, Suzuki & Uritani (1976)
Initiation of pollen germination	Pinus	Katsumata, Takahashi & Ejiri (1978)
Activation of antiviral factor	Nicotiana	Rast, Skrivanova & Bachofen (1973)
		Sela, Hauscher & Mozes (1978)
		Gat-Edelbaum, Altman & Sela (1983)
Lengthening periodicity of leaf movement	Trifolium	Bollig, Mayer, Mayer & Engelmann (1978)
Regulation of H$_2$O transport	Vicia	Karimova & Gusev (1980)
Activation and inhibition of	Lemna	Kato, Uno, Ishikawa & Fujii (1983)
soluble protein kinases		
Depression of glucose effect on	Asparagus	Tassi, Restivo, Puglisi & Cacco (1984)
synthesis of glutamate dehydrogenase		
and acid phosphatase		

142

Table 4.8. *Effect of light on cAMP concentration*

Tissue	cAMP level	Report
Phaseolus		
Light-grown shoot	4.6	Brown & Newton
Dark-grown shoot	0.6 pmol/mg dry wt	(unpublished observation)
Dark-grown shoot, illuminated 18 h	15.2	
Avena		
Etioplasts	67 pmol/mg protein	Wellburn, Ashby & Wellburn (1973)
Etio-chloroplast, illuminated 1 h	350	
Zea		
Dark-grown shoot	0.1–1.8	
Light-grown shoot	0.2–2.4	Bachofen (1973)
Dark-grown root	2.8–4.2 pmol/mg dry wt	
Light-grown root	0.1–0.5	

ase, calmodulin has also been shown to be present in plant tissues (Anderson, Charbonneau, Jones, McCann & Cormier, 1980). This poses the question as to whether a cyclic nucleotide-mediated control mechanism is functional in plants and, if so, what form it takes. Is it likely to be similar to that of animal tissues, in which the cyclic nucleotide mediates the effects of non-maintenance hormones? If this is the case, we would need to be looking at the short-term physiological responses of plants. Such examples as the seismonastic movements of *Mimosa pudica*, the onset of seed germination, and the various light-related responses are the type of phenomena in which cyclic-nucleotide mediation may be involved. There are, however, some fundamental differences between the phytohormones and animal non-maintenance hormones. Those animal hormones the effect of which is mediated by cAMP, do not penetrate the cell but trigger the adenylate cyclase system on the cell membranes. It is the newly synthesized cyclic nucleotides, released inside the cell, which produce the specific response. More commonly, phytohormones produce a wider range of effects and are able to penetrate cells. Although, for these reasons, a close similarity between plant and animal hormonal mechanisms would seem to be precluded, some phytohormonal mechanisms may, nevertheless, be mediated by cAMP.

Another possibility is that it is not a phytohormone which controls cAMP concentration but rather the calmodulin-Ca^{2+} complex. One of the important roles of the calmodulin-Ca^{2+} complex in animal tissues is to regulate cyclic nucleotide metabolism and thereby couple the two second messengers, Ca^{2+} and cAMP.

In comparing the higher plant with mammalian systems, it is pertinent to consider also regulation of cell proliferation. It has been suggested that in mammalian cells cAMP controls the expression of cell maturation and cell proliferation (Cho-Chung, 1980); evidence that phosphodiesterase inhibitors inhibit cell elongation and postgerminative growth (Levi, Sparvoli & Galli, 1981, 1982) in *Haplopappus* and that both these inhibitors and dibutyryl cAMP induce a block in the cell cycle in G1 (Levi, Chiatante & Sparvoli, 1984) in the same organism indicates that a parallel involvement may be present in plants.

In discussing the possible role of cAMP in higher plants, it may well be more relevant to consider possible similarity with microbial systems. The 'primary messenger' role of cAMP in bacteria is related to the nutritional requirements of an individual cell. A similarly functioning system in a multicellular, autotrophic plant would therefore seem feasible. The existence in plant tissues of an intercellular cAMP communication system, possibly signalling nutritional status, has been previously suggested by Becker & Ziegler (1973) and by Bachofen (1973). These views are supported by the recent studies of Tassi, Restivo, Puglisi & Cacco (1984), in which cAMP was shown to derepress the effect of glucose on the synthesis of glutamate dehydrogenase and acid phosphatase in single-cell cultures of *Asparagus*. The endosymbiont theory of the origin of chloroplasts and mitochondria, which suggests that these organelles had their evolutionary origin in endosymbiotic prokaryotes, points also to the possibility of cyclic nucleotides functioning in an intracellular signalling system. Amongst the roles of such a system in higher plants could be regulation of the permeability of the organelle membrane, or even that of the plasmalemma, to essential metabolites.

Consideration of the known functions of cyclic nucleotides in microorganisms also indicates that in the tissues of higher plants these compounds may be involved in regulating gene expression. In bacteria, according to their metabolic requirements, cAMP controls repression or synthesis of inducible enzymes. Involvement of cAMP in regulating nucleic acid synthesis in higher plants has been suggested by the work of Tarantowicz-Marek & Kleczkowski (1975), and of Kapoor & Sachar (1976). Lanzani, Giannattasio, Manzocchi, Bollini, Soffientini & Macchia (1974) have presented evidence that cyclic GMP is also implicated in this area of plant metabolism, stimulating polypeptide synthesis in a cell-free system obtained from wheat embryos.

Substantial evidence has now been amassed to show that, in animal tissues, cyclic GMP is a unique component in a complex network of regulation (Goldberg & Haddox, 1977). This involves lipid metabolism, the visual process, lymphocyte proliferation, antibody production, leukocyte chemotaxis and cell-mediated toxicity, and there is also evidence for cyclic GMP involvement in regulation of the synthesis or expression of specific mRNA molecules in both prokaryotic and eukaryotic systems, including microbial and animal systems. A variety of other cellular functions of animal tissues have been shown to be affected by exposure of the intact cells to exogenous cyclic GMP. Numerous agents, including neurotransmitters and the calmodulin-Ca^{2+} complex have been shown to alter the steady-state concentrations of cellular cyclic GMP. (For a review of this field, see Goldberg & Haddox, 1977.) Although cAMP and cyclic GMP play a number of roles independently of one another, there are also biological systems in which they play opposing roles (Yin-Yang hypothesis, Nawrath, 1976; Goldberg *et al.* 1975). This needs to be considered when planning and appraising experiments designed to investigate the effects of cAMP on various plant tissues. Thus, in order to make interpretations of analytical results, it is advisable to determine the concentrations of both nucleotides.

Although a regulatory role for cyclic GMP in plant tissues has not yet been firmly established, the occurrence of this cyclic nucleotide in plants has been the subject of a number of reports. Using a radioimmunoassay, Lundeen *et al.* (1973) found this nucleotide in excised tobacco pith cultures, and Haddox, Stephenson & Goldberg (1974) discerned its presence in meristematic and elongating regions of bean root. Intriguingly high concentrations of cyclic GMP (10–35 nmol g dry wt) have been found in fruits of *Evodia rutaecarpa* and *E. officinalis* (Cyong, Takahashi, Hanabusa & Otsuka, 1982). More recently, Janistyn (1983) has used a GC/MS technique to show the presence of cyclic GMP in a plant tissue. Newton, Kingston, Evans, Younis & Brown (1984) have now unequivocally confirmed identification of the compound (from *Phaseolus vulgaris*) as guanosine 3′:5′-cyclic monophosphate, using NMR and FAB-mass spectrometry with MIKES scanning. They report a concentration of 33 pmol/10 g fresh wt of leaves. In addition, these latter authors have demonstrated the presence in *Phaseolus* tissues of a guanylate cyclase and a cyclic GMP phosphodiesterase.

It is only by a clearer understanding of the properties, subcellular location, and regulation of adenylate and guanylate cyclase and cyclic nucleotide phosphodiesterases, and of the distribution, properties and metabolic effects of the activated cAMP-binding protein, that we will begin to answer the questions posed in this chapter. In particular, as interest in Ca^{2+} metabolism and calmodulin in plant tissues develops, we can expect further elucidation of the involvement of cyclic nucleotides in the metabolism of higher plants.

The authors thank Pergamon Press Ltd for permission to reproduce Figs. 4, 5, and 6 and Table 3 from papers published by them in *Phytochemistry*.

4.8   References

Al-Azzawi, M. J. & Hall, J. L. (1976).Cytochemical localization of adenyl cyclase activity in maize root tips. *Plant Sci. Lett.* **6**: 285–9.

Amrhein, N. (1974a). Evidence against the occurrence of cyclic AMP in higher plants. *Planta* **118**: 241–58.

Amrhein, N. (1974b). Cyclic nucleotide phosphodiesterases in plants. *Z. pflanzenphysiol.* **72**: 249–61.

Amrhein, N. (1977). The current status of cyclic AMP in higher plants. *Ann. Rev. Physiol.* **28**: 123–32.

Anderson, J. A., Charbonneau, H., Jones, H. P., McCann, R. O. & Cormier, M. J. (1980). Characterization of the plant nicotinamide adenine dinucleotide kinase activator protein and its identification as calmodulin. *Biochemistry* **19**: 3113–20.

146   R. P. Newton and E. G. Brown

Anderson, W. B. & Pastan, I. (1973). The cyclic AMP receptor of *E. coli:* immunological studies in extracts of *E. coli* and other organisms. *Biochim. Biophys. Acta* **320:** 570–7.

Ashton, A. R. & Polya, G. M. (1975). Higher plant cyclic nucleotide phosphodiesterases. *Biochem. J.* **149:** 329–39.

Ashton, A. R. & Polya, G. M. (1977). Adenosine 3':5'-cyclic monophosphate in higher plants. *Biochem. J.* **165:** 27–32.

Ashton, A. R. & Polya, G. M. (1978). Cyclic AMP in axenic Rye grass endosperm cell culture. *Plant Physiol.* **61:** 718–22.

Azhar, S. & Krishna Murti, C. R. (1971). Effect of indole-3-acetic acid on the synthesis of cyclic 3':5' adenosine phosphate by Bengal gram seeds. *Biochem. Biophys. Res. Commun.* **43:** 58–64.

Bachofen, R. (1973). Distribution of cyclic AMP in maize seedlings. *Plant. Sci. Lett.* **1:** 447–50.

Barton, K. A., Veerbeek, R., Ellis, R. & Khan, A. A. (1973). Abscisic acid inhibition of GA₃ and cyclic AMP induced α-amylase synthesis. *Physiol. Plantarum* **29:** 186–93.

Basile, D. V., Wood, H. N. & Braun, A. C. (1973). Programming of cells for death under defined experimental conditions: relevance to the tumour problem. *Proc. Nat. Acad. Sci. U.S.A.* **70:** 3055–9.

Becker, D. & Ziegler, H. (1973). Cyclisches Adenosin-3':5'-monophosphat in pflanzlichen Leitbahnen. *Planta* **110:** 85–9.

Bianco, J. & Bulard, C. (1974). Cyclic AMP and germination of lettuce seeds. *Z. Pflanzenphysiol.* **74:** 160–4.

Bollig, I., Mayer, K., Mayer, W. & Englemann, W. (1978). Effect of cyclic AMP, theophylline, imidazole and 4-(3,4-dimethoxylbenzyl)-2-imidazolidone on the leaf movement rhythm of *Trifolium repens* – a test of the cyclic AMP hypothesis of circadian rhythms. *Planta* **141:** 225–30.

Bonnafous, J. C., Olive, J. L., Borgna, J. L. & Mousseron-Cadet, M. (1975). L'AMP cyclique dans les graines et les plantules d'orge et la contamination bactérienne ou fongique. *Biochimie* **57:** 661–3.

Bressan, R. A., Handa, A. K., Cherniak, J. & Filner, P. (1980a). Synthesis and release of adenosine 3':5'-cyclic monophosphate by *Chlamydomonas reinhardtii*. *Phytochem.* **19:** 2089–93.

Bressan, R. A., Handa, A. K., Quader, H. & Filner, P. (1980b). Synthesis and release of adenosine 3':5'-cyclic monophosphate by *Ochromonas malhamensis*. *Plant Physiol.* **65:** 165–70.

Brewin, N. J. & Northcote, D. H. (1973a). Variations in the amounts of 3':5'-cyclic AMP in plant tissues. *J. Exp. Bot.* **24:** 881–8.

Brewin, N. J. & Northcote, D. H. (1973b). Partial purification of a cyclic AMP phosphodiesterase from soybean callus. Isolation of a non-dialysable inhibitor. *Biochim. Biophys. Acta* **320:** 104–22.

Brown, B. L., Albano, J. D. M., Elkins, R. P. & Sgherzi, A. M. (1971). A simple and sensitive saturation assay method for the measurement of adenosine 3':5' cyclic monophosphate. *Biochem. J.* **121:** 561–2.

Brown, E. G., Al-Najafi, T. & Newton, R. P. (1975). Partial purification of adenosine 3':5'-cyclic monophosphate phosphodiesterase from *Phaseolus vulgaris* L.: Associated activator and inhibitors. *Biochem. Soc. Trans.* **3:** 393–5.

Brown, E. G., Al-Najafi, T. & Newton, R. P. (1977). Cyclic nucleotide phosphodiesterase activity on *Phaseolus vulgaris*. *Phytochem.* **16:** 1333–7.

Brown, E. G., Al-Najafi, T. & Newton, R. P. (1979). Adenosine 3':5'-cyclic monophosphate, adenylate cyclase and a cyclic AMP binding protein in *Phaseolus vulgaris*. *Phytochem.* **18:** 9–14.

Brown, E. G., Edwards, M. J., Newton, R. P. & Smith, C. J. (1979). Plurality of cyclic nucleotide phosphodiesterase in *Spinacea oleracea*: subcellular distribution, partial purification and properties. *Phytochem.* **18**: 1943–8.

Brown, E. G., Edwards, M. J., Newton, R. P. & Smith, C. J. (1980). The cyclic nucleotide phosphodiesterase of spinach chloroplasts and microsomes. *Phytochem.* **19**: 23–30.

Brown, E. G. & Newton, R. P. (1973). Occurrence of adenosine 3':5'-cyclic monophosphate in plant tissues. *Phytochem.* **12**: 2683–5.

Brown, E. G. & Newton, R. P. (1981). Cyclic AMP and higher plants. *Phytochem.* **20**: 2453–63.

Brown, E. G., Newton, R. P. & Smith, C. J. (1980). A cyclic AMP binding protein from barley seedlings. *Phytochem.* **19**: 2263–6.

Carratu, G., Manzocchi, L. A., Lanzani, G. A. & Giannattasio, M. (1974). Soluble and ribosome associated protein kinases from wheat embryos. *Plant Sci. Lett.* **3**: 313–21.

Chataway, F. W., Wheeler, P. R. & O'Reilly, J. (1981). Involvement of cyclic AMP in the germination of blastospores of *Candida albicans*. *J. Gen. Microbiol.* **123**: 233–40.

Cho-Chung (1980). On the mechanism of cyclic AMP mediated growth arrest of solid tumors. *Adv. in Cyclic Nucleotide Res.* **12**: 111–23.

Chopra, R. N. & Bhatla, S. C. (1981). Involvement of cyclic 3',5'-adenosine monophosphate in sex induction in the moss *Bryum argenteum*. *Z. Pflanzenphysiol.* **103**: 393–402.

Cyong, J-C., Takahashi, M., Hanabusa, K. & Otsuka, Y. (1982). Guanosine 3',5'-monophosphate in fruits of *Evodia rutaecarpa* and *E. officinalis*. *Phytochem.* **21**: 777–8.

Dandekar, S. & Modi, V. V. (1980). Involvement of cyclic AMP in carotenogensis and cell differentiation in *Blakeslee trispora*. *Biochim. Biophys. Acta* **628**: 398–406.

DeNicola, M. G., Amico, V. & Piatteli, M. (1975). Effects of light and kinetin on amarathin synthesis induced by cAMP. *Phytochem.* **14**: 989–91.

Donnelly, T. E., Kuo, J. F., Miyamoto, E. & Greengard, P. (1973a). Protein kinase modulator from lobster tail muscle. II. Effects of the modulator on holoenzyme and catalytic submit of guanosine 3',5'-monophosphate-dependent and adenosine 3',5'-monophosphate-dependent protein kinases. *J. Biol. Chem.* **248**: 199–203.

Donnelly, T. E., Kuo, J. F., Reyes, P. L., Liu, Y-P. & Greengard, P. (1973b). Protein kinase modulator from lobster tail muscle. I. Stimulatory and inhibitory effects of the modulator on the phosphorylation of substrate proteins by guanosine 3',5'-monophosphate-dependent and adenosine 3',5'-monophosphate-dependent protein kinase. *J. Biol. Chem.* **248**: 190–8.

Drlica, K. A., Gardner, J. M., Kadoc, C. I., Vijay, I. K. & Troy, F. A. (1974). Cyclic AMP levels in normal and transformed cells of higher plants. *Biochem. Biophys. Res. Commun.* **56**: 753–8.

Duffus, C. M. & Duffus, J. H. (1969). A possible role for cyclic AMP in gibberellic acid triggered release of α-amylase in barley endosperm slices. *Experientia* **25**: 281.

Earle, K. M. & Galsky, A. G. (1971). The action of cyclic AMP on GA₃ controlled responses. II. *Plant and Cell Physiology* **12**: 727–32.

Edlich, W. & Gräser, H. (1978). Presence of cyclic adenosine 3',5'-monophosphate in primary shoots of *Zea mays* L. *Biochem. Physiol. Pflanzen.* **173**: 114–22.

Elliott, D. C. & Murray, A. W. (1975). Evidence against an involvement of cyclic nucleotides in the induction of betacyanin synthesis by cytokinins. *Biochem. J.* **146**: 333–7.

Endress, R. (1978). Allosteric regulation of phosphodiesterase from *Portulaca callus* by cyclic GMP and papavarin. *Phytochem.* **18**: 15–21.

Endress, R. (1983). Cooperative interaction of cyclic AMP with its binding protein from *Portulaca grandiflora*. *Phytochem.* **22**: 2147–54.

Forte, J. W. & Orlowski, M. (1980). Profile of cyclic AMP-binding proteins during the conversion of yeasts to hyphae in the fungus *Mucor*. *Exp. Mycol.* **4**: 78–86.

148   *R. P. Newton and E. G. Brown*

Francko, D. A. (1983). Cyclic AMP in photosynthetic organisms: recent developments. *Adv. Cyclic Nucleotide Res.* **15:** 97–119.

Francko, D. A. & Wetzel, R. G. (1980). Cyclic AMP production and extracellular release from green and blue-green algae. *Physiol. Plant.* **49:** 65–7.

Francko, D. A. & Wetzel, R. G. (1981). Synthesis and release of cyclic AMP by aquatic macrophytes. *Physiol. Plant.* **52:** 33–6.

Galsky, A. G. & Lippincott, J. L. (1969). Promotion and inhibition of α-amylase production in barley endosperm by cyclic 3′,5′-adenosine monophosphate and adenosine diphosphate. *Plant Cell Physiol.* **10:** 607–20.

Gat-Edelbaum, O., Altman, A. & Sela, I. (1983). Polyinosinic and polycytidylic acid in association with cyclic nucleotides activates the antiviral factor (AVF) in plant tissues. *J. Gen. Virol.* **64:** 211–14.

Giannattasio, M., Carrutu, G. & Tucci, G. F. (1974). Presence of a cyclic AMP binding protein in Jerusalem artichoke tubers. *FEBS Lett.* **49:** 249–53.

Giannattasio, M. & Macchia, V. (1973). Adenylate cyclase and cyclic 3′,5′-AMP-diesterase in Jerusalem artichoke tubers. *Plant Sci. Lett.* **1:** 259–64.

Giannattasio, M., Mandato, E. & Macchia, V. (1974). Content of 3′,5′-cyclic AMP and cyclic AMP phosphodiesterase in dormant and activated tissues of Jerusalem artichoke tubers. *Biochem. Biophys. Res. Commun.* **57:** 365–71.

Giannattasio, M., Sica, G. & Macchia, V. (1974). Cyclic AMP phosphodiesterase from dormant tubers of Jerusalem artichoke. *Phytochem.* **13:** 2729–33.

Gilbert, M. L. & Galsky, A. G. (1972). The action of cyclic AMP on GA₃ controlled reponses. *Plant Physiol.* **13:** 867–73.

Gilman, A. G. & Murad, F. (1974). Assay of cyclic nucleotides by receptor protein binding displacement. In *Methods in Enzymology*, vol. 38, ed. J. G. Hardman & B. W. O'Malley, pp. 49–58. New York: Academic Press.

Goldberg, N. D. & Haddox, M. K. (1977). Cyclic GMP metabolism and involvement in biological regulation. *Annual Rev. Biochem.* **46:** 823–96.

Goldberg, N. D., Haddox, M. K., Nicol, S. E., Glass, D. B., Sandford, C. H., Kuehl, F. & Estensen, R. (1975). Biologic regulation through opposing influences of cyclic GMP and cyclic AMP: the Yin Yang hypothesis. *Adv. Cyclic Nucleotide Res.* **5:** 307–30.

Gomes, S. L., Mennucci, L. & deCosta Maia, J. C. (1980). Induction of germination of *Blastocladiella emersonii* by cAMP. *Cell Differentiation* **9:** 169–79.

Habaguchi, K. (1977). A possible function of cyclic AMP in the induction of polyphenol oxidase preceding root formation in cultured carrot root callus. *Plant Cell Physiol.* **18:** 191–7.

Haddox, M. K., Stevenson, J. H. & Goldberg, N. D. (1974). Cyclic GMP in meristematic and elongating regions of bean root. *Fed. Proc.* **33:** 522.

Hall, K. A. & Galsky, A. G. (1974). The action of cyclic AMP on GA₃ controlled responses. IV. *Plant and Cell Physiol.* **14:** 565–71.

Hanabusa, K., Cyong, J. & Takahashi, M. (1981). High level of cyclic AMP in the *Jujube* plum. *Planta Medica* **42:** 380–4.

Handa, A. K., Bressan, R. A., Quader, H. & Filner, P. (1981). Association of formation and release of cyclic AMP with glucose depletion and onset of chlorophyll synthesis in *Poterioochromonas malhamensis*. *Plant Physiol.* **68:** 460–3.

Hartung, W. (1973). Die Wirtung von cyclischem Adenosin-3′,5′ monophosphat auf das Streckungswachstum von *Avena*-koleoptilzylindern. *Z. Pflanzenphysiol.* **67:** 380–2.

Higgins, T. J. V., Goodwin, P. B. & Carr, D. J. (1974). The induction of nitrate reductase in mung bean seedlings. *Austr. J, Plant Physiol.* **1:** 1–8.

Hintermann, R. & Parish, R. W. (1979). Determination of adenylate cyclase activity in a variety of organisms: Evidence against the occurrence of the enzyme in higher plants. *Planta* **146:** 459–61.

Hixson, C. S. & Krebs, E. G. (1980). Characterization of a cyclic AMP binding protein from bakers' yeast. Identification as a regulatory subunit of cyclic AMP-dependent protein kinase. *J. Biol. Chem.* **255:** 2137–45.

Holm, R. E. (1972). Enhancement of hypocotyl elongation in rootless *Soybean* seedlings by cyclic AMP. *Plant Physiol. (Suppl.)* **49:** 30.

Holm, R. E. & Miller, M. R. (1972). Hormonal control of weed seed germination. *J. Weed Sci.* **20:** 209–11.

Hood, E. E., Armour, S., Ownby, J. D., Handa, A. K. & Bressan, R. A. (1979). Effect of nitrogen starvation on levels of cyclic AMP in *Anabaena variabilis. Biochim. Biophys. Acta* **588:** 193–200.

Isherwood, F. A. & Ring, S. G. (1977). Adenosine cyclic monophosphate in potato tubers during storage at +10° and +2°. *Phytochem.* **16:** 309–10.

Janistyn, B. (1972). IES-gesteigerte Adenylcyclase-Aktivatät im Homogenat der Maiskoleoptile. *Z. Naturforsch.* Teil B **27:** 872.

Janistyn, B. (1978). Short time effect of cyclic AMP on corn (*Zea mays*) coleoptile growth. *Z. Naturforsch.* **33c:** 801–2.

Janistyn, B. (1981). Gas chromatographic mass- and infrared-spectrometric identification of cyclic adenosine 3′,5′-monophosphate (cAMP) in maize seedlings (*Zea mays*). *Z. Naturforsch.* **36:** 193–6.

Janistyn, B. (1983). Gas chromatographic–mass spectroscopic identification and quantification of cyclic guanosine-3′,5′-monophosphate in maize seedlings. *Planta* **159:** 382–6.

Janistyn, B. & Drumm, H. (1975). Phytochrome-mediated rapid changes of cyclic AMP in mustard seedlings (*Sinapis alba* L.). *Planta* **125:** 81–5.

Johnson, R. A., Hardman, J. G., Broadus, A E. & Sutherland, E. W. (1970). Analysis of adenosine 3′,5′-monophosphate with luciferase luminescence. *Analyt. Biochem.* **35:** 91–7.

Johnson, L. P., McLeod, J. K., Parker, C. W. & Letham, D. S. (1981). The quantitation of adenosine 3′,5′-cyclic monophosphate in cultured tobacco tissue by mass spectrometry. *FEBS Lett.* **124:** 119–21.

Jost, J-P. & Rickenberg, H. V. (1971). Cyclic AMP. *Ann. Rev. Biochem.* **40:** 741–4.

Kamisaka, S. & Masuda, Y. (1971). Stimulation of gibberellin-induced germination in lettuce seeds by cyclic 3′,5′-adenosine monophosphate. *Plant and Cell Physiol.* **12:** 1003–5.

Kamisaka, S., Sakurai, N. & Masuda, Y. (1973). Auxin induced growth of Jerusalem artichoke tubers: role of cyclic AMP, auxin, cytokinin and gibberellic acid. *Plant and Cell Physiol.* **14:** 183–93.

Kamisaka, S., Sano, H. & Katsumi, M. (1972). Effects of cyclic AMP and gibberellic acid on lettuce hypocotyl elongation and mechanical properties of its cell wall. *Plant and Cell Physiol.* **13:** 167–73.

Kapoor, H. C. & Sachar, R. C. (1976). Stimulation of ribonuclease activity and its isoenzymes in germinating seeds of cowpea by gibberellic acid and cyclic AMP. *Experientia* **32:** 558.

Karimova, F. G. & Gusev, N. A. (1980). Effect of adenosine 3′,5′-monophosphate on water exchange of plant cells. *Fiziologiya Rastenii* **27:** 766–72.

Kato, R., Uno, I., Ishikawa, T. & Fujii, T. (1983). Effects of cyclic AMP on the activity of soluble protein kinases in *Lemna pauciostata. Plant and Cell Physiol.* **24:** 841–8.

Katsumata, T., Takahashi, N. & Ejiri, S. (1978). Changes of cyclic AMP level and adenylate cyclase activity during germination of pine pollen. *Agric. Biol. Chem.* **42:** 2161–2.

Keates, R. A. B. (1973*a*). Evidence that cyclic AMP does not mediate the action of gibberellic acid. *Nature* **244:** 355–7.

Keates, R. A. B. (1973*b*). Cyclic nucleotide-independent protein kinase from pea shoots. *Biochem. Biophys. Res. Commun.* **54:** 655–61.

150   R. P. Newton and E. G. Brown

Kessler, B. & Kaplan, B. (1972). Cyclic purine nucleotides: induction of gibberellic acid biosynthesis in barley endosperm. *Physiol. Plantarum* **27**: 424–31.

Kessler, B. & Levenstein, R. (1974). Adenosine 3',5'-cyclic monophosphate in higher plants. Assay, distribution, and age-dependency. *Biochim. Biophys. Acta* **343**: 156–66.

Kuo, J-F. & Greengard, P. (1972). An assay method for cyclic AMP and cyclic GMP based upon their abilities to activate cyclic AMP dependent and cyclic GMP dependent protein kinases. *Adv. Cyc. Nuc. Res.* **4**: 4–51.

Kuo, J-F., Wyatt, G. R. & Greengard, P. (1971). Cyclic nucleotide-dependent protein kinases. IX. Partial purification and some properties of guanosine 3',5'-monophosphate-dependent and adenosine 3',5'-monophosphate-dependent protein kinases from various tissues and species of *Arthropoda. J. Biol. Chem.* **246**: 7159–67.

Langan, T. A. (1973). Protein kinases and protein kinase substrates. *Adv. Cyc. Nuc. Res.* **3**: 99–153.

Lanzani, G. A., Giannattasio, M., Manzocchi, L. A., Bollini, R., Soffientini, A. N. & Macchia, V. (1974). The influence of cyclic GMP on polypeptide synthesis in a cell-free system derived from wheat embryos. *Biochem. Biophys. Res. Commun.* **58**: 172–7.

Lawson, A. M., Stillwell, R. N., Tacker, M. M., Tsuboyama, K. & McCloskey, J. A. (1971). Mass spectrometry of nucleic acid components. Trimethylsilyl derivatives of nucleotides. *J. Am. Chem. Soc.* **93**: 1014–23.

Levi, M., Chiatante, D. & Sparvoli, E. (1984). Evidence of a possible physiological role for cAMP in plants. *Adv. Cyc. Nuc. Res.* **17a**: 181.

Levi, M., Sparvoli, E. & Galli, M. G. (1981). Interference of the phosphodiesterase inhibitor aminophylline with the plant cell cycle. *European Journal of Cell Biol.* **25**: 71–5.

Levi, M., Sparvoli, E. & Galli, M. G. (1982). Inhibition of elongation and H+ and K+ transport by the phosphodiesterase inhibitor aminophylline in plant tissue. *Planta* **156**: 369–3.

Lieberman, M. & Kunishi, A. T. (1969). Cyclic nucleotide phosphodiesterase in pea seedlings. *11th Int. Cong. Botany, Seattle.*

Lin, P. P-C. (1974). Cyclic nucleotides in higher plants. *Adv. Cyc. Nuc. Res.* **4**: 439–61.

Lin, P. P-C. & Varner, J. E. (1972). Cyclic nucleotide phosphodiesterase in pea seedlings. *Biochim. Biophys. Acta* **276**: 454–74.

Lundeen, C. V., Wood, H. N. & Braun, A. C. (1973). Intracellular levels of cyclic nucleotides during cell enlargement and cell division in excised tobacco pith tissues. *Differentiation* **1**: 255–60.

Makman, R. S. & Sutherland, E. W. (1965). Adenosine 3',5'-phosphate in *Escherichia coli. J. Biol. Chem.* **240**: 1309–14.

Malik, C. P., Chabra, N. & Vermani, S. (1976). Cyclic AMP induced elongation of pollen tubes in *Tradescantia paludosa. Biochem. Physiol. Pflanzen* **169**: 311.

Marriott, K. M. & Northcote, D. H. (1977). Influence of abscisic acid, cyclic AMP, and gibberellic acid on induction of isocitrate lyase activity in endosperm of germinating castor bean seeds. *J. Exp. Bot.* **28**: 219–25.

Maslowski, P., Maslowski, H. & Urbanski, T. (1974). Correlation between changes in ion-stimulated ATP-ase activity and protein content in *Phaseolus vulgaris* cotyledon tissue during germination. *Pflanzenphysiol.* **73**: 119–24.

Miller, J. J. & Galsky, A. G. (1974). Radioimmunological evidence for the presence of cyclic AMP in *Hordeum* seeds. *Phytochem.* **13**: 1295–6.

Narayanan, A., Vermeersch, J. & Pradet, A. (1970). Dosage enzymatique de l'acide adenosine 3',5' monophosphate cyclique dans les semences de laitue, variété Reine de mai. *C.R. Acad. Sci.* **271**: 2406–7.

Nawrath, H. (1976). Cyclic AMP and cyclic GMP may play opposing roles in influencing force of contraction in mammalian myocardium. *Nature* **262**: 509–11.

Newton, R. P., Gibbs, N., Moyse, C. D., Wiebers, J. C. & Brown, E. G. (1980). Mass spectrometric identification of adenosine 3',5'-cyclic monophosphate isolated from a higher plant tissue. *Phytochem.* **19:** 1909–11.

Newton, R. P., Kingston, E. E., Evans, D. E., Younis, L. M. & Brown, E. G. (1984). Occurrence of guanosine 3',5'-cyclic monophosphate (cyclic GMP) and associated enzyme systems in *Phaseolus vulgaris. Phytochem.* **23:** 1367–72.

Nickels, M. W. & Galsky, A. G. (1970). Effects of 3',5'-cyclic AMP and 5'-ADP on protease induction in barley half seeds and growth promotion in dwarf peas. *Plant Physiol.* **46:** suppl. 102.

Nickels, M. W., Schaeffer, G. M. & Galsky, A. G. (1971). The action of cyclic AMP on GA_3 controlled responses. I. *Plant and Cell Physiol.* **12:** 717–25.

Niimi, M., Niimi, K., Tokinaga, J. & Nakayama, H. (1980). Changes in cyclic nucleotide levels and dimorphic transition in *Candida albicans. J. Bacteriol.* **142:** 1010–4.

Nimmo, H. G. & Cohen, P. (1977). Hormonal control of protein phosphorylation. *Adv. Cyc. Nuc. Res.* **8:** 145–265.

Ockerse, R. & Mumford, L. M. (1972). The relation of auxin-induced pea stem growth to cyclic AMP treatment. *Naturwissenschaften* **59:** 166–7.

Ogumi, I., Suzuki, K. & Uritani, J. (1976). Terpenoid induction in sweet potato roots by cyclic AMP. *Agric. Biol. Chem.* **40:** 1251.

Oota, Y. & Kondo, T. (1974). Removal by cyclic AMP of the induction of duckweed flowering due to ammonia and water treatment. *Plant Cell Physiol.* **15:** 403.

Orlowski, M. (1980). Cyclic AMP and germination of sporangiospores from the fungus *Mucor. Arch. Mikrobiol.* **126:** 133–40.

Pall, M. L. (1981). Adenosine 3',5'-phosphate in fungi. *Microbiol. Revs.* **45:** 462–80.

Pastan, I. & Perlman, R. L. (1972). Regulation of gene transcription in *Escherichia coli* by cyclic AMP. *Adv. Cyc. Nuc. Res.* **1:** 11–16.

Pollard, C. J. (1970). Influence of gibberellic acid on the incorporation of 8-^{14}C adenine into adenosine 3',5'-cyclic phosphate in barley aleurone layers. *Biochim. Biophys. Acta* **201:** 511–2.

Pollard, C. J. (1972). Rapid gibberellin responses and the action of adenosine 3',5'-monophosphate in aleurone layers. *Biochim. Biophys. Acta* **252:** 553–8.

Pollard, C. J. & Singh, B. N. (1979). Levels of adenosine 3',5'-cyclic phosphate in plants. *Abstracts of 11th International Congress of Biochemistry, Toronto 1979*, p. 487.

Polya, G. M. (1975). Purification and characterization of a cyclic nucleotide regulated 5'-nucleotidase from potato. *Biochim. Biophys. Acta* **384:** 443–57.

Polya, G. M. & Ashton, A. R. (1974). Inhibition of wheat seedling 5'(3') ribonucleotide phosphodiesterase by adenosine 3',5'-cyclic monophosphate. *Plant Sci. Lett.* **1:** 349–57.

Polya, G. M., Ashton, A. R. & Sia, J. P. H. (1976). Occurrence in plants of an adenosine 3',5'-cyclic monophosphate-like compound and adenosine 3',5'-cyclic monophosphate binding activity. *Proc. Austr. Biochem. Soc.* **9:** 32.

Polya, G. M. & Bowman, J. A. (1981). Resolution and properties of two high affinity cyclic adenosine 3',5'-monophosphate-binding proteins from wheat germ. *Plant. Physiol.* **68:** 577–84.

Polya, G. M. & Sia, J. P. H. (1976). Properties of an adenosine 3',5' cyclic monophosphate-binding protein from wheat embryo. *Plant. Sci. Lett.* **7:** 43–50.

Posner, H. B. (1973). Reversal of sucrose inhibition of *Lemna* flowering by adenine derivatives. *Plant and Cell Physiol. Tokyo* **14:** 1199–1200.

Potempa, L. A. & Galsky, A. G. (1973). Confirmation of the presence of adenosine 3',5'-cyclic monophosphate in lettuce seeds. *C.R. Acad. Sci.* **275:** 1987–95.

Rall, T. W., Sutherland, E. W. & Berthet, J.(1957). The relation of epinephrine and glucagon to liver phosphorylase. *J. Biol. Chem.* **224:** 463–75.

Rao, V. S. & Khan, A. A. (1975). Enhancement of polyribosome formation by gibberellic acid and 3',5'-adenosine monophosphate in barley embryos. *Biochem. Biophys. Res. Commun.* **62:** 25–30.

Rast, D., Skřivanová, R. & Bachofen, R. (1973). Replacement of light by dibutyryl-cAMP and cAMP in betacyanin synthesis. *Phytochem.* **12:** 2669–72.

Raymond, P., Narayanan, A. & Pradet, A. (1973). Evidence for the presence of 3′,5′-cyclic AMP in plant tissues. *Biochem. Biophys. Res. Commun.* **53:** 1115–21.

Roy, T., Ghose, B. & Sircar, S. M. (1973). Adenosine 3′,5′-cyclic monophosphate promotion and abscisic acid inhibition of α-amylase activity in the seeds of rice. *J. Exp. Bot.* **24:** 1064–73.

Safran, E. M. & Galsky, A. G. (1974). The action of cyclic AMP in GA₃ controlled responses. VI. Characteristics of the promotion of light inhibited seed germination in *Phacelia tanacetifolia* by GA₃ and cyclic 3′,5′-adenosine monophosphate. *Plant Cell Physiol.* **15:** 527–32.

Salomon, D. & Mascarenhas, J. P. (1971). Auxin induced synthesis of cyclic 3′,5′-adenosine monophosphate in *Avena* coleoptiles. *Life Sci.* **10:** 879–85.

Salomon, D. & Mascarenhas, J. P. (1972). The time course of synthesis of cyclic AMP in *Avena* coleoptile sections in response to auxin. *Plant, Physiol.* **49:** 5–30.

Sela, I., Hauschner, A. & Mozes, R. (1978). The mechanism of stimulation of the antiviral factor in *Nicotiana* leaves. *Virology* **89:** 1–6.

Sen, S. & Pilet, P. E. (1981). Adenosine 3′,5′-cyclic monophosphate levels in maize roots. *Experienta* **37:** 1279.

Sen, S. & Stevenson, S. (1978). Adenosine 3′,5′-cyclic monophosphate levels in normal and indolyl acetic acid induced tumours in bean (*Phaseolus vulgaris* L.) embryos. *Cytobios* **21:** 143–9.

Shimoyama, M., Sakamoto, M., Nasu, N., Shigehisa, S. & Ueda, I. (1972). Identification of the 3′,5′-cyclic AMP phosphodiesterase inhibitor in potato: feedback control by inorganic phosphate. *Biochem. Biophys. Res. Commun.* **48:** 235–41.

Smith, C. J., Brown, E. G., Newton, R. P., Al-Najafi, T. & Edwards, M. J. (1977). Adenylate cyclase activity in higher plants. *Biochem. Soc. Trans.* **5:** 1976–7.

Smith, C. J., Brown, E. G., Newton, R. P. & Edwards, M. J. (1979). Isolation of an adenosine 3′,5′-cyclic monophosphate binding protein from the tissues of higher plants. *Biochem. Soc. Trans.* **6:** 1268–9.

Srivastava, A. K., Azhar, S. & Krishna Murti, C. R. (1972). A possible role of cyclic 3′,5′-adenosine monophosphate in the germination of *Cicer arietinum* seeds. *Phytochem.* **14:** 903–7.

Steiner, A. L., Wehman, R. E., Parker, C. W. & Kipnis, D. M. (1972). Radioimmunoassay for the measurement of cyclic nucleotides. *Adv. Cyc. Nuc. Res.* **2:** 51–61.

Tarantowicz-Marek, E. & Kleckowski, K. (1975). Effect of adenosine 3′,5′-cyclic monophosphate on the pattern of RNA synthesis in protoplasts isolated from gibberellin-sensitive maize seedlings. *Plant Sci. Lett.* **5:** 417–25.

Tarantowicz-Marek, E. & Kleczkowski, K. (1978). Effect of gibberellic acid on the adenosine 3′,5′-cyclic monophosphate content in dwarf maize shoots. *Plant Sci. Lett.* **13:** 121–4.

Tassi, F., Restivo, F. M., Puglisi, P. P. & Cacco, G. (1984). Effect of glucose on glutamate dehydrogenase and acid phosphatase and its reversal by cyclic adenosine monophosphate in single cell cultures of *Asparagus officinalis*. *Physiol. Plant* **60:** 61–5.

Truelsen, T. A., Langhe, E. D. & Verbeek-Wyndaele, R. (1974). Adenosine 3′,5′-cyclic monophosphate induced growth promotion in sunflower callus tissue. *Arch. Int. Physiol. Biochim.* **82:** 109–14.

Tu, J. C. (1978). Biochemical and histochemical investigation of diurnal variation in adenosine 3′,5′-cyclic monophosphate concentration and adenylate cyclase activity in white dutch clover. *Protoplasma* **99:** 139–46.

Uno, I. & Ishikawa, T. (1981). An adenosine 3′,5′-monophosphate-receptor protein and protein kinase in *Coprinus macrorhizus*. *J. Biochem. Tokyo* **89:** 1275–81.

Vandepeute, J., Huffaker, R. C. & Alvarez, R. (1973). Cyclic nucleotide phosphodiesterase activity in barley seeds. *Plant Physiol.* **52:** 278–82.

Venere, R. J. (1972). Dissertation Abstr. 1369B.

Weintraub, R. L. & Lawson, V. R. (1972). Mechanism of phytochrome-mediated effects of light on cell growth. *VI Internat. Congr. Photobiol.:* 161.

Wellburn, A. R., Ashby, J. P. & Wellburn, F. A. M. (1973). Occurrence and biosynthesis of adenosine 3′,5′ cyclic monophosphate in isolated *Avena* etioplasts. *Biochim. Biophys. Acta* **320:** 363–71.

Wells, J. N. & Hardman, J. G. (1977). Cyclic nucleotide phosphodiesterases. *Adv. Cyc. Nuc. Res.* **8:** 119–43.

Wertman, K. F. & Paznokas, J. L. (1981). Effects of cyclic nucleotides upon the germination of *Mucor racemosus* sporangiospores. *Exp. Mycol.* **5:** 314–22.

Wilson, T., Moustafa, E. & Renwick, A. E. C. (1978). Isolation, characterization and distribution of adenosine 3′,5′ cyclic monophosphate from *Pinus radiata*. *Biochem. J.* **175:** 931–6.

Wood, H. N., Lin, M. C. & Braun, A. C. (1972). The inhibition of plant and animal adenosine 3′,5′ cyclic monophosphate phosphodiesterase by a cell division promoting substance from tissues of higher plant species. *Proc. Natl. Acad. Sci. USA* **69:** 403–6.

Yunghams, W. N. & Marré, D. J. (1977). Adenylate cyclase activity, not found in soybean hypocotyl and onion meristem. *Plant Physiol.* **60:** 144–51.

5

Receptors for cell communication in Dictyostelium

Peter C. Newell

5.1 Introduction

Organisms or cells that communicate by chemical signals need outwardly facing cell-surface receptors that avidly and specifically bind the signal molecules in order to discriminate between similar compounds in soups of competing molecules.

Such receptors play an important role during all phases of the life cycle of the cellular slime moulds. During growth, the amoebae perceive and move towards patches of multiplying bacteria. Such bacteria emit folates, and slime moulds can sense the bacterial food source using highly specific folate receptors on their cell surface membranes. (Pan, Hall & Bonner, 1972; Wurster & Butz, 1980; Van Driel, 1981; De Wit, 1982; Tillinghast & Newell, 1984). When the food source runs out and the amoebae begin to starve they develop different receptors that recognize other (in some cases related) types of small communicating molecules ('acrasins') that direct the amoebae to collect together to form a multicellular aggregate and to begin differentiation into spores and stalk cells. The molecules used for this process, and the receptors involved, differ from species to species (Fig. 5.1). The particular compound emitted by the aggregation centres of the four *Dictyostelium* species *D. discoideum*, *D. rosarium*, *D. purpureum*, and *D. mucoroides* is cyclic AMP (cAMP) (Konijn *et al.* 1969; Konijn, 1972). Of these species, *D. discoideum* has been by far the best studied and will be considered in most detail below.

Other genera and some other *Dictyostelium* species use radically different types of molecules for communication during aggregation. For example, *D. minutum* (a primitive slime mould in terms of its simple aggregation system which lacks signal relay) uses a derivative of the food-seeking attractant folate, modified in the glutamic acid moiety

(Kakebeeke, Mato & Konijn, 1978; De Wit & Konijn, 1983). The species *D. lacteum* is attracted to a pterin derivative during aggregation (Mato, Van Haastert, Krens & Konijn, 1977; Van Haastert, De Wit, Grijpma & Konijn, 1982) and the members of the more advanced genus *Polysphondylium* (which produce multiple branched fruiting bodies) are attracted by a small dipeptide of glutamate and ornithine called 'glorin' which somewhat resembles peptides that cause chemotaxis of leukocytes (Shimomura, Suthers & Bonner, 1982; Fig. 5.1).

5.2   Cell-surface cAMP receptors
5.2.1   *The excitability of aggregating amoebae*
Aggregation is triggered by starvation. After several hours without food amoebae develop the ability to communicate with each other

Fig. 5.1. Molecular structures of folate, pterin, cAMP and glorin which are used by various members of the cellular slime moulds as chemoattractants. Glorin is a dipeptide of glutamate and ornithine that has an amino group and a carboxyl group of the glutamate blocked respectively by a propionyl group and an ethyl ester (indicated by the curved broken lines) and an amino group on the ornithine blocked by formation of a lactam ring.

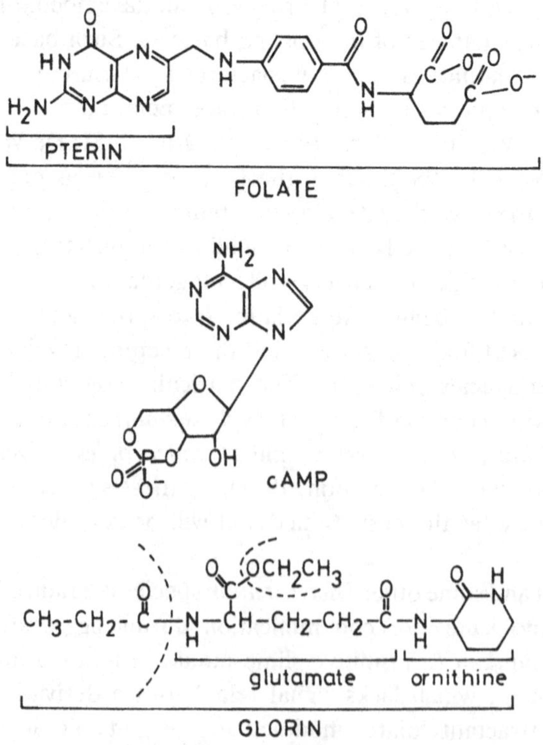

using long-range signalling which induces chemotactic movement and aggregation. During this process they exhibit clearly visible changes in their light-scattering properties. Using darkfield optics, lawns of amoebae synchronously aggregating in Petri dishes can be seen to produce outwardly expanding light and dark bands as they move towards the centres (Fig. 5.2). Analogous fluctuations in their optical density may also be observed and quantified in oxygenated suspensions in spectrophotometer cuvettes (Gerisch & Hess, 1974).

In the *Dictyostelium* species that produce cAMP as the signalling molecule for aggregation, cAMP receptors are formed on the external face of their plasma membranes and at the same time phosphodiesterases are produced both bound to the plasma membrane and excreted into the surrounding medium. When cAMP binds to the cell-surface cAMP receptors, adenylate cyclase is transiently activated to increase the production of cAMP. The system is now in an excitable state in which the activities of the signal-producing system (the adenylate cyclase) and the signal-destructive system (the phosphodiesterase) are in an unstable equilibrium (Goldbeter & Segel, 1980). (Fig. 5.3). In this state these enzyme activities oscillate spontaneously in a few of the cells (the 'initiating cells') such that the adenylate cyclase goes into a self-stimulating rhythmic activation cycle that liberates cAMP in micromolar quantities every 5–10 min (Stage C, Fig. 5.3). These initiating cells become the signalling centres for aggregation. Surrounding cells that have not reached this self-perpetuating state of excitement but are nevertheless highly excitable (State B, Fig. 5.3) are able to secrete a pulse of cAMP in response to a pulse of cAMP received, and this has the effect of transmitting a signal from the original initiating cell to amoebae further away. The signal is not relayed inwards because the amoebae which have just relayed are refractory for further stimulation for about 2.5 min. Besides relaying the signal, the amoebae also respond by moving for about 100 s towards the signalling centre (forming the bright bands in Fig. 5.2) and then wait for at least twice this time for the next signal to arrive. It is by this discontinuous but rhythmic movement that amoebae aggregate and form multicellular organisms containing up to 100 000 cells.

5.2.2   *Binding and dissociation properties of the cAMP receptor*

Several techniques have been used for assay of cAMP binding to the cell surface receptors. Initially Malchow & Gerisch (1974) measured the loss of [^{3}H]-cAMP from the supernatant after separation of amoebae from the medium by centrifugation in a microfuge. The phos-

Fig. 5.2. Aggregating fields of *Dictyostelium discoideum* (strain
HL100) on an agar surface seen using darkfield optics. The signalling
centres which appear as bright dots are emitting pulses of cAMP; the
signal is relayed outwards from the initiating cells as the amoebae
themselves move inwards. Bands of amoebae moving towards the
centres appear bright and the stationary amoebae dark. The frequency
of the rhythmic pulses is not constant but slowly increases with time
so that the dark and light bands appear closer together towards the
centre of the aggregation field. Commonly concentric fields become
broken at some point by an inhomogeneity in the responding cells
and give rise to spiral patterns. A double spiral wave in the process
of being formed from a concentric wave may be seen in the lower
middle portion of the photograph. The bar marker represents 5 mm.

phodiesterase inhibitors (such as theophylline) used for mammalian systems do not inhibit the slime mould phosphodiesterases and Malchow & Gerisch used a vast excess (0.5 mM) of cyclic GMP to compete with the cAMP for phosphodiesterase enzyme which is a thousandfold less specific for cAMP than the cell-surface cAMP receptor. Later studies used dithiothreitol to inhibit the phosphodiesterases (Henderson, 1975; Green & Newell, 1975) and cell separation from the medium using Millipore membrane filtration (Green & Newell, 1975), Nucelopore filters (which give much lower backgrounds due to lower [³H]-cAMP retention) (Mullens & Newell, 1978) or centrifugation through silicone oil (Klein & Juliani, 1977). More recently, Van Haastert & Kien (1983) have des-

Fig. 5.3. The partitioned adenylate cyclase (σ) – phosphodiesterase (k) activity plane. Domain A represents the state where the cells would be capable of a steady low level of cAMP secretion. Domain B corresponds to the ability to relay an incoming cAMP signal and domain C represents the ability to secrete cAMP autonomously in an oscillatory manner. In domain D the cells secrete a high steady-stage level of cAMP. Domain E represents a domain with multiple steady states. During the course of starvation the parameter state is represented by a point that moves slowly along the path indicated by the arrow from A to C. The path crosses successively the no-relay, relay, and oscillatory domains; upon ceasing to oscillate, the system may either return to the excitability domain or settle to a high, steady cAMP secretion. (From Goldbeter & Segel, 1980.)

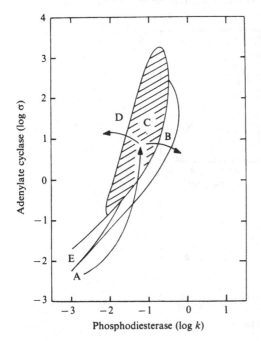

cribed an assay that uses addition of a tenfold excess of saturated ammonium sulphate to stabilize the [³H]-cAMP–receptor complex which may then be removed by centrifugation.

Compared to the binding of many hormones to their receptors on the surfaces of mammalian cells, the kinetics of binding and dissociation of cAMP to the cell-surface receptor in *Dictyostelium* are very rapid. Experiments using a technique involving rapid dilution of [³H]-cAMP-labelled amoebae revealed that 80–90% of the dissociation of cAMP from the receptors was first order with a half-life of about 4 s at 0°C (Fig. 5.4; Mullens & Newell, 1978) or about 1 s at 22°C (Van Haastert

Fig. 5.4. Effect of the presence of excess unlabelled cAMP on the rate of dissociation of [³H]-labelled cAMP from the surface of *D. discoideum* amoebae. Aggregation-competent amoebae were incubated for 1 min at 0°C with 1 nM [³H]-cAMP and then diluted 83-fold with ice-cold buffer (●) or ice-cold buffer containing excess (100 μM) unlabelled cAMP (○) for the times indicated. The dissociation is biphasic due to the relatively fast (H plus L) receptors dissociating before the slowly dissociating (S) class. The rate of dissociation during its fast part may be seen to be increased by the presence of excess cAMP. The inset shows a semilogarithmic plot of the dissociation data excluding the slowly dissociating S fraction. The triangles show the rate of association in the absence (▲) and presence (△) of 100 μM cAMP as a test of the extent of reassociation under the conditions used. (From Mullens & Newell, 1978.)

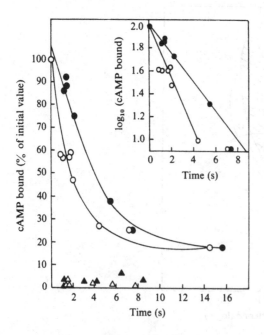

& De Wit, 1984). Association of cAMP to the receptors is a second-order process dependent on the cAMP concentration. Measurements using 30 nM cAMP indicated that maximum binding occurred after approximately 6 s. However, equilibrium was not reached until 30–40 s due to relatively slow receptor transitions (Van Haastert & De Wit, 1984).

Scatchard plots of equilibrium binding were found by Green & Newell (1975) to be strongly curvilinear (concave upwards) and were analysed in terms of two classses of receptors: low affinity ($K_d = 150$ nM) ('L') receptors) present at approximately 10^5/cell and high affinity ($K_d = 10$ nM) ('H' receptors) present at only 10^4/cell (Fig. 5.5). Similar plots have since been observed with other species of *Dictyostelium* such as *D. purpureum* and *D. mucoroides* (Mullens & Newell, 1978). It was also found by dissociation experiments that the rate of dissociation was faster in the presence of excess unbound cAMP (Mullens & Newell, 1978). Implying a change in affinity with a transition between H and L receptors dependent on the presence of cAMP (Fig. 5.4). Such variable affinity, which has also been noticed with some cell-surface hormone receptors in animal systems, explains the curvilinear Scatchard plots and may be a system for sensing (and reacting to) both large and small cAMP stimuli. These two forms have recently been analysed by Van Haastert & De Wit (1984) who have found that in the absence of cAMP about 40% of the receptors are present in the H form and 60% in the L form. During the cAMP-binding reaction the number of H receptors decreases to a minimum of about 10% with a simultaneous rise in L receptors to about 90%, so that the equilibrium mixture of H and L receptors normally measured in equilibrium studies (such as in Fig. 5.5) is not reached until 45 s. The receptor transition shows first-order kinetics with a half-life of about 9 s. It takes place at 0°C and is seen in cells preincubated in the presence of sodium azide to deplete ATP content, implying that phosphorylation is not the driving force. It is of interest that drugs can affect the H–L equilibrium as shown by Scatchard analysis of [^{3}H]-cAMP binding in the presence of the drug. For example, at 0.1–0.5 mM the phenothiazine antipsychotics and the tricyclic antidepressants (both of which have large hydrophobic triple-ring structures) convert all the receptors to the H form (Newell & Hardwicke unpublished), and in the presence of 10 mM caffeine nearly all the sites are in the low-affinity form (Van Haastert & De Wit, 1984). From a consideration of hormonal binding systems it seems likely that some modulating protein is involved whose occupancy affects the affinity of the cAMP receptor.

A third class of receptors has also been demonstrated using dissociation experiments in which it was found that a fraction of the bound

162   *P. C. Newell*

[³H]-cAMP did not dissociate rapidly compared to the H and L receptors (Mullens & Newell, 1978; Mullens, Ashley & Newell, unpublished; Fig. 5.4). The appearance of these slowly dissociable ('S') receptors which constitute 10–20% of total cAMP-binding activity is seen within a few seconds of cAMP addition at 0°C and, therefore, is probably not attributable to 'down regulation' in which receptors are removed from the membrane over a period of minutes at 22°C. The properties of these S receptors have recently been investigated extensively by Van Haastert & De Wit (1984) who found evidence that they are a stable receptor type that are distinct from the H or L receptors and not freely interconvertible with them. Their number and affinity did not appear to change during incubation with cAMP. Scatchard plots indicate their K_d to be

Fig. 5.5. Scatchard plot of [³H]-cAMP binding showing high affinity (H), low affinity (L) and slowly dissociating (S) classes of receptor binding. Cells were incubated at 22°C in the presence of different [³H]-cAMP concentrations for 45 s (○) then excess unlabelled cAMP was added and the remaining [³]-cAMP still bound was measured 8 s later (●). (Note that for quantitative assessment of binding to either L or H receptors from such data, a correction must be made for the minor contribution to the observed binding resulting from binding to the other two classes; see Green & Newell, 1975.) The inset shows a Scatchard plot of the S fraction. The broken line represents the calculated plot allowing for about 33% dissociation of the S receptors during incubation in the presence of excess unlabelled cAMP. (From the data of Van Haastert & De Wit, 1984.)

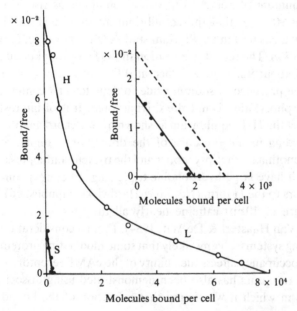

approximately 12.5 nM with about 3×10^3 receptors per cell (Fig. 5.5) and from dissociation experiments they were shown to have a half-life of about 15 s. Their function is at present unknown but it is conceivable that they are the modulating proteins that affect the transition of the H and L receptors. It is interesting in this connection that the half-maximal H-to-L transition occurs at 12.5 nM which is the K_d for the S receptors (Van Haastert & De Wit, 1984).

In addition to all these receptor transitions, futher complexity has been found at very low concentrations of cAMP (less than 1.5 nM). At these concentrations, Scatchard plots of [^{3}H]-cAMP binding are often steeply curvilinear with the concavity downwards (that is, reversing the curvature seen at higher ligand concentrations) (Coukell, 1981). This phenomenon, which is normally interpreted for other protein–ligand interactions as positive cooperativity, was found in all five strains of *D. discoideum* examined as well as in a strain of *D. purpureum*. It was seen at 0°C and 22°C and with whole cells and plasma membrane preparations, but while it was seen routinely with aggregating amoebae, post-aggregative amoebae (packed into cell aggregates) often failed to show the effect. Because, in the *D. discoideum* strains studied, the lowest concentration of cAMP pulses that can appreciably stimulate membrane differentiation was found to be 0.15–1.5 nM, and as similar concentrations have been reported to trigger minimal chemotactic and relay responses, Coukell speculated that the purpose of the apparently positive cooperative behaviour of the cAMP receptors might be to generate a steep cellular-response threshold.

'Down regulation' or 'desensitization' of receptors, which was first observed for hormone binding to cells of higher organisms (Gavin *et al.* 1974; Lesniak & Roth, 1976) has also been observed for the cAMP receptors in *D. discoideum* by Klein & Juliani (1977) who found that binding of [^{3}H]-cAMP was greatly decreased by preincubation of amoebae at 22°C with unlabelled cAMP. This decrease in binding was shown to be due to loss of binding sites rather than to a change in their affinity, and the loss and subsequent reappearance (in the absence of excess ligand) was not dependent on protein synthesis. The concentration of cAMP needed to evoke the response was high, 10 uM–1 mM, but this may have been due to the presence of active phosphodiesterases that rapidly reduced the effective concentration of added ligand. In later experiments Klein (1979) employed a strain of *Dictyostelium* that lacked the phosphodiesterase enzymes. With this strain much lower concentrations of cAMP were found to be effective and significant effects at 10 nM–100 nM could be seen.

5.2.3   *Rate-receptor versus occupation-receptor mechanism*

One of the actions of pulses of cAMP during aggregation of D. *discoideum* is to induce the synthesis of phosphodiesterase. Using this induction as an assay, Van Haastert, Van der Meer & Konijn (1981) tested the ability of a series of cAMP derivatives to mimic cAMP in their action. Not surprisingly the derivatives showed variously reduced abilities to induce phosphodiesterase. What was more surprising, however, was that the activities of the derivatives when applied at concentrations sufficiently high to saturate the receptors were inversely related to their affinity. This finding is hard to explain on the basis of a receptor that transmits into the cell interior a graded response which is in proportion to the number of receptors occupied. If the receptor response is, however, in proportion to the frequency of associations (a rate-receptor mechanism) then the results of Van Haastert, Van der Meer & Konijn would be more easily explicable. A derivative with a lower affinity would require much higher concentrations than cAMP for saturation of the receptor and because of its lower affinity it would dissociate more rapidly and have to reassociate again to maintain receptor saturation. If the activity of the receptor were proportional to these associations then such a derivative would be capable of considerably higher stimulation compared to cAMP at saturating levels, as was found experimentally. Such a rate-receptor mechanism, Van Haastert, Van der Meer & Konijn (1981) suggest, would be more suited to the rapid requirements for signal transduction. Similar rate receptors have been proposed for other fast-responding systems such as the gustatory stimulation of insects (Heck & Ericson, 1973) and ileum contraction by acetylcholine in guinea pigs (Paton, 1961).

Other workers have expressed reservations about the interpretation of the data of Van Haastert, Van der Meer & Konijn and have suggested that phosphodiesterase induction (the response measured) may not require a cell-surface receptor at all. Using the N^6-(aminohexyl) derivative of cAMP ('hexyl cyclic AMP') Juliani, Brusca & Klein (1981) found that pulses of this derivative could induce phosphodiesterase formation very effectively yet clearly showed no ability to bind to the cell-surface cAMP receptors and could not bring about cAMP-induced chemotaxis. Obviously the simplest interpretation of this apparently anomalous result is that induction of the phosphodiesterase proceeds by a receptor-independent mechanism, thereby apparently casting doubt on the evidence for the rate-receptor model. However, more recent work with this analogue by Van Haastert, Bijleveld & Konijn (1982) provides a reasonable explanation of the anomaly. They found that one of the actions of hexyl cAMP is to cause transient inhibition of both extracellular and cell-surface phosphodiesterase enzyme activity. As a consequence of this effect, there is a build up of the small amounts of cAMP continuously released by the cells which then bind to the cell-surface cAMP receptors, and hence indirectly bring about induction of phosphodiesterase activity. So the rate-receptor model lives on.

5.2.4   Inhibition by adenosine

Early work of Bradley, Sussman & Ennis (1956) revealed that adenine decreased the number of centres formed by starving populations on agar surfaces. When this effect was investigated more recently by Newell & Ross (1982), it was found that adenine (and the more soluble nucleoside adenosine) differentially inhibited the ability of starving lawns to form autonomously oscillating aggregation centres while not apparently affecting the relay of signals that were formed (Fig. 5.6). Such inhibition, which leads to the formation of very large aggregation territories was correlated with inhibition of the cell-surface cAMP receptors. At an adenosine concentration of 5 mM, these receptors were inhibited by 90%. (Higher concentrations of adenosine did not inhibit further.) The inhibition was found to be non-competitive with cAMP over the range 2–400 nM cAMP. Because of the non-competitive inhibition, it was concluded that adenosine produces its effects indirectly by binding to receptor sites other than the cAMP receptors (Newell & Ross, 1982; Van Haastert, 1983). To confirm this notion, Newell (1982) and Van Haastert (1983) investigated the binding of [^{3}H]-labelled adenosine to whole cells and found that receptors existed on the surface of these amoebae that rapidly and specifically bound adenosine with an affinity that was consistent with the concentration needed for inhibition of cAMP receptors and initiation of aggregation.

The function of the adenosine receptors is at present unknown. In higher systems purinergic receptors have important regulatory roles (reviewed by Burnstock & Brown, 1981) and some of the higher cell adenosine-binding receptors have been located on the plasma membrane (Fain & Malbon, 1979; Londos, Cooper, Schegel & Rodbell, 1978). In some higher systems adenosine has been found to inhibit adenylate cyclase (Baer & Drummond, 1979). Whether adenosine (or some more active, naturally occurring compound) has physiological effects during aggregation of *Dictyostelium* via the interaction of adenosine and cAMP receptors remains to be resolved.

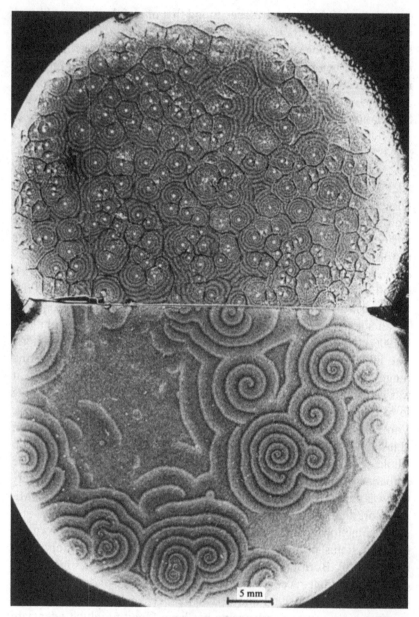

Fig. 5.6 The inhibition by adenosine of autonomous signal generation in starving amoebae, leading to the formation of large aggregation territories. Aliquots of starving amoebae were incubated on non-nutrient agar in 50-mm Petri dishes in the presence or absence of 5 mM adenosine for 17 h at 7°C. Agars from the two conditions were then spliced together and incubated at 22°C in the same 90-mm Petri dish to allow rapid signal propagation and aggregation to occur under

5.2.5   *Molecular configuration of the cAMP receptor*

By testing approximately 50 purine nucleotide derivatives for their ability to bind to the receptor and to induce chemotaxis of *D. discoideum*, Mato & Konijn (1977), Mato, Jastorff, Morr & Konijn (1978) and Van Haastert & Kien (1983) determined the parts of the cAMP molecule that were essential for activity and presumably were involved with binding to the receptor protein. From experiments testing adenine derivatives that were induced to adopt preferentially the *syn* rather than *anti* conformation by substitution at C-8, they concluded that the active form of cAMP that bound to the receptor was probably the *anti* form (Fig. 5.7). This is an interesting finding as it differs from similar studies on binding of cAMP to the regulatory subunit of protein kinase enzymes (in various types of cells, including *Dictyostelium*; De Wit, Arents & Van Driel, 1982), for which substitution at C-8 generally enhances binding as it occurs in the *syn* conformation. Chemotaxis of *D. discoideum* and binding of cAMP to the cell-surface receptors are comparable to cAMP binding to the regulatory subunit of protein kinase in that an unsubstituted 3′ hydroxyl group on the ribose ring is required, but, in contrast to protein kinase, *Dictyostelium discoideum* cAMP binding is much less sensitive to bulky substitution at the 2′ position. Based on such comparisons of binding and chemotactic activity of pairs of cAMP analogues, a model has been proposed in which the cAMP binds to the *D. discoideum* cell in the *anti* configuration via hydrogen bonds with the receptor at N^6H_2 and $0^{3′}$ (Fig. 5.8). The adenine moiety is bound in a hydrophobic cleft (shown stippled) while the phosphate moiety is probably located in a narrow cave without being bound by any specific electrostatic forces (Mato *et al.* 1978; Van Haastert & Kien, 1983).

Caption for fig. 6.5 (*cont.*)

identical environmental conditions. Numerous aggregation territories were formed under these conditions on the control agar (upper portion of picture) whereas amoebae on adenosine-containing agar (lower portion of picture) initiate centres more rarely and form large spreading territories. The frequency of signalling (indicated by the interval between bands), which appears lower in the presence of adenosine in the picture shown, increases in time to similar values on both control agar and on agar containing adenosine. The photographs were taken from an experiment using a streamer mutant (hence the wide bright bands) although similar effects of adenosine on territory size are seen with the wild type and other mutant strains. The bar marker represents 5 mm.

5.2.6   *Di-equatorial phosphate and receptor activation*

The bonding required not only for cAMP binding but also for activation of the receptor was studied by Van Haastert & Kien (1983) by comparing pairs of cAMP analogues in their ability to induce formation of cyclic GMP in aggregating amoebae (a response which peaks within 10 s of cAMP binding to the cell-surface receptors and is thought to be connected with chemotaxis – see below). Generally a close correlation existed between cAMP binding affinity and the ability to induce cyclic GMP formation. However, two pairs of stereoisomeric analogues, differing only in the position of $-S$ or $-N(CH_3)_2$ groups attached to the phosphate atom, showed similar binding but very different activation ability. From these studies only the phosphate group appears to be implicated specifically in activation of the receptor and from quantum-mechanical calculations of Van Ool & Buck (1982) and specific properties of the active molecules, the model of activation shown in Fig. 5.9 is proposed by Van Haastert & Kien (1983). In this model, after cAMP is bound to the receptor, an active group in the receptor cleft performs a nucleophilic attack on the phosphorus atom of cAMP and a covalent

Fig. 5.7. Molecular structures of the *anti* and *syn* forms of cAMP. (From Van Haastert & Kien, 1983.)

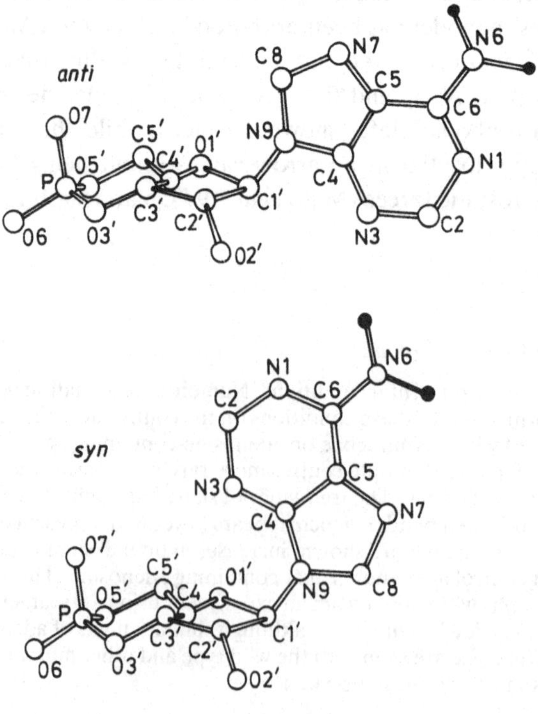

bond is formed. Formation of this bond results in a pentacovalent phosphorus atom with the cyclo phosphate ring in the energetically favourable diequatorial position (Fig. 5.9) and compounds that bind but do not activate cannot easily adopt this configuration. In some way as yet unknown this change induces the receptor to adopt the activated conformation that triggers the chemotactic responses.

5.2.7   *Isolation of the cAMP receptor*

Using the commercially available light-sensitive 8-azido [P^{32}]-cAMP derivative it would seem feasible to label the cAMP receptor on whole cells and hence to identify it in subsequent gel electrophoresis. However, when this technique has been used it has not proved to be straightforward. For example, while soluble cAMP-binding proteins have been found to be labelled efficiently, Cooper, Chambers & Scanlon (1980) could not detect any labelling of cell-surface components. One of the difficulties arises from the fact that the azido compound used is a derivative of cAMP at the 8-position of the adenine moiety. As mentioned above, such derivatives exist predominantly in the *syn* form and hence have only very low activity (approximately one hundredth that of cAMP) as measured by their chemotactic-inducing ability and

Fig. 5.8. Model of the binding of cAMP to the cell-surface receptor of *D. discoideum*. The cAMP binds to the receptor in the *anti* configuration by two hydrogen bonds at N^6H_2 and $O^{3'}$ respectively. The adenine moiety is bound in a hydrophobic cleft of the receptor (shown stippled). 'A' represents a hydrogen bond acceptor and 'D' a hydrogen bond donator. (From Van Haastert & Kien, 1983.)

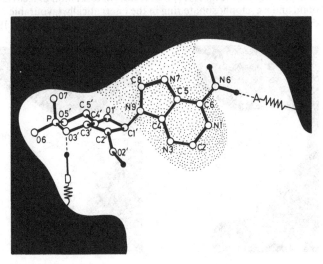

their affinity for the cAMP receptor. Wallace & Frazier (1979) also showed that the azido cAMP derivative was very rapidly broken down to the inactive AMP derivative unless the phosphodiesterase enzymes were inhibited. They showed that the most efficient phosphodiesterase inhibitor, dithiothreitol, could not be used because this compound reduced the azido group to an inactive amino group. Using the alternative phosphodiesterase inhibitor (or more strictly 'active site occupier') cyclic GMP, in 250-fold excess, Wallace & Frazier were able to label whole cells apparently successfully and found a band that was prominently labelled on one-dimensional SDS gels of cell extracts. They demonstrated that this band, which had an M_r of 40 000, was not labelled in the presence of competing unlabelled cAMP and was formed on a developmental time scale that correlated with the appearance of the cAMP receptors in starving suspensions of amoebae. Was this 40 000 M_r protein the cell-surface cAMP receptor? At the time it was considered that this was indeed the receptor, but it was significant (for reasons not fully appreciated at that time) that this protein could not be labelled using cell ghosts or purified plasma membranes.

Using a similar labelling technique that also incorporated a number of important controls, Juliani & Klein (1981) found a band on one-dimensional SDS gels that was labelled by $8\text{-}N_3\text{-}[^{32}P]\text{-cAMP}$ and had an apparent M_r of 45 000 (or under certain conditions 47 000). Labelling

Fig. 5.9. Model of the activation of the cell-surface cAMP receptor. The cAMP initially binds to the receptor as shown in Fig. 5.8, and then an active group (N^-) performs a nucleophilic attack on the phosphorus atom (Fig. 5.9A) that results in formation of a covalent bond and a cyclophosphate ring in the energetically favourable diequatorial position (Fig. 5.9B). (From Van Haastert & kien, 1983.)

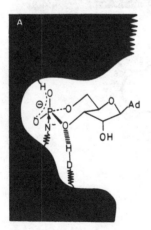
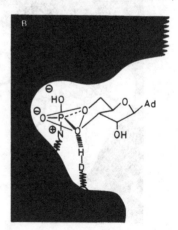

of this band was specifically inhibited by unlabelled cAMP and was seen only in gels prepared from developing amoebae. Developmental mutants lacking cAMP receptors failed to show this band, but in those mutants where cAMP receptors could be induced by exogenous pulses of cAMP, the band was labelled. In another control, the labelled band was shown not to be the membrane-bound phosphodiesterase enzyme by its presence in gels prepared from the mutant HPX235 that lacks phosphodiesterase activity. So was the cAMP receptor a 45 000 or 47 000 M_r protein?

It now seems possible from more recent experiments of Meyers-Hutchins & Frazier (1984) that neither the 40 000 M_r nor the 45 000/47 000 M_r proteins identified as cAMP-binding proteins are in fact the cell-surface cAMP receptors. Using a different approach, Meyers-Hutchins & Frazier first isolated plasma membranes of aggregating *D. discoideum* amoebae and solubilized these with a non-ionic detergent (Emulphogene BC-720). Purification of fractions showing cAMP-binding activity by DEAE-Sephadex and decyl-agarose, followed by non-denaturing PAGE, finally revealed a very acidic glycoprotein of M_r 70 000 that was purified 22 000-fold. Detailed studies of the formation and properties of this glycoprotein strongly suggest that it is (really) the cell-surface cAMP receptor. Not only was it derived from the cell surface (as the starting material was plasma membranes) but it was found only in plasma membranes from developing cells. Its nucleotide specificity and protease sensitivity were similar to those of the cell-surface cAMP receptor and it showed the characteristic rapid rates of association and dissociation. In the light of this work it seems likely that previous studies with azido analogue labelling of whole cells (rather than using plasma membranes) probably caused some cell rupture or that the compound was internalized and labelled the regulatory subunit of the cytoplasmic protein kinase (which favours the *syn* form and has an M_r of 40 000 (de Gunzberg & Veron, 1982; De Wit *et al.* 1982) and the soluble intracellular adenosine/cAMP-binding protein observed by de Gunzberg, Hohman, Part & Veron (1983) which has a native M_r of 190 000 made up of four 47 000 M_r subunits.

What, however, is not yet clear is whether the 70 000 M_r protein represents a receptor that spans the membrane and binds both cAMP and some internal effector protein or whether it is just the cAMP-binding moiety of a larger membrane-spanning protein complex.

5.2.8   *Coupling of receptors to chemotaxis*

Amoebae respond very rapidly to a stimulus of cAMP. From calculations based on the speed of movement of signal waves (as shown in Fig. 5.2) across lawns of starving amoebae (Cohen & Robertson, 1971) it seems necessary on theoretical grounds for the *start* of the relay response to occur within about 15 s of the incoming signal, although in practice it is not normally detected until about 30 s. However, whichever of these figures is correct, relay is clearly not as fast as chemotactic movement responses and unlike the latter it appears to continue with a steady accumulation of cAMP over a period of 60 to 100 s (Roos, Nanjundiah, Malchow & Gerisch, 1975; Devreotes, Derstine & Steck, 1979). The rapidity of chemotactic responses is shown by the observation, for example, that amoebae can extend a pseudopodium in the direction of a capillary containing cAMP within 5 s of cAMP-receptor binding (Gerisch *et al.* 1975; Swanson & Taylor, 1982). Cellular events that have been observed to occur over such a rapid timescale in response to a pulse of cAMP are: (1) cyclic GMP accumulation; (2) increased influx of Ca^{2+}; (3) actin accumulation in the cytoskeleton; (4) acidity increase of the medium; (5) protein methylation; (6) phospholipid methylation; and (7) myosin heavy chain kinase inhibition. Let us consider these in more detail.

(1) Cyclic GMP accumulation
Within 2 s of cAMP binding to amoebal receptors a detectable increase in cellular cyclic GMP occurs that peaks at 10 s before declining rapidly over the next 15 s (Mato *et al.* 1977; Wurster, Schubiger, Wick & Gerisch, 1977; Mato & Malchow, 1978). The rise is due to activation of guanylate cyclase (Mato & Malchow, 1978). The rapid return of cyclic GMP to basal levels is due to a highly specific cyclic GMP phosphodiesterase (Mato *et al.* 1977; Dicou & Brachet, 1980). This interesting enzyme shows positive cooperativity with a cyclic GMP-binding site for activation that is distinct from the catalytic cyclic GMP site and which probably accounts for the sharp decline in cyclic GMP concentration observed after its rapid accumulation (Bulgakov & Van Haastert, 1983).

   Although a temporal correlation between chemotactic movement and cyclic GMP formation is not in itself evidence for any causal connection between these events, the correlation is a very strong one. All active chemoattractants tested in several species trigger the cyclic GMP response, including cAMP and folate in *D. discoideum*, pterins in *D. lacteum* (Mato & Konijn, 1977) and glorin in *P. violaceum* (Wurster, Bozzaro & Gerisch, 1978). A further correlation is connected with the cyclic GMP phosphodiesterase. In mutants of complementation group *stmF* (streamer F mutants) (Ross & Newell, 1979) the movement period is prolonged for up to 500 s rather than the normal 100 s and wide (bright) movement bands are seen under darkfield illumination (Fig. 5.10). Such

Fig. 5.10. Aggregating fields of parental strain XP55 (top) and
streamer F mutant (bottom) on agar observed with darkfield optics.
The moving bands of amoebae appear bright and the stationary
amoebae dark under these conditions. Note the wide movement bands
with the streamer mutant compared to XP55. The bar marker
represents 5 mm. (From Ross & Newell, 1981.)

5 mm

mutants seem normal in every respect tested except that their cyclic GMP accumulation in response to pulses of cAMP is abnormally large and (like their movement) is long lasting (Fig. 5.11) (Ross & Newell, 1981). It has been found that their specific defect is the lack of the cyclic GMP-specific phosphodiesterase enzyme (Van Haastert, Van Lookeren Campagne & Ross, 1982; Coukell, Cameron, Pitre & Mee, 1984) and what little (if any) activity of this enzyme remains cannot be activated by cyclic GMP. Receptor binding, movement and cyclic GMP seem likely, therefore, to be related but the molecular connection is as yet obscure.

(2) Increased influx of Ca^{2+} ions

In many systems Ca^{2+} has been implicated as a secondary signalling molecule. The large difference in concentration between inside and outside of cells and the presence of powerful Ca^{2+} pumps makes the transient movement of Ca^{2+} ions an efficient device to connect membrane receptors with cytosolic metabolism (Hazelbauer, 1978). With *D. discoideum*, the stimulation of amoebal suspensions with cAMP or folate induces a transient influx of Ca^{2+} from the medium that is detectable within

Fig. 5.11. Formation of intracellular cyclic GMP in starving amoebae of parental strain XP55 (○), and a streamer F mutant (●) in response to a 50-nM pulse of cAMP at time zero. Note the rapid formation and degradation of the cyclic GMP peak in the parental strain and the slow degradation of the large peak of cyclic GMP formed in the streamer mutant. (From Ross & Newell, 1981.)

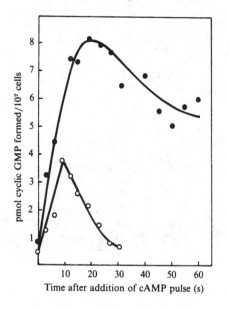

6s and which peaks at 30s, the cellular increase being estimated as roughly 60–100 μM (Wick, Malchow & Gerisch, 1978; Bumann, Wurster & Malchow, 1984). This is followed by a slower efflux over a period of several minutes (Chi & Francis, 1971; Wick *et al.* 1978).

Whether such uptake from the medium is the important aspect of Ca^{2+} movement is uncertain, as the cellular slime moulds contain large internal stores of sequestered Ca^{2+}, equivalent to 2 mM if uniformly distributed throughout the cell (Wick *et al.* 1978) and these have to be greatly depleted before the Ca^{2+} chelator EGTA has any inhibitory effect at blocking chemotaxis or cyclic GMP formation in response to exogenous cAMP pulses (Europe-Finner, McClue & Newell, 1984). Recent studies with the intracellular Ca^{2+} blocker TMB-8 confirm the notion that it is intracellular Ca^{2+} liberation that is important. It was found that TMB-8 completely blocked aggregation and cyclic GMP formation in a Ca^{2+}-reversible manner. That this result was due to an effect on internal Ca^{2+} stores rather than on uptake of external Ca^{2+} was shown by the inability of high (7 mM) external EGTA concentrations to mimick the inhibition of TMB-8, (Europe-Finner & Newell, 1984). The finding of Brenner & Thoms (1984) that the Ca^{2+} ionophore A23187 can enhance the cyclic GMP response in the absence of extracellular Ca^{2+} (presumably by redistributing intracellular Ca^{2+}) also gives support to the importance of internal Ca^{2+} stores being involved in chemotactic responses. Meanwhile, the role of the observed influx of external Ca^{2+} remains to be explained.

(3) Actin accumulation

Actin may exist in cells as monomers of M_r 43 000 G-actin or as filamentous (polymeric) F-actin. Recent experiments have revealed that pulsing whole cellular slime mould amoebae with specific chemoattractants causes rapid accumulation of F-actin in the triton-insoluble cytoskeleton isolated from the pulsed amoebae. With *D. discoideum* amoebae, pulses of cAMP or folate produced a sharp peak of actin accumulation at 3–5 s, followed by a rapid fall (or trough) often to below initial levels and then further peaks of accumulation at roughly 20–25 s and 60–70 s (Fig. 5.12; McRobbie & Newell, 1983). Similar actin accumulation is also found to be a characteristic response to pulses of cAMP in *D. mucoroides*, to monapterin in *D. lacteum* and to glorin in *P. violaceum* (McRobbie & Newell, 1984). Recent work has revealed that the actin involved in the first peak shown in Fig. 5.12 is almost exclusively the most acidic (and plentiful) isoform designated A_1 (McRobbie & Newell, 1985a). The first actin peak is very similar in timing to the formation of a pseudopodium in response to a cAMP-containing capillary placed near an amoeba, and the subsequent trough (Fig. 5.12) correlates well with the 'cringe' (or rapid rounding up of the amoebae) and the (probably equivalent) first peak of the change in light scattering seen in oxygenated suspensions (McRobbie & Newell, 1985b). The actin response is obviously very strongly implicated in chemotactic movement and represents one of the fastest observed responses, yet it must lie near the end of the transduction chain. The relationship of this response to the binding of chemoattractant to the receptors is presently unknown, but Ca^{2+} could well be involved.

(4) Acidity increase in the medium

In unbuffered amoebal suspensions of aggregation-competent *D. discoideum*, exogenous pulses of cAMP elicit a rapid increase in acidity of the medium (Malchow, Nanjundiah & Gerisch, 1978; Malchow, Nanjundiah, Wurster, Eckstein & Gerisch, 1978). The measured decrease of the extracellular pH was biphasic, the first peak occurring at 30 s and the second at 2.5 min after stimulation. The origin and significance of these proton changes are completely unknown, but trivial artefacts are thought to be unlikely considering the magnitude of the release: between 10 and 30 times the number of protons are accumulated in the medium compared to the number of molecules of cAMP used in the stimulating pulse.

Fig. 5.12. Time course of the accumulation of actin associated with the Triton-insoluble cytoskeleton in response to chemotactic stimulation of *D. discoideum* amoebae with cAMP or folate at time zero. (From McRobbie & Newell, 1983.)

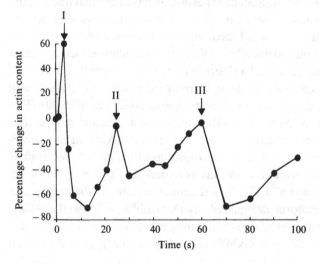

(5) Protein methylation

Methylation of a protein of M_r 120 000 was reported by Mato & Marín-Cao (1979) to occur in amoebae of *D. discoideum* in response to a pulse of cAMP with a peak of methylation being reached within 15–30 s. Van Waarde (1982, 1983), using acid urea PAGE rather than the usual alkaline PAGE (to prevent labilization of the carboxyl methyl groups from proteins) found at least four proteins that were rapidly methylated with M_r values of: 110 000; 46 000; 28 000 and 16 000. The methylation responses were in most cases multiphasic and showed several peaks over a 2-min period. However, all showed a peak of labelling at 15 s with the 46 000 M_r protein being labelled maximally at 3 s after stimulation. The link between these responses and cAMP-receptor binding is completely unknown and the significance of methylation for chemotaxis is not understood. However, an indication that methylation does have some significance is in the finding of Van Waarde & Van Haastert (1984) that transmethylation inhibitors such as cycloleucine decrease chemotactic sensitivity and delay aggregation at concentrations that do not adversely affect cAMP binding or cAMP phosphodiesterase activity.

(6) Phospholipid methylation

The role of the phospholipid methylation that occurs in response to a cAMP pulse seems even more obscure at present than protein methylation. First reports by Mato & Marín Cao (1979) were of phospholipid demethylation peaking within 2 min in response to cAMP in *D. discoideum* amoebae. Using a different approach, Alemany, García Gil & Mato (1980) investigated the effects of physiological concentrations of cyclic GMP (such as are formed in response to cAMP) on amoebal homogenates and found a 2-fold increase in phospholipid methylation, with phosphatidyl-*N*-monoethylethanolamine being formed preferentially. The authors also found that the membrane-permeable derivative of cyclic GMP, 8-bromo-cyclic GMP had similar effects when used on whole cells and concluded that chemotaxis involves membrane methylation via the increase in cyclic GMP produced as a response to cAMP–receptor binding.

(7) Myosin heavy-chain kinase inhibition

The phosphorylated form of myosin seems less able to assemble into thick filaments than the dephosphorylated form (Kuczmarski & Spudich, 1980). Phosphorylation apparently occurs at two threonine residues on the heavy chains of the myosin and strongly inhibits actin-activated myosin ATPase, suggesting a role for phosphorylation in control of contraction (Maruta, Baltes, Dieter, Marmé & Gerisch, 1983). Changes in the

state of myosin phosphorylation following chemotactic stimulation of
D. discoideum amoebae with cAMP pulses were found by Rahmsdorf,
Malchow & Gerisch (1978) and Malchow, Böhme & Rahmsdorf (1981)
who employed [^{32}P]-ATP incorporation to measure the concentration
of dephosphorylated myosin heavy chain. Accumulation of dephosphorylated myosin, apparently due to inhibition of myosin heavy-chain
kinase, began 5–10 s after cAMP binding to the cell-surface receptors.
Calmodulin and Ca^{2+} may be involved in this response as in vitro experiments of Malchow et al. (1981) showed that Ca^{2+} inhibited heavy-chain
kinase and this effect was enhanced by addition of calmodulin.

When the timing of the various responses is considered and their apparent interrelations, the most likely candidate for the primary response
to receptor activation is movement of Ca^{2+} ions (probably released from
internal stores). These ions could have direct effects on actin-binding
proteins and myosin kinases and have further indirect responses via
stimulation of cyclic GMP formation. Whether all responses stem from
such a primary event or from a more complex simultaneous action of
the receptor on several intermediary messengers has yet to be determined.

5.3   Comparison of folate and cAMP receptors

Compared to the cAMP receptors, relatively little is known
about other cell-surface receptors in the cellular slime moulds. Of these,
cell-surface folate receptors on D. discoideum are probably the best
studied although even in this case there is not complete agreement about
the specificity and best method of assay.

The original work of Wurster & Butz (1980) and Van Driel (1981)
was complicated by the slow deamination of folate to 2-deamino-2-hyd-
roxyfolate (DAFA) as the folate deaminase was not inhibited. This work
did indicate, however, the presence of 6×10^4 (Wurster & Butz) and
$1–2 \times 10^5$ receptors per cell (Van Driel) in the growth phase, the number
decreasing rapidly during aggregation. Like cAMP binding, folate binding and dissociation were found to be fast, with equilibrium being reached
in less than 30 s. Later studies used non-degradable derivatives of folate
such as methotrexate and aminopterin (Nandini-Kishore & Frazier,
1981) and, more recently, the deaminase has been inhibited with 8-azagu-
anine (De Wit, 1982). Using 8-azaguanine with growth phase cells, De
Wit found evidence for receptors that bound both DAFA and folate
(2×10^5 per cell) as well as highly specific folate receptors (1×10^5 per
cell). However, Tillinghast & Newell, (1984) found no satisfactory

evidence for such specific receptors. Highly purified DAFA appeared to compete off all bound folate from its receptors and these were present at 4.5×10^4 per cell. The binding ability of DAFA clearly needs further explanation as it has the curious property of binding to the folate receptor while not having the activity of folate as a chemotactic agent nor competing strongly with folate in chemotactic assays (Van Haastert, De Wit & Konijn, 1982).

Studies with Triton X-100 insoluble cytoskeletons have shown that, like the cAMP receptors (Galvin, Stockhausen, Meyers-Hutchins & Frazier, 1984), folate receptors are present attached to the cytoskeletons (Tillinghast & Newell, 1984). However, knowledge of the molecular identity of the folate receptors is as yet less advanced than of cAMP receptors described in earlier sections. Using Triton X-100-solubilized plasma membranes Van Driel (1981) isolated potential folate receptors by passing the extracts down folate–Sepharose affinity chomatography columns and specific elution with folate. Of seven proteins isolated in this way and identified by SDS PAGE, only one was found to be clearly amphiphilic (as required of an integral membrane protein) when tested by the phase-separation method of Bordier (1981) and this represents a promising candidate for the receptor. However, two other proteins showed some amphiphilic properties and rigorous identification of the protein acting as the folate receptor must await further studies.

Folate resembles cAMP in triggering the events of chemotaxis such as formation of cyclic GMP and accumulation of actin. This suggests that the two receptors at some point activate a common part of the signal-transduction chain that regulates the internal chemotactic machinery. From the work of Van Haastert (1983) the link may before guanylate cyclase as stimulation of cells with folate and cAMP simultaneously did not cause an additive cyclic GMP response. That both receptors regulate Ca^{2+} ion movements would be a reasonable hypothesis, and this is supported by the finding that the stimulation of cyclic GMP formation by both folate and cAMP are similarly blocked by TMB-8 (Europe-Finner, Tillinghast, McRobbie & Newell, in press).

5.4   Further outlook

A general understanding of cAMP-receptor binding and some of the cellular responses related to it is progressing steadily, but our appreciation of receptor activation and the triggering of the initial primary response is only rudimentary. Major problems to be tackled include elucidating how the changes noted in the phosphate group of the bound cAMP molecule bring about conformational changes in the receptor protein and how this change (directly or indirectly) could cause release of Ca^{2+} in the cytosol. The effects of Ca^{2+} on actin-binding proteins and myosin kinase may soon be comprehended, but what physically connects and coordinates the actin and myosin responses? Until the molecular mechanism of force generation in non-muscle cells is more adequately understood, the question may be hard to answer. Meanwhile, exploring the roles of cyclic GMP and protein methylation may add significant pieces to a pattern whose outline is dimly there to see.

I wish to thank Hank Tillinghast, Stuart McRobbie and Peter Van Haastert for their helpful comments and suggestions and Mr Frank Caddick for all his help in preparing the figures. The author's research is supported by the Science and Engineering Research Council.

5.5   References

Alemany, S., García Gil, M. G., Mato, J. M. (1980). Regulation by cyclic GMP of phospholipid methylation during chemotaxis in *Dictyostelium discoideum*. *Proc. Natl. Acad. Sci. U.S.A.* **77:** 6996–9.

Baer, H. P. & Drummond, G. I. (1979). *Physiological and Regulatory Functions of Adenosine and Adenine Nucleotides.* New York: Raven Press.

Bordier, C. (1981). Phase separation of integral membrane proteins in Triton X-114 solution. *J. Biol. Chem.* **256:** 1604–7.

Bradley, S. G., Sussman, M. & Ennis, H. L. (1956). Environmental factors affecting the aggregation of the cellular slime mold *Dictyostelium discoideum. J. Protzool.* **3:** 33–8.

Brenner, M. & Thoms, S. (1984). Caffeine blocks activation of cyclic AMP synthesis in *Dictyostelium discoideum. Dev. Biol.* **101:** 136–46.

Bulgakov, R. & Van Haastert, P. J. M. (1983). Isolation and partial characterisation of a cyclic GMP-dependent cyclic GMP-specific phosphodiesterase from *Dictyostelium discoideum. Biochim. Biophys. Acta* **756:** 56–66.

Bumann, J., Wurster, B. & Malchow, D. (1984). Attractant-induced changes and oscillations of the extracellular Ca^{++} concentration in suspensions of differentiating *Dictyostelium* cells. *J. Cell Biol.* **98:** 173–8.

Burnstock, G. & Brown, C. M. (1981). An introduction to purinergic receptors. In *Receptors and Recognition*, vol. *B12*, ed. G. Burnstock, pp. 1–45. London: Chapman and Hall.

Chi, Y. Y. & Francis, D. (1971). Cyclic AMP and calcium exchange in a cellular slime mold. *J. Cell Physiol.* **77:** 169–74.

Cohen, M. H. & Robertson, A. (1971). Wave propagation in the early stages of aggregation of cellular slime molds. *J. Theoret. Biol.* **31:** 101–18.

Cooper, S., Chambers, D. A. & Scanlon, S. (1980). Identification and characterization of the adenosine 3′, 5′-cyclic monophosphate binding proteins appearing during the development of *Dictyostelium discoideum. Biochim. Biophys. Acta* **629:** 235–42.

Coukell, M. B. (1981). Apparent positive cooperativity at a surface cAMP receptor in *Dictyostelium. Differentiation* **20:** 29–35.

Coukell, M. B., Cameron, A. M., Pitre, C. M. & Mee, J. D. (1984). Developmental regulation and properties of the cGMP-specific phosphodiesterase in *Dictyostelium discoideum. Dev. Biol.* **103:** 246–57.

Devreotes, P. N., Derstine, P. L. & Steck, T. L. (1979). Cyclic 3′, 5′ AMP relay in *Dictyostelium discoideum. J. Cell Biol.* **80:** 291–9.

De Wit, R. J. W. (1982). Two distinct types of cell surface folic acid binding proteins in *Dictyostelium discoideum. FEBS Letters* **150**: 445–8.

De Wit, R. J. W., Arents, J. C. & van Driel, R. (1982) Ligand binding properties of the cytoplasmic cAMP-binding protein of *Dictyostelium discoideum. FEBS Letters* **145**: 150–4.

De Wit, R. J. W. & Konijn, T. M. (1983) Identification of the acrasin of *Dictyostelium minutum* as a derivative of folic acid. *Cell Differentiation* **12**: 205–10.

Dicou, E. & Brachet, P. (1980). A separate phosphodiesterase for the hydrolysis of cyclic guanosine 3′, 5′-monophosphate in growing *Dictyostelium discoideum* amoebae. *Eur. J. Biochem.* **109**: 507–14.

Europe-Finner, G. N., McClue, S. J. & Newell, P. C. (1984). Inhibition of aggregation in *Dictyostelium* by EGTA-induced depletion of calcium. *FEMS Microbiol. Letters* **21**: 21–5.

Europe-Finner, G. N. & Newell, P. C. (1984). Inhibition of cyclic GMP formation and aggregation in *Dictyostelium* by the intracellular Ca^{2+} antagonist TMB-8. *FEBS Letters* **171**: 315–19.

Fain, J. N. & Malbon, C. C. (1979). Regulation of adenylate cyclase by adenosine. *Molec. Cell Biochem.* **25**: 143–69.

Galvin, N. J., Stockhausen, D., Meyers-Hutchins, B. L. & Frazier, W. A. (1984). Association of the cyclic AMP chemotaxis receptor with the detergent-insoluble cytoskeleton of *Dictyostelium discoideum. J. Cell Biol.* **98**: 584–95.

Gavin, J. R. III, Roth, J., Neville, D. M. Jr., De Meyts, P. & Buell, D. N. (1974). Insulin-dependent regulation of insulin receptor concentrations: a direct demonstration in cell culture. *Proc. Natl. Acad. Sci. U.S.A.* **71**: 84–8.

Gerisch, G. & Hess, B. (1974). Cyclic AMP-controlled oscillations in suspended *Dictyostelium* cells: their relation to morphogenetic cell interactions. *Proc. Natl. Acad. Sci. U.S.A.* **71**: 2118–22.

Gerisch, G., Malchow, D., Huesgen, A., Nanjundiah, V., Roos, W. & Wick, U. (1975). Cyclic AMP reception and cell recognition in *Dictyostelium discoideum*. In *ICN-UCLA Symposium on Developmental Biology*, ed. D. MacMahon & F. Fox. New York: Benjamin.

Goldbeter, A. & Segel, L. A. (1980). Control of developmental transitions in the cAMP signalling system of *Dictyostelium discoideum. Differentiation* **17**: 127–35.

Green, A. A. & Newell, P. C. (1975). Evidence for the existence of two types of cAMP binding sites in aggregating cells of *Dictyostelium discoideum. Cell* **6**: 129–36.

de Gunzburg, J., Hohman, R., Part, D. & Veron, M. (1983). Evidence that a cAMP binding protein from *Dictyostelium discoideum* carries *S*-adenosyl-L-homocysteine hydrolase activity. *Biochimie* **65**: 33–41.

de Gunzburg, J. & Veron, M. (1982). A cAMP-dependent protein kinase is present in differentiating *Dictyostelium discoideum* cells. *EMBO J.* **1**: 1063–8.

Hazelbauer, G. L. (1978). Taxis and behavior. *Receptors and Recognition*, series B, vol. **5**: 341 pp. London: Chapman & Hall.

Heck, G. L. & Ericson, R. P. (1973). A rate theory of gustatory stimulation. *Behav. Biol.* **8**: 687–712.

Henderson, E. J. (1975). The cyclic adenosine 3′, 5′-monophosphate receptor of *Dictyostelium discoideum. J. Biol. Chem.* **250**: 4730–6.

Juliani, M. H., Brusca, J. & Klein, C. (1981). cAMP regulation of cell differentiation in *Dictyostelium discoideum* and the role of the cAMP receptor. *Dev. Biol.* **83**: 114–21.

Juliani, M. H. & Klein, C. (1981). Photoaffinity labelling of the cell surface adenosine 3′, 5′-monophosphate receptor of *Dictyostelium discoideum* and its modification in down-regulated cells. *J. Biol. Chem.* **256**: 613–19.

Kakebeeke, P. I. J., Mato, J. M. & Konijn, T. M. (1978). Purification and preliminary characterization of an aggregation-sensitive chemoattractant of *Dictyostelium minutum*. *J. Bacteriol.* **133:** 403–5.

Klein, C. (1979). Slowly dissociating form of the cell-surface cAMP receptor of *Dictyostelium discoideum. J. Biol. Chem.* **254:** 12 573–8.

Klein, C. & Juliani, M. H. (1977). cAMP-induced changes in cAMP binding sites on *Dictyostelium discoideum* amoebae. *Cell* **10,** 329–35.

Konijn, T. M. (1972). Cyclic AMP as a first messenger. *Advances in Cyclic Nucleotide Research* **1:** 17–31.

Konijn, T. M., Van de Meene, J. G. C., Chang, Y. Y., Barkley, D. S. & Bonner, J. T. (1969). Identification of adenosine-3′,5′-monophosphate as the bacterial attractant for myxamoebae of *Dictyostelium discoideum. J. Bacteriol.* **99:** 510–12.

Kuczmarski, E. R. & Spudich, J. A. (1980). Regulation of myosin self-assembly: phosphorylation of *Dictyostelium* heavy chain inhibits formation of thick filaments. *Proc. Natl. Acad. Sci. U.S.A.* **77:** 7292–6.

Lesniak, M. A. & Roth, J. (1976). Regulation of receptor concentration by homologous hormone. *J. Biol. Chem.* **251:** 3720–9.

Londos, C., Cooper, D. M. F., Schlegel, W. & Rodbell, M. (1978). Adenosine analogs inhibit adipocyte adenylate cyclase by a GTP-dependent process: basis for action of adenosine and methyl xanthines on cAMP production and lipolysis. *Proc. Natl. Acad. Sci. U.S.A.* **75:** 5362–6.

Malchow, D., Böhme, R. & Rahmsdorf, H. J. (1981). Regulation of phosphorylation of myosin heavy chain during the chemotactic response of *Dictyostelium* cells. *Eur. J. Biochem.* **117:** 213–18.

Malchow, D. & Gerisch, G. (1974). Short-term binding and hydrolysis of cyclic 3′:5′-adenosine monophosphate by aggregating *Dictyostelium* cells. *Proc. Natl. Acad. Sci. U.S.A.* **71:** 2423–7.

Malchow, D., Nanjundiah, V. & Gerisch, G. (1978). pH oscillations in cell suspensions of *Dictyostelium discoideum*: their relation to cyclic-AMP signals. *J. Cell Sci.* **30:** 319–30.

Malchow, D., Nanjundiah, V., Wurster, B., Eckstein, F & Gerisch, G. (1978). cAMP-induced pH changes in *Dictyostelium discoideum* and their control by Ca^{2+}. *Biochim. Biophys. Acta* **538:** 473–80.

Maruta, H., Baltes, W., Dieter, P., Marmé, D. & Gerisch, G. (1983). Myosin heavy chain kinase inactivated by Ca^{2+} / calmodulin from aggregating cells of *Dictyostelium discoideum. EMBO J.* **2:** 535–42.

Mato, J. M., Jastorff, B., Morr, M. & Konijn, T. M. (1978). Model for cAMP chemoreceptor interaction in *Dictyostelium discoideum. Biochim, Biophys. Acta* **544:** 309–14.

Mato, J. M. & Konijn, T. M. (1977). Chemotactic activity of cAMP and AMP derivatives with substitutions in phosphate moiety in *Dictyostelium discoideum. FEBS Letters* **75:** 173–6.

Mato, J. M. & Malchow, D. (1978). Guanylate cyclase activation in response to chemotactic stimulation in *Dictyostelium discoideum. FEBS Letters* **90:** 119–22.

Mato, J. M. & Marín-Cao, D. (1979). Protein and phospholipid methylation during chemotaxis in *Dictyostelium discoideum* and its relationship to Ca^{2+} movements. *Proc. Natl. Acad. Sci. U.S.A.* **76:** 6106–9.

Mato, J. M., Van Haastert, P. J. M., Krens, F. A. & Konijn, T. M. (1977). An acrasin-like attractant from yeast extract specific for *Dictyostelium lacteum. Dev. Biol.* **57:** 450–3.

Mato, J. M., Van Haastert, P. J. M., Krens, F. A., Rhijnsburger, E. H., Dobbe, F. C. P. M. & Konijn, T. (1977). Cyclic AMP and folic acid mediated cyclic GMP accumulation in *Dictyostelium discoideum. FEBS Letters* **79:** 331–6.

McRobbie, S. J. & Newell, P. C. (1983). Changes in actin associated with the cytoskeleton following chemotactic stimulation of *Dictyostelium discoideum*. *Biochem. Biophys. Res. Commun.* **115**: 351–9.

McRobbie, S. J. & Newell, P. C. (1984). Chemoattractant-mediated changes in cytoskeletal actin of cellular slime moulds. *J. Cell Sci.* **68**: 139–51.

McRobbie, S. J. & Newell, P. C. (1985a). Cytoskeletal accumulation of a specific iso-actin during chemotaxis of *Dictyostelium*. *FEBS Letters* **181**: 100–2.

McRobbie, S. J. & Newell, P. C. (1985b). Effects of cytochalasin B on cell movements and chemoattractant-elicited actin changes of *Dictyostelium*. *Exp. Cell Res.* (In Press).

Meyers-Hutchins, B. L. & Frazier, W. A. (1984). Purification and characterisation of a membrane-associated cAMP-binding protein from developing *Dictyostelium discoideum*. *J. Biol. Chem.* **259**: 4379–88.

Mullens, I. A. & Newell, P. C. (1978). cAMP binding to cell surface receptors of *Dictyostelium*. *Differentiation* **10**: 171–6.

Nandini-Kishore, S. G. & Frazier, W. A. (1981). [³H]Methotrexate as a ligand for the folate receptor of *Dictyostelium discoideum*. *Proc. Natl. Acad. Sci. U.S.A.* **78**: 7299–303.

Newell, P. C. (1982). Cell surface binding of adenosine to *Dictyostelium* and inhibition of pusatile signalling. *FEMS Microbiol. Letters* **13**: 417–21.

Newell, P. C. & Ross, F. M. (1982). Inhibition by adenosine of aggregation centre initiation and cyclic AMP binding in *Dictyostelium*. *J. Gen. Microbiol.* **128**: 2715–24.

Pan, P., Hall, E. M. & Bonner, J. T. (1972). Folic acid as second chemostatic substance in the cellular slime moulds. *Nature, New Biology* **237**: 181–2.

Paton, W. D. M. (1961). A theory of drug action based on the rate of drug–receptor combination. *Proc. R. Soc. London. Series B.* **154**: 21–69.

Rahmsdorf, H. J., Malchow, D. & Gerisch, G. (1978). Cyclic AMP-induced phosphorylation in *Dictyostelium* of a polypeptide comigrating with myosin heavy chains. *FEBS Letters* **88**: 322–6.

Roos, W., Nanjundiah, V., Malchow, D. & Gerisch, G. (1975). Amplification of cyclic-AMP signals in aggregating cells of *Dictyostelium discoideum*. *FEBS Letters* **53**: 139–42.

Ross, F. M. & Newell, P. C. (1979). Genetics of aggregation pattern mutations in the cellular slime mould *Dictyostelium discoideum*. *J. Gen. Microbiol.* **115**: 289–300.

Ross, F. M. & Newell, P. C. (1981). Streamers: chemotactic mutants of *Dictyostelium* with altered cyclic GMP metabolism. *J. Gen. Microbiol.* **127**: 339–50.

Shimomura, O., Suthers, H. L. B. & Bonner, J. T. (1982). Chemical identity of the acrasin of the cellular slime mold *Polysphondylium violaceum*. *Proc. Natl. Acad. Sci. U.S.A.* **79**: 7376–9.

Swanson, J. A. & Taylor, D. L. (1982). Local and spatially coordinated movements in *Dictyostelium discoideum* during chemotaxis. *Cell* **28**: 225–32.

Tillinghast, H. S. Jr & Newell, P. C. (1984). Retention of folate receptors on the cytoskeleton of *Dictyostelium* during development. *FEBS Letters* **176**: 325–30.

Van Driel, R. (1981). Binding of the chemoattractant folic acid by *Dictyostelium discoideum* cells. *Eur. J. Biochem.* **115**: 391–5.

Van Haastert, P. J. M. (1983). Binding of cAMP and adenosine derivatives to *Dictyostelium discoideum* cells. *J. Biol. Chem.* **258**: 9643–8.

Van Haastert, P. J. M., Bijleveld, W. & Konijn, T. M. (1982). Phosphodiesterase induction in *Dictyostelium discoideum* by inhibition of extracellular phosphodiesterase activity. *Dev. Biol.* **94**: 240–5.

Van Haastert, P. J. M. & De Wit, R. J. W. (1984). The cell surface cAMP receptor of *Dictyostelium discoideum*: demonstration of receptor heterogeneity and affinity modulation by nonequilibrium experiments. *J. Biol. Chem.* **259**: 13321–8.

Van Haastert, P. J. M., De Wit, R. J. W. & Konijn, T. M. (1982). Antagonists and chemoattractants reveal separate receptors for cAMP, folic acid and pterin in *Dictyostelium*. *Exp. Cell Res.* **140**: 453–6.

Van Haastert, P. J. M., De Wit, R. J. W. Grijpma, Y. & Konijn, T. M. (1982).

Identification of a pterin as the acrosin of the cellular slime mold *Dictyostelium lacteum*. *Proc. Natl. Acad. Sci. U.S.A.* **79**: 6270–4.

Van Haastert, P. J. M. & Kien, E. (1983). Binding of cAMP derivatives to *Dictyostelium discoideum* cells: activation mechanism of the cell surface receptor. *J. Biol. Chem.* **258**: 9636–42.

Van Haastert, P. J. M., Van der Meer, R. C. & Konijn, T. M. (1981). Evidence that the rate of association of cAMP to its chemotactic receptor induces phosphodiesterase activity in *Dictyostelium discoideum*. *J. Bacteriol.* **147**: 170–5.

Van Haastert, P. J. M., Van Lookeren Campagne, M. M. & Ross, F. M. (1982). Chemotactic mutants of *Dictyostelium discoideum* with altered cyclic GMP-phosphodiesterase activity. *FEBS Letters* **147**: 149–52.

Van Ool, P. J. J. M. & Buck, H. M. (1982). The mechanisms of action of cAMP: a quantum chemical study. *Eur. J. Biochem.* **121**: 329–4.

Van Waarde, A. (1982). Rapid, transient methylation of four proteins in aggregative amoebae of *Dictyostelium discoideum* as a response to stimulation with cyclic AMP. *FEBS Letters* **149**: 266–70.

Van Waarde, A. (1983). Cyclic AMP, folic acid and pterin-mediated protein carboxymethylation in cellular slime molds. *FEBS Letters* **161**: 45–50.

Van Waarde, A. & Van Haastert, P. J. M. (1984). Transmethylation inhibitors decrease chemotactic sensitivity and delay cell aggregation in *Dictyostelium discoideum*. *J. Bacteriol.* **157**: 368–74.

Wallace, L. J. & Frazier, W. A. (1979). Photoaffinity labelling of cAMP binding and AMP binding proteins of differentiating *Dictyostelium discoideum* cells. *Proc. Natl. Acad. Sci. U.S.A.* **76**: 4250–4.

Wick, U., Malchow, D. & Gerisch, G. (1978). cAMP stimulated Ca^{2+} influx into aggregating cells of *Dictyostelium discoideum*. *Cell Biology Int. Reports* **2**: 71–9.

Wurster, B., Bozzaro, S. & Gerisch, G. (1978). Cyclic GMP regulation and responses of *Polysphondylium violaceum* to chemoattractants. *Cell Biology Int. Rep.* **2**: 61–9.

Wurster, B. & Butz, U. (1980). Reversible binding of the chemoattractant folic acid to cells. *Eur. J. Biochem.* **109**: 613–18.

Wurster, B., Schubiger, K., Wick, U. & Gerisch, G. (1977). Cyclic GMP in *Dictyostelium discoideum*. *FEBS Letters* **76**: 141–4.

6

Receptors in slime mould cell adhesion

C. M. Chadwick, J. E. Ellison and D. R. Garrod

6.1 Introduction: the cellular slime moulds

The cellular slime moulds or Acrasiales are unusual organisms
of uncertain phylogenetic relationships. They are classified among the
lower fungi as Mycetozoa together with other groups including the true
slime moulds or Myxomycetales and the parasitic Plasmodiophorales
(Bonner, 1967). As the name Mycetozoa implies, the cellular slime
moulds have features in common with both fungi and animals. The resting stage of their life cycle is a fruiting body or sorocarp consisting of
a stalk, usually cellular in nature, and spores (Fig. 6.1). It is only at
this stage of the life cycle that cells become enclosed by solid cell walls.
They are unusual among fungi in that the stalk sheath consists of cellulose. At all other stages of the life cycle the cells are naked, motile
and amoeboid.

Apart from possessing cell walls at the sorocarp stage, the cellular
slime moulds resemble plants in few characteristics. One plant-like feature which may be particularly pertinent in the present context is that
their glycoproteins contain no sialic acid (Jermyn, Kilpatrick, Schmidt
& Sterling, 1977; Gilkes, Laroy & Weeks, 1979; Sharon & Lis, 1979).
Otherwise, slime mould amoebae appear more like the cells of animals.
Indeed the very phenomenon we are considering, cell adhesion, is characteristic of the behaviour of animal cells which are either brought into
contact by aggregation or movement, or find themselves as neighbours
due to cell division. In both animals and slime moulds, adhesion occurs
between molecules in the plasma membranes of adjacent cells, whereas
in plants, because the cells are surrounded by walls, such plasma membrane interactions do not as a rule take place. Whereas animal cells
have many structural and chemical specializations of their plasma membranes for adhesion (Garrod & Nicol, 1981), these are absent from plant

cells unless the cell wall itself is regarded as such a structure. There are some situations where formation of an adhesion between plant cells is a functional requirement. For example in the processes of mating and fertilization, adhesion between unlike gametes is required (see Chapter 8). The cells concerned usually adhere cell-wall to cell-wall before plasma membrane contact is established; they would therefore not be expected to have receptors for adhesion on the outer surface of their plasma membranes.

Fig. 6.1. The life cycle of *Dictyostelium discoideum*. Beginning at top left-hand side: vegetative amoebae feeding on bacteria; aggregation centre with aggregation streams (enlargement shows cells in aggregation stream); migrating grex, slug or pseudoplasmodial stage; early culmination stage (from side view); fruiting body (from side view); crushed fruiting body showing spores and stalk. The cells are 10–15 μ in diameter, the grexes of the order of one millimetre in length and the spores about 5–7 μ in length.

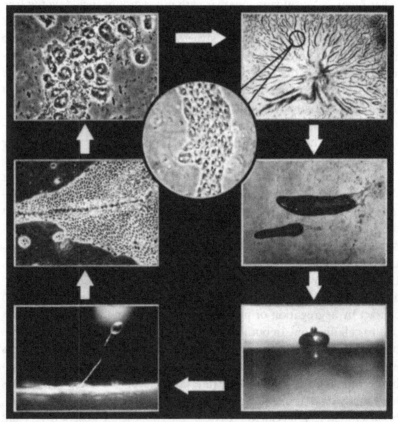

The relevance of adhesion in this book is that it is one of a whole range of phenomena which may be included within the scope of the rather loose term 'cellular recognition'. This includes any process which involves binding between a cell and (a) another cell (b) a bacterium or inert particle or (c) a molecule. Examples of (a) are gamete interactions, immune adherence of killer cells to target cells, tissue cell adhesion and cell adhesion in slime moulds. Examples of (b) are phagocytosis and the uptake of bacteria into root nodules (see Chapter 11). Examples of (c) are lectin–cell interactions, the uptake of blood proteins by the liver, the binding of antigens to lymphocytes, antibodies to cells and hormone–receptor interactions. To describe a process as recognition implies that it is dependent on some type of complementary molecular interaction i.e. between a ligand and receptor. In some cases, such as those mentioned under (c) above, the molecular interactions have been well demonstrated. In the present context some cell adhesion molecules have now been identified. Examples are the neural and liver cell-adhesion molecules (N-CAM and L-CAM) (Edelman, 1983) and those of the epithelial intercellular junctions known as desmosomes (Cowin, Mattey & Garrod, 1984). A good deal is known also about the aggregation factor mechanisms of sponges (Jumblatt, Schlup & Burger, 1980; Muller, 1982) which in some cases appear to involve protein–carbohydrate interactions. The latter, however, may turn out to be untypical of animal cell adhesion systems.

6.2   The functions of cell adhesion in animals

The morphogenesis which occurs during animal development is extensively dependent upon cell movements. These are of various types: the bending of cell sheets or rolling of them into tubes, invagination or evagination of cell sheets or cell masses and migration of cells either singly or in masses (Gustafson & Wolpert, 1967). The control of these movements and the positioning of cells as a result of them is believed to be by means of cell adhesion both to other cells and to the extracellular matrix (Trinkaus, 1969; Garrod, 1973, 1981; Garrod & Nicol, 1981). Once the correct positioning of cells has been achieved, it must be maintained throughout adult life. In cancer the integrity of tissues is not maintained. Cells become invasive and metastatic, spreading into other tissues and other parts of the body. A primary reason for the acquisition of these properties by neoplastic cells may be an alteration in their adhesive and motile properties. In certain respects controlled cell motility is still an important normal process in the adult, for example in the migration and circulation of cells in the immune system and during wound healing.

6.3   The functions of cell adhesion in slime moulds

The cellular slime moulds, in particular *Dictyostelium discoideum*, provide a simple developmental system in which cell adhesion plays an important role in morphogenesis (Shaffer, 1962, 1964; Gerisch, 1968; Garrod & Ashworth, 1973; Garrod, Swan, Nicol & Forman, 1978; Rosen & Barondes, 1978; Barondes, Springer & Cooper, 1982). Their life cycle (Fig. 6.1) is divisible into a vegetative phase in which free-living amoebae (soil microorganisms in nature) feed on bacteria while increasing in number by division. The onset of the developmental phase occurs at aggregation which is a chemotactic process (see Chapter 5) but which is also much dependent upon cell adhesion. The multicellular mass resulting from aggregation elongates vertically to form a polarized structure which is surrounded by a thin, polysaccharide slime sheath. This moves into a horizontal position and migrates over the substratum as the migrating pseudoplasmodium, slug or grex. Next, and finally, the grex returns to the vertical position and forms a sorocarp containing a stalk and spores. Precursors of these two cell types, called prespore and prestalk cells, are present in the migrating grex at the back and front respectively, (Raper, 1940; Bonner, 1944; Takeuchi, 1963; Gregg, 1965). Differentiation of the precursors begins at the onset of elongation of the cell mass (Hayashi & Takeuchi, 1976; Forman & Garrod, 1977a,b).

The functions of cell adhesion in slime moulds vary according to the stage of the life cycle and appear to be as follows:

6.3.1   Vegetative cells

These are solitary amoeboid cells which feed and divide. They need to adhere to the substratum in order to move and they need to adhere to their food microorganisms such as bacteria and yeasts in order to phagocytose them (Vogel, Thilo, Schwarz & Steinhart, 1980). They tend to repel each other actively, showing negative chemotaxis (Samuel, 1961; Shaffer, 1962). Nevertheless, vegetative *D. discoideum* cells are cohesive, those of the axenic strain, Ax-2, grown in suspension being more so than wild-type cells grown on bacteria (Garrod, 1972; Swan & Garrod, 1975). It seems likely that the properties of phagocytosis and cellular adhesiveness are at least partly related at the molecular level (Vogel *et al.* 1980; Chadwick, Ellison & Garrod, 1984) (see below).

6.3.2 Aggregating cells

The onset of the multicellular or social phase of the life cycle is accompanied by the acquisition of new, stage-specific, cohesive properties. Aggregating cells are elongated and polarized (Bonner, 1947). When a cell joins an aggregation stream its leading edge attaches to the tail of a cell already in the stream (Schaffer, 1964). Thus by means of mutual adhesion they follow each other in streams into the aggregation centre, an aspect of behaviour termed 'contact following' by Shaffer (1965). The new cohesive properties of aggregation stage cells subserve this contact following behaviour. Whereas the cohesion of vegetative cells is inhibited by EDTA, that of aggregating cells is only partially so. Dehaan (1959) and, particularly, Gerisch (1961) showed that the end-to-end contacts of aggregating cells persisted in the presence of EDTA. Much work has been done subsequently on the cohesive mechanism of aggregating or 'aggregation-competent' cells (see below).

Aggregating cells and, indeed, cells at later stages of the life cycle, retain their ability to adhere to the substratum. Aggregating cells must move over the substratum in order to reach the aggregation centre so that adhesion to the substratum has functional importance. Later, the grex becomes surrounded by a thin slime sheath which then forms the substratum for cell movement. Dissociated cells can, however, adhere to other substrata if dissociated.

Another intriguing possibility is that intercellular contact acts as a trigger for differentiation. Thus when slime mould cells aggregate the establishment of specific intercellular contacts may switch on those cellular events which are involved in later development. This has been suggested by several groups (Newell, Longlands & Sussman, 1971; Newell, Franke & Sussman, 1972; Garrod & Forman, 1977; Forman & Garrod, 1977b; Garrod et al. 1978; Grabel & Loomis, 1977; Lodish & Alton, 1977; Burger & Clark, 1983). Recently it has been shown that only mutants which can establish EDTA-resistant cohesions will accumulate late stage messenger RNA species: mutants which show chemotactic aggregation but do not form these cohesions do not accumulate these messengers (Blumberg et al. 1982). Further, when an aggregate is dissociated, stage-specific mRNA and enzyme species are lost, a phenomenon known as erasure (Sussman, 1976; Finney, Varnum & Soll, 1979). Sussman's group have in fact isolated a mutant (FR-17) that will continue to differentiate as isolated cells when disaggregated (Wilcox & Sussman, 1978). FR-17 has not yet been characterized fully, but should prove important in the analysis of contact-mediated regulation of gene expression.

To conclude, we wrote in a previous paper: 'It seems probable that the cohesion of cells triggers a sequence of differentiative changes possibly in a manner analogous to the changes brought about by the binding of hormones to cell surfaces. Thus the interaction of receptors on apposed cell surfaces may trigger cytoplasmic events leading to differentiation' (Garrod *et al.* 1978). However, Gross *et al.* (1981) proposed that there are no cell interactions dependent on contact *per se*. They have shown that formation of mature stalk cells from isolated amoebae requires only cAMP and the lipid-like differentiation-inducing factor (DIF). In addition sporogenous mutants can form spores in the absence of contact under appropriate conditions.

6.3.3   The aggregate and the migrating grex

After aggregation, the cells still possess the capacity for contact following (Garrod, 1969, 1974*b*) so that end-to-end adhesion is important.

At these stages an entirely new aspect of development begins, that of pattern formation involving prestalk and prespore cells (see above). Cell adhesion may play an important role in this. There is now increasing evidence that the formation of the anterior-posterior prestalk-prespore pattern may arise by cell sorting-out (Takeuchi, 1969; Bonner, Sieja & Hall, 1971; Leach, Ashworth & Garrod, 1973; Maeda & Maeda, 1974; Sharpe, Treffry & Watts, 1982). Using a prespore-specific antibody prepared according to Takeuchi (1963), Garrod & Forman (1977) and Forman & Garrod (1977*a*,*b*) found that prespore cells developed in an apparent spatially random fashion within spherical aggregates of slime mould cells maintained in suspension for 12–18 h. Similar results were found by Sternfeld & Bonner (1977) using PAS-staining as a cell marker, and more recently, Tasaka, Noce & Takeuchi (1983) have observed a random differentiation of prestalk cells prior to aggregation. It now seems that differentiation into prespore and prestalk cells may precede sorting out, and that no positional information is needed to select either of the alternative pathways of differentiation. (See, however, Krefft *et al.*, 1984). On the other hand, it is still unclear whether sorting-out of prespore and prestalk cells is brought about by differential cell adhesion (Steinberg, 1964) or by chemotaxis. Thus Garrod *et al.* (1978) concluded that sorting out is probably caused by a difference in the relative strengths of adhesion of prestalk and prespore cells, while Durston & Vork (1979) and Sternfeld & David (1981) have shown that chemotaxis is also important.

6.4   Molecular basis of slime mould cell cohesion

There is much evidence that cohesion in the cellular slime mould is complex, involving several different molecules not all of which are present at any particular stage of development. It is therefore useful to consider them in chronological order.

6.4.1   Vegetative cells

Working with *D. discoideum*, V12-M2, Beug, Gerisch, Kempf, Riedel & Cremer (1970) prepared univalent Fab fragments of antibodies raised against crude particulate fractions of vegetative amoebae. (Fab fragments retain the antigen-binding specificity of the parent immunoglobulin molecules and are therefore capable of blocking receptor sites. However, because they are univalent they do not cause patching, capping or agglutination and are thus ideal for the study of cell cohesion. They were first used for this purpose by Gerisch and his colleagues with *Dictyostelium* but have since been used for the study of cell cohesion in animal systems.) The Fab fragments inhibited cohesion of vegetative cells and these workers suggested the name contact sites B (CSB) for the cohesive molecules involved. As we shall see, CSB persist on the amoebae at least until the aggregation stage. A similar approach has led to the isolation of a membrane antigen (contact sites A(CSA)) which appears to be involved in cohesion of aggregation stage cells. This will be considered in detail in the next section.

Our approach to vegetative cell cohesion arose because of two discoveries. Firstly, as axenic cells progressed from the log phase to the stationary phase of growth they become non-cohesive. Secondly, a low molecular weight inhibitor of vegetative cell cohesion accumulated in stationary-phase medium (Swan & Garrod, 1975; Swan, Garrod & Morris, 1977; Garrod *et al.* 1978). Jaffé, Swan & Garrod (1979) showed that the low molecular weight inhibitor was specific for vegetative cell cohesion. Like EDTA, it completely inhibited the cohesion of vegetative cells, but only partially inhibited that of aggregation-stage cells. The inhibitor was not a chelating agent, however, because its activity was not prevented by calcium. It is resistant to boiling, to proleolytic enzymes and periodate oxidation and there is some evidence that it binds reversibly to the cell surface (Swan *et al.* 1977; Ellison, unpublished). If this inhibitor can be identified it may provide an important clue to the molecular mechanism of vegetative cell cohesion. It was also found that although stationary phase cells were not mutually cohesive, they would adhere to log phase vegetative cells. One interpretation of these facts was that vegetative cell cohesion may involve a ligand-receptor mechanism (Jaffé et al. 1979).

The effects of isolated plasma membranes on vegetative cohesion in D. discoideum have also been studied (Jaffé & Garrod, 1979). Membranes from log phase cells gave complete inhibition of log phase cell cohesion but only partial inhibition of aggregation-competent cell cohesion. Those isolated from stationary cells gave moderate inhibition of vegetative cohesion. Also, membranes from aggregation-competent cells completely inhibited the cohesion of log-phase cells but at higher concentrations than those of log-phase membranes.

It was hoped that isolated plasma membranes would provide a starting point for the isolation of cohesion molecules. Since a purified cohesion molecule may display an affinity for its receptor, an extract of membranes in which it is present might therefore block cell cohesion. We have found that proteins may be extracted from the plasma membrane of D. discoideum using the chaotrope lithium diiodosalicylate (LIS) (Jaffé, 1980). These tend to form large aggregates akin to micelles since they are hydrophobic and insoluble membrane proteins (Chadwick and Garrod, unpublished). Although such extracts inhibit the cohesion of D. discoideum they are also toxic to the cells and eventually cause lysis. The extracts have similar effects upon erythrocytes and it is therefore likely that some solubilized membrane proteins can disrupt the plasma membrane of cells. We feel therefore, that if extracts of cells are employed in the above way, it is essential to establish that the viability of the cells is unaffected.

Vogel et al. (1980) have provided evidence for two receptor systems that mediate cohesion in vegetative cells of D. discoideum. They obtained mutant strains that were defective in phagocytosis. Mutant cells were unable to ingest the bacterium E. coli B/r when glucose was added to the medium. They also did not form EDTA-sensitive cohesions (mediated by CSB) or adhere to substrata. The mutants were able to ingest strongly hydrophobic particles but not hydrophilic ones. Vogel et al. (1980) suggested that the cells carried a lectin type of receptor specific for glucose residues (present on E. coli B/r) and furthermore that a nonspecific receptor was present which allowed the binding of particles by hydrophobic interaction. The mutant cells were also defective in cell adhesion at the vegetative stage suggesting that the nonspecific receptor activity might be involved in this as well as in phagocytosis.

In this laboratory we have raised several antisera against vegetative cells of D. discoideum strain Ax-2, which yield cohesion-inhibiting Fab' fragments. (Fab' is a slightly larger univalent antibody fragment than Fab, being derived from IgG with pepsin rather than papain treatment.)

They affect vegetative cohesion but not the developmentally regulated system of cohesion acquired by aggregation-competent cells. We have used these antisera to identify the contact site of vegetative cells. Immunoprecipitation and absorption studies led to the conclusion that the contact sites activity resides in a molecule of M_r 126 000 which is exposed on the cell surface, as demonstrated by its ready iodination with lactoperoxidase. An antibody raised against this molecule, purified by gel electrophoresis, also yielded cohesion-inhibiting Fab' (Chadwick & Garrod, 1983). The molecule was later found to be glycosylated and hence is now referred to as gp126 (Chadwick *et al.* 1984).

The antibody raised against purified gp126 is monospecific since it immunoprecipitates only gp126 from the surface of vegetative amoebae (Chadwick *et al.* 1984). Its Fab' has been shown to inhibit cell cohesion, cell–substratum adhesion and phagocytosis of both bacteria (*E. coli* B/r) and latex particles (Chadwick *et al.* 1984; Garrod, 1984). Furthermore, immunoprecipitation studies demonstrated that there was significantly less gp126 on the surface of both wild type cells and axenic cells grown on bacteria than on the latter grown axenically. These observations may suggest that gp126 is internalized during phagocytosis and may account for the low cohesiveness of vegetative wild type cells (Garrod, 1972) compared with vegetative axenic cells (Swan & Garrod, 1975). Anti-gp126 Fab' also inhibits cell–substratum adhesion and phagocytosis of bacteria and latex beads (Chadwick *et al.* 1984; Garrod, 1984). Can gp126 be the non-specific receptor of Vogel *et al.* (1980)? Duffy & Vogel (1984) think this unlikely because cells treated with anti-gp126 Fab' should be able to phagocytose *E. coli* B/r by means of the glucose receptor, unless glucose and nonspecific receptor activity reside in the same molecule.

Duffy & Vogel (1984) have demonstrated by complementation analysis of phagocytosis-deficient mutants that there are at least two recessive loci on different chromosomes (linkage groups IV and VIII) governing the expression of non-specific receptor properties. They conclude that the nonspecific 'receptor' is unlikely to be a single molecule.

Chadwick *et al.* (1984) have also investigated the effect of anti-gp126 Fab' on development of *D. discoideum*. It was found that in the presence of excess Fab' development proceeded normally in every respect except that much smaller aggregates were formed at the aggregation stage and, hence, the fruiting bodies were much smaller. Although many factors may influence the size of aggregates formed at the aggregation stage, we suggested that the effect of the Fab' was in reducing cellular adhesiveness by blocking one of the cellular adhesion mechanisms. From this result we also believe that gp126 cannot be involved in 'triggering' development (Chadwick *et al.* 1984). Instead the developmental trigger appears to be starvation, because cultures of cells which were fed with bacteria were delayed in development whereas parallel cultures which were treated with anti-gp126 Fab' to prevent them phagocytosing entered development immediately (Chadwick, Ellison and Garrod, unpublished observations).

An important question which will depend upon further characterization of the molecules involved is whether gp126 is a receptor for a non-specific property of cells, bacteria and substrata. A cautionary aspect which has not yet been entirely eliminated is that Fab' such as gp126 may be causing a general masking of the cell surface rather than specifically blocking a receptor molecule.

Marin, Goyette-Boulay & Rothman (1980) using *D. discoideum*, strain A3, have suggested that CSB cohesion may involve a ligand-receptor mechanism based on glucose or a related molecule. We find, however, that glucose and its analogues cause only slight inhibition of cell cohesion even at very high concentrations (100 mM) (Swan, 1978).

Using polyacrylamide gels derivatized with various sugars, Bozzaro & Roseman (1982) obtained the additional evidence that growth-phase cells adhere to glucose, maltose, cellobiose, *N*-acetylglucosamine and mannose. They observed a decline in *N*-acetylglucosamine-binding ability as development progressed to the aggregation stage and also that neither vegetative nor aggregation-competent cells had any affinity for lactose or galactose which might be interpreted as evidence against the involvement of the lectin, discoidin (see below), in cell adhesion.

The behaviour of cells on carbohydrate-derivatized gels is curious and interesting. Instead of undergoing normal development, they stop at the aggregation stage and proceed to undergo cycles of disaggregation and reaggregation. The pattern of gene expression found under these conditions has been studied by Bozzaro, Perlo, Ceccarelli & Mangiorotti (1984).

What is the significance of these studies for slime mould cell adhesion? Our own view is that they clearly demonstrate the presence of carbohydrate receptors on the cell surface and these are probably primarily involved in phagocytosis under normal conditions, being receptors for carbohydrates on the surface of food microorganisms. These receptors may also be involved in cell–cell adhesion, but at present there is no evidence to suggest that they are.

In the related species *Polysphondylium pallidum*, the molecule responsible for vegetative cohesion has also been identified with the aid of cohesion-blocking Fab fragments. Steinemann, Hinterman & Parish (1979) prepared a butanol extract of plasma membranes from *P. pallidum*. This contained only two antigens of molecular weight 110 000 and 71 000 respectively. The latter absorbed the cohesion-blocking effect of Fab fragments suggesting that this molecule is involved vegetative cohesion. They also found that the molecule was present on aggregation-competent cells although it disappeared once aggregation had taken place. It contained the sugars glucosamine and fucose (Hinterman & Parish, 1979) and was therefore a glycoprotein.

Vegetative cohesion of *P. pallidum* has also been studied by Bozzaro *et al* (1981) who found that a butanol extract of cells absorbed the cohesion-blocking activity of a Fab. The factor that absorbed this activity was purified by isolectric focussing. It had a molecular weight of 64 000 and has been called contact site 1 (CS1). An interesting property of CS1 is its apparent ability to dimerize and form a molecule of 130 000 molecular weight. This may indicate that the molecule can form inter-cellular bonds by dimerization. An additional, minor component of the butanol extract was also found to absorb the cohesion-blocking activity. It was distinct from CS1 after isolectric focusing and had a molecular weight of 58 000. This molecule, CS2, may not be a glycoprotein since it was unstained by Schiff's reagent. Bozzaro and colleagues were unsure whether these molecules were identical to those described by Steinemann *et al.* (1979). They felt, however, that CS2 was not a precursor of CS1 since peptide maps of each were dissimilar and the amino acid composition differed. The similarity in molecular weight between CS1 (130 000 daltons) and the CSB molecule from *D. discoideum* (126 000 daltons) should be noted.

Toda *et al.* (1984) have carried out a detailed study of the *P. pallidum* vegetative contact site molecule (which we shall now refer to as gp64) using both polyclonal and monoclonal antibodies. Firstly they showed that it could be resolved into two bands on polyacrylamide gels, with molecular weights of 66 kD and 60 kD (referred to as gp66 and gp60, respectively). Polyclonal antibodies raised against these also reacted with many other membrane proteins in immunoblotting studies, and almost all of this activity could be absorbed with purified gp64. Monoclonal antibodies against gp64 showed similar cross-reactivity.

Fab from several polyclonal anti-gp64 antibodies and from one mono-clonal antibody completely blocked adhesion of aggregation-competent cells. The activity could be entirely absorbed by a carbohydrate fraction deprived from the purified glycoprotein. It was concluded that both poly-clonal and monoclonal antibodies react with the carbohydrate portion of the molecule and that the relevant carbohydrate moieties are shared with other membrane proteins. These results show how anticarbohydrate antibodies may have a general nonspecific inhibitory effect on cell cohesion and emphasize the care needed in interpreting results derived from Fab inhibition studies.

6.4.2   *Aggregating cells*

Gerisch (1968) reported that a new cell-surface antigen appeared at the aggregation stage in *D. discoideum*, and its appearance paralleled the development of EDTA-insensitive, end-to-end cell cohesion. Beug *et al.* (1970) and Beug, Katz & Gerisch (1973) produced Fab against aggregation-stage cells and this inhibited their cohesion completely. Absorption with vegetative cells produced a Fab which specifically inhibited end-to-end cohesion of aggregating cells. The Fab had no effect on chemotaxis or cell motility. The developmentally regulated molecules presumed to be responsible for end-to-end cohesion were named contact sites A (CSA).

Since this first work a number of other observations of fundamental importance have emerged. Beug, Katz, Stein & Gerisch (1973) found that 3×10^5 anti-CSA Fab molecules per cell completely blocked CSA-mediated cohesion. They estimated that not more than 2% of the cell surface had been covered by the Fab molecules. Furthermore, 8 times the quantity of Fab fragments directed against a carbohydrate fraction of the plasma membrane could bind to the cell surface without blocking cohesion. Thus it was the specific blocking of contact sites by anti-CSA Fab which inhibited cohesion and not nonspecific coating of the cell surface.

A further interesting point is that Beug *et al.* were able to label the whole cell surface including the cell ends with fluorescent anti-carbohydrate Fab and yet still observe a cohesion-blocking effect of anti-CSA Fab. Steric hindrance by the anti-carbohydrate Fab had thus not affected the inhibitory qualities of anti-CSA Fab. It therefore seemed that CSA must project from the cell surface into the intercellular space to a distance greater than the projection of the derivatized Fab from the cell surface. This was confirmed by measuring the distance between the plasma membrane and ferritin molecules when the binding of Fabs to cells was detected by immunoelectromicroscopy (Gerish, Beug, Malchow, Schwarz & Stein, 1974). Rather curiously, the same study showed that anti-CSA Fab bound over the entire surface and not just at the ends of the cells.

The available evidence therefore indicated that CSA was a large molecule that was part of 'or anchored to' the plasma membrane. The next step was to isolate the molecule. Membrane proteins were extracted with sodium deoxycholate and the solubilized material was found to absorb the cohesion-blocking activity of the anti-CSA Fab. The solubilized material resolved into various constituents on Sephadex G200. CSA was eluted from the Sephadex column in fraction separate from that of cAMP phosphodiesterase (Heusgen & Gerisch, 1975).

The purification of CSA from *D. discoideum* has since been completed. Butanol/water extracts were made from plasma membranes and the CSA antigen could be detected in the water phase. Using ion-exchange chromatography and sucrose gradient centrifugation of glycoprotein was purified that had a molecular weight of 80 000. This molecule (gp80) contained protein and carbohydrate in a ratio of 3 : 1. The sugars mannose, N-acetylglucosamine and glucose were present in equal amounts but there was less fucose. N-acetylgalactosamine was absent and there was a small quantity of galactose. The polypeptide moiety contained much proline and an abundance of threonine and serine (Muller, Gerisch, Fromme, Mayer & Tsugita, 1979). Schmidt & Loomis (1982) has shown that [^{32}P] becomes incorporated into gp80 in the form of serine phosphate. Recently Ochai, Schwarz, Merkl, Wagle & Gerisch (1982) have raised monoclonal antibodies against purified gp80. Ferritin immunoelectron microscopy using the monoclonals has shown that gp80 is distributed in clusters over the entire surface of aggregation-stage cells. If the cells were allowed to develop to the slug stage the monoclonals recognized an additional cell-surface molecule of 106 000 daltons. Polyclonal antisera to aggregation competent cells have been found to cause dissociation of slug cells (Beug, Gerisch & Muller, 1971) suggesting either that gp80 persists to the slug stage of development or that the 106 000 dalton protein is a further cohesion molecule. Other studies on late-stage cohesion (Chadwick, Collodi and Sussman, manuscript in preparation) show that about half of the cohesiveness of 16-h structures is due to CSA and that it must therefore persist after the aggregation stage. The 106 000 dalton protein reported by Ochai *et al.* (1982) was therefore probably a cross-reactant with their antiserum.

Ochai *et al.* (1982) have studied further the properties of their monoclonal antisera. They find that membranes of tunicamycin-treated cells contain molecules of 66 000 and 53 000 daltons which react with anti-gp80 monoclonals. They interpret their results as showing that gp80 carries two carbohydrate chains. However, since EDTA-stable adhesion is not blocked by monoclonals that react specifically with the glycosylated protein, they feel that the carbohydrate moieties are not involved directly in adhesion.

Hirano, Yamada & Miyazaki (1984) report that an antibody raised against a 68kD glycoprotein shows cross-reactivity with both gp80 and another component of 56 kD. This rabbit antiserum, in the presence of Fab from goat anti-rabbit IgG, inhibits cell cohesion.

Stadler, Gerisch, Bauer, Suchanek & Huttner (1983) have suggested an explanation for the mechanism of cohesion employed by gp80, and also for the interesting observation that gp80-mediated cohesiveness can rapidly be turned off and on by the extracellular presence and absence of cAMP. They find that sulphate is incorporated into gp80 where, being highly electronegative, it might bind to positive charges on the opposing cell surface. Regulation of adhesiveness might thus be accomplished by direct sulphation, by an alteration of configuration such as that produced by a protein kinase or by interaction with the cytoskeleton. The reversible inhibition of gp80-mediated cohesiveness has also been studied by McConaghy, Saxe, Williams & Sussman (1984), who have coined the term 'modulation' for the abrupt change. They found that modulation occurred when cells were treated with diffusible metabolites that affect the cAMP relay.

Stadler, Bauer & Gerisch (1984) have now demonstrated the lipoprotein nature of gp80 by showing it to become acylated by palmitic acid. This property is one shared by a variety of integral membrane proteins from other sources.

Monoclonal antibody raised against the lectins, discoidins I and II, from *D. discoideum* show cross-reactivity not only with gp80 but also the β-galactosidase and the *lac* repressor from *E. coli*. (The latter are both carbohydrate-binding proteins.) The two *E. coli* proteins and discoidin share an amino acid sequence (-Ser-X-X-Ile-His(Pro)-Pro(His)-Leu-Thr-) which may be responsible for the cross-reactivity. One possible inference from this result is that gp80 may have carbohydrate-binding activity, though it is not yet known whether the monclonal antibody-binding site lies within carbohydrate-binding sites of the other proteins.

Elsewhere, gp80-mediated cohesion has been studied by several laboratories and there are now a further three that have produced monoclonal antibodies. Klein has obtained a monoclonal antibody that blocks adhesion and which immunoprecipitates a molecule of 80 000 daltons from aggregation-stage cells (Loomis, personal communication). Springer & Barondes (1983) have obtained monoclonal antibodies that block cell cohesion and which fall into two categories, those specific for aggregating cells and a second family of antibodies active against both vegetative and aggregation stages. Loomis and co-workers have studied the developmental regulation of molecules that react with an anti-gp80 monoclonal and find that gp80 can be detected in AX3 cells from about 10 h (bacterially grown cells) or 4 h (axenically grown cells) of development. It persists on the cells surface and is present even on spores (Murray, Yee & Loomis, 1981).

Finally, Loomis has also obtained mutants that do not react with an anti-gp80 monoclonal antibody. The mutants nevertheless are able to cohere because the portion of gp80 to which the antibody binds is probably not involed directly in adhesion. The mutants have been crossed with marker strains by established techniques of parasexual genetics and the pattern of segregation indicates that the mutation (modB) can be ascribed to linkage group VI (Murray, Wheeler, Jongens & Loomis, 1984.)

As in *D. discoideum*, developing cells of *P. pallidum* were found to acquire contact sites in addition to those present at the vegetative stage. Anti-*P. pallidum* Fab was found to dissociate aggregates of developing cells of this species but did not affect aggregates of *D. discoideum*. Anti-*D. discoideum* Fab was found to display a similar specificity for cells against which it had been raised (Bozzaro & Gerisch, 1978). Contact sites of *D. discoideum* and *P. pallidum* are therefore serologically distinct. On the face of it this would seem to support Gerisch's view that contact sites should be species specific (Gerisch, 1977). However we would point out, firstly, that serological specificity does not necessarily denote adhesive specificity, and, secondly, that there is abundant evidence that cells of different slime mould species adhere to each other (Nicol & Garrod, 1978; Sternfeld, 1979; Bozzaro & Gerisch, 1978; Springer & Barondes, 1978).

To conclude this section we should mention some additional facts. Firstly, Bozzaro & Gerisch (1979) have reported the presence of a low molecular weight factor in a soluble extract of vegetative axenic cells which inhibits cohesion of aggregation stage cells. Although it blocked the formation of EDTA-stable contacts, no indication has been provided that it interacts directly with CSA. Secondly, Laroy & Weeks (1982) have prepared an octyl glucoside extract which partially inhibits the cohesion of aggregation stage cells. They claim that this pronase-sensitive activity might represent an alternative cohesion mechanism to CSA. This is preliminary work and further evidence is awaited. Finally, Brodie, Klein & Swierkosz (1983) have used monoclonal antibodies to detect an antigen which appears to be involved in aggregation stage adhesion.

This is a doublet of 69 000 and 73 000. Its appearance is not controlled by pulses of cAMP.

6.4.3   Post-aggregation stages
gp95

One approach to the study of cohesiveness in cellular slime moulds has been to use surface labelling techniques to look for changes in plasma membrane proteins which correlate with changes in cohesiveness. Using this approach Parish, Schmidlin & Parish (1979) and Parish & Schmidlin (1979) have found that the synthesis of gp80 ceases after aggregation and that it subsequently disappears from the plasma membrane (see however, Ochai et al. 1982 and above). They also discovered that the plasma membranes of grex cells contained a glycoprotein of molecular weight 95 000 that was absent in vegetative amoebae and aggregating cells. It was synthesized from the tip stage until late culmination since radioactive acetate and glucosamine were incorporated into the molecule during this period (Parish & Schmidlin, 1979). The glycoprotein was therefore a strong candidate for a cohesion molecule and in later experiments this was confirmed. Steinemann & Parish (1980) raised an antibody to plasma membranes at the grex stage of development. Fab fragments were made from this and were found to block the reaggregation of dissociated grex cells. When slices of polyacrylamide gel that contained fixed gp95 were incubated with the inhibitory Fab, the cohesion-blocking effect of the latter was removed.

Kunzli & Parish (1983) have since shown that gp95 is located predominantly on prespore cells of the migrating slug, an interesting point because differential cell adhesiveness could be important in formation of the prespore–prestalk pattern.

Cohesiveness in cells of this developmental stage have also been studied by Wilcox & Sussman (1981a). They obtained a mutant of D. discoideum, strain JC5, which at 22°C showed development that was identical to the parent strain. At 27°C a temperature-sensitive mutation was expressed in which the normal aggreagtes at 11 h of development dissociated into a lawn of single cells. The defect was shown to be a phenomenon peculiar to late stage cells because incubation at 27°C before 11 h had no affect on aggregation or development.

The mutation seemed to be important until at least 14 h of development because cultures placed at 27°C could be made to regress at this time. Wilcox & Sussman prepared ghosts of cells cultured at 22°C and found that they had normal cohesiveness, even at non-permissive temperatures. Furthermore, ghosts from noncohesive cells were noncohesive at both normal and non-permissive temperatures. This indicated that the temperature-sensitive defect lay in the inability of the mutant cells to produce a factor at the cell surface that was required for cohesion. No defect of CSA nor any change in the lectin discoidin were observed. Wilcox & Sussman therefore suggested that a molecule must assume the function of CSA in later development.

In later work, Wilcox & Sussman (1981*b*) prepared cohesion-blocking Fab against grex-stage cells. It was observed that JC5 cells cultured at the permissive temperature were able to absorb the cohesion-blocking activity from the Fab. However, no such effect was found when the Fab was incubated with cells that had been cultured at 27°C. Wilcox & Sussman suggested that the antigen at the cell surface might be gp95.

More recently Saxe & Sussman (1982) purified a wheat germ agglutinin-binding fraction from late-stage plasma membranes of *D. discoideum*, which induced noncohesive cells of strain JC5 to cohere. The active fraction could be obtained from wild type cells and from JC5 grown at 22°C, but not from JC5 cultured at 27°C. As yet however, there is no indication that the inducing activity is due to gp95.

In the studies described above, however, the mutant JC5, although defective in late-stage cell adhesion, was difficult to analyse because JC5 was itself a derivative of FR-17, which has an accelerated development, will continue to differentiate even when disaggregated, and forms an upside-down fruit. Moreover, none of the normal structures encountered during wild-type morphogenesis could be observed. Clearly, the mutant JC5 was of little value in elucidating the function of the late-stage cohesive system in normal morphogenesis. However, the late-stage temperature-sensitive mutation (CoA) has since been transferred to a wild-type background by crossing JC5 with DdB, and selecting for haploid segregants. Mutant JC36 has thus been obtained and has the CoA mutation, but with an entirely normal morphology at 22°C. At the restrictive temperature, however, the defect in cohesion can be observed at 14 h and is fully apparent by 16 h after which no further development occurs and the cell aggregates slowly regress in shape. Under conditions of growth which produce migrating slugs, cells are deposited in the form of a trail behind the slug until, at 27°C, it eventually disappears. The late-stage structures do not fall apart completely and show a considerable amount of cohesiveness at 27°C, due to residual aggregation-stage (gp80-mediated) cohesion which persists until late in development (C. M. Chadwick, R. P. Collodi & M. Sussman, manuscripts in preparation). The CoA mutation is currently being mapped and should soon be ascribed to a particular linkage group.

gp150

Using an approach similar to that of Parish and colleagues, Geltosky, Siu & Lerner (1976) found that the content of a concanavalin A-binding surface glycoprotein of 150 000 molecular weight increased between 6 and 18 h of development. Later, Geltosky, Weseman, Bakke & Lerner (1979) purified this protein and raised antibodies to it. Using a fluorescein-conjugated antibody they were able to quantitate the development of the glycoprotein (gp150) with a fluorescence activated cell sorter. A maximum level was attained after 10 h and maintained until the final stages of differentiation. They also found that the Fab prepared from the antibody prevented the cohesion of dissociated 15 h cells. In further work, Geltosky, Birdwell, Weseman & Lerner (1980) studied the distribution of gp150 within aggregates of developing *D. discoideum* by immunohaemocyanin labelling and scanning electron-microscopy. They found that gp150 was concentrated at sites of contact between cells.

Lam, Pickering, Geltosky & Sui (1981), have provided evidence that gp150 is differentially expressed in prestalk and prespore cells. Culmination-stage grexes were separated into populations of prespore and prestalk cells by discontinuous gradient centrifugation. They found that 75% of prespore cells would aggregate after 15 min of roller culture. The prespore cells formed large clumps that were resistant to dissociation by 30 mM EDTA. By contrast, prestalk cells were less cohesive. Only 60% of these cells formed aggregates in 10 mM EDTA and complete inhibition of aggregation was observed in 20 mM EDTA. Furthermore the aggregates themselves were very small. This suggested that prespore cells had acquired an extra mechanism for cohesion. The behaviour of the two populations of cells when treated with anti-gp150 Fab was then examined. At 1 mg/ml Fab more than half the prestalk cells were able to form small aggregates but prespore cells were dissociated completely under these conditions. Thus, prespore cells had aquired an extra mechanism of cohesion and this was probably gp150.

Given that gp150 is concentrated more on prestalk cells it comes as no surprise to find that Siu, Roches & Lam (1983) have shown it to be involved in the cell-sorting process. They observed that Fab directed against gp150 inhibited the sorting of prespore and prestalk cells while Fab directed against lectins, contact sites A or vegetative cells had no effect.

On the other hand, the entire gp150 study has been thrown in doubt by a report from Loomis, Murray, Yee & Yongens (1983), who have shown that anti gp150 antibodies cross-react with gp80. Thus if gp80 persists throughout development many of the properties ascribed to gp150 could in fact stem from the former molecule.

6.5   Slime mould lectins

These molecules were discovered by Barondes & Rosen and their colleagues (Rosen, Kafka, Simpson & Barondes, 1973) in a search for possible mechanisms of cohesion. Cell cohesion may involve some type of ligand–receptor interaction and there are glycoproteins on the cell surface. Why not look, therefore, for a molecule with carbohydrate-binding activity, i.e. a lectin?

Rosen *et al.* (1973) obtained soluble extracts from *D. discoideum* (wild type, strain NC-4) grexes which agglutinated formalin-fixed sheep red blood cells. The extractable agglutinating activity was developmentally regulated, increasing dramatically in parallel with the acquisition by the cells of EDTA-insensitive cohesion. Furthermore, the haemagglutinating activity was specifically blocked by galactose-like sugars such as *N*-acetylgalactosamine and lactose. This enabled the lectin to be purified by affinity chromatography on the galactose-containing polymer Sepharose 4B. Two separate fractions which had slightly different properties were resolved. These fractions (discoidins I and II) have since been shown to consist of tetramers of 4 polypeptides. In the case of discoidin I these are of 26 000 daltons and for discoidin II, 24 000 daltons. The lectins therefore both have a molecular weight of approximately 100 000 but the composition and carbohydrate-binding specificities are slightly different (Frazier, Rosen, Reitherman & Barondes, 1975). Recently, Berger & Armant (1981) have shown that discoidins I and II contain regions of polypeptide that are unique to each. They also contain common sequences and Berger & Armant (1981) have therefore drawn an analogy with other families of recognition proteins such as immunoglobulins.

Other slime moulds also contain lectins. The best characterized of these is pallidin which has been purified from extracts of *P. pallidum*. The purified lectin differs from the discoidins in consisting of three closely related isolectins composed of different combinations of three subunits of molecular weight 25 000–27 000. Its binding specificity is unlike that of discoidin in that pallidin binds lactose but has little affinity for *N*-acetyl-D-galactosamine (Rosen, Simpson, Rose & Barondes, 1974). The lectins of other slime moulds have not been studied to the same extent.

A lectin should have a receptor and receptors have indeed been found for slime mould lectins. Reitherman, Rosen, Frazier & Barondes (1975) found that receptors for discoidins I and II were present at the cell surface and furthermore these were developmentally regulated. Receptors for pallidin were also found to be present on *P. pallidum* cells. The receptors seemed to show some specificity for lectins from their own species. Madley, Herries & Hames (1981) demonstrated that there are common surface receptors for discoidins I and II, while Bartles & Frazier (1982) showed two types of receptor for discoidin I on fixed *D. discoideum* cells. The latter receptors are dependent on carbohydrate binding (C-receptors) and ionic binding (I-receptors). Bartles, Santaro & Frazier (1982) showed that in living cells a certain proportion of exogenous discoidin I became much more tightly bound than that which was associated with either of the above receptors. Reitherman *et al.* (1975) showed that the binding of the slime mould lectins to their receptors had an affinity one order of magnitude greater than the binding to receptors on erythrocytes. Chang, Rosen & Barondes (1977) have found that the receptor for pallidin is distributed over the whole cell surface.

Recently, Drake & Rosen (1982) have demonstrated that the pallidin receptor will agglutinate *P. pallidum* cells which have been treated with dinitrophenol and have suggested that the receptor is analogous to a sponge aggregation factor. Ray & Lerner (1982) have isolated a membrane protein of 80 000 molecular weight from *D. discoideum* which they claim is the discoidin I receptor. Instead of agglutinating the cells, it causes inhibition of cohesion of aggregation-competent cells in the presence of EDTA. It is distinct from CSA which has the same molecular weight.

The cellular location of slime mould lectins has also been described and early studies suggested that a proportion was present at the cell surface. Slime mould amoebae at the aggregation stage were found to form rosettes with erythrocytes and this effect could be blocked by those sugars known to interact with the lectins (Rosen *et al.* 1973, 1974; and Chang *et al.* 1977). Further evidence that discoidin occurs at the cell surface was provided by Chang, Reitherman, Rosen & Barondes (1975) who raised an antibody to discoidin and found that this bound to the cell surface. The binding was observed for developing cells but not for vegetative cells. This result was in contrast to that found for cells of *P. pallidum*, since the latter had a certain amount of pallidin at the surface of vegetative amoebae (Chang *et al.* 1977). Further evidence of the surface location has come from the fact that the surface of *D. discoideum* cells may be radio-iodinated and radioactive discoidin observed. If anti-discoidin is used to immunoprecipitate protein, the antibody is able to recover radio-iodinated discoidin (Siu *et al.* 1976).

However, slime mould lectins now appear to be principally cytoplasmic. Pallidin is found within the cell where it is localized at the endoplasmic reticulum (Chang *et al.* 1977). For *Dictyostelium purpureum* the distribution of purpurin has been assayed. Springer, Haywood & Barondes (1980) have found that as little as 2% of this lectin is normally found at the cell surface. They also made the interesting observation that anti-purpurin IgG could bring about the transfer of purpurin to the cell surface. More recently, Springer & Barondes (1982) have shown that other reagents can have the same effect provided that they are bivalent.

Bartles *et al.* (1982) have determined that there are only 1.3×10^3 lectin tetramers on the surface of each cell and 6×10^3 per cell in the surrounding medium at the time when the cells are maximally cohesive and conclude that the lectin is therefore very unlikely to be involved in cohesion. Since there are 5×10^6 tetramers inside each cell, Bartles *et al.* propose that the molecule has some type of internal regulatory function.

Erdos & Whitaker (1983) attempted to refine the techniques employed in localization of discoidin and found that all was localized within the cell. They concluded that other reports of discoidin on the cell surface stemmed from cell lysis by the experimental techniques employed. They suggested further that lectins have only an intracellular role.

Some studies on a mutant deficient in the complement of active lectin seemed to provide the only real evidence for the role of lectins in cell cohesion. Ray, Shinnick & Lerner, (1979) obtained a mutant of *D. discoideum*, HJR 1, that formed only loose mounds at the aggregation stage and failed to develop thereafter. The soluble fraction of the mutant failed to agglutinate sheep erythrocytes in contrast to that of wild type NC4 cells. It was known that the 26 kD subunit of discoidin I was responsible for such agglutination rather than the 24 kD subunit of discoidin II. Hence, it was proposed that the mutation was in the 26 kD subunit of discoidin (carbohydrate-binding protein of CBP26). The mutant cells had levels of CBP26 comparable with that of NC4 cells (determined by radio-immunoassay), but the lectin did not bind to galactose. Hence, the mutant was characterized by a structural defect in CBP26. Revertants of HJR1 showed normal cohesiveness and development. Since the revertants also had normal haemagglutination activity the ability to cohere could be correlated with the recovery of active CBP26.

More recently Alexander, Shinnick & Lerner (1983) have described two other mutants (SA-31 and SA-219) which express vastly reduced amounts of the two discoidins (less than 1% and 1–2% respectively) compared to the wild type. These mutants lack lectin activity, lectin protein and lectin mRNA. Nevertheless they are able to undergo morphogenesis and differentiation which differs from that of the wild type only in being very slightly accelerated. These results strongly indicate not only that the lectins are not involved in cell adhesion but further that they are not required for development.

The above review shows that the slime mould lectins have many of the properties one might expect of a cell cohesion molecule. However, the evidence that these lectins are involved in cell cohesion is not convincing. Thus neither the lectins themselves nor their carbohydrate inhibitors affect cell cohesion except under highly unphysiological conditions (Rosen, Haywood & Barondes, 1978; Rosen & Barondes, 1978; Marin *et al.* 1980). Furthermore, anti-lectin Fabs are similarly without effect (Rosen *et al.* 1976; Rosen, Chang & Barondes, 1977). A recent report (Cano & Pestaña, 1984) that a discoidin Fab inhibits cell adhesion must be treated cautiously because of the demonstration of cross-reactivity of anti-discoidin antibodies and contact sites A (Stadler, Bauer, Westphal & Gerisch, 1984b). A final point is that discoidin is not developmentally regulated in axenic strains even though EDTA-insensitive cohesion is (Ochai *et al.* 1982).

In conclusion, we believe that slime mould lectins are most interesting molecules whose true function has yet to be established. (It has now been suggested that discoidin may be involved in cell–substratum adhesion. See Springer, Cooper & Barondes, 1984).

6.6   The interplay of contact sites

Evidence for the multiplicity of contact sites in cellular slime moulds is now overwhelming. They seem to be produced at various phases of the life cycle and all are developmentally regulated. The function of the various contact sites is still a little unclear but there are two explanations for the multiplicity of mechanisms for cohesion. In the first case a particular contact site could be produced simply to replace a molecule employed earlier in development, but it is difficult to see what advantage this might bring. The second explanation is that additional contact sites are produced in response to a different type of level of cohesiveness required as a consequence of differentiation and morphogenesis.

This is probably the better explanation, especially in view of the evidence that gp95 and gp150 are associated mainly with prespore cells. The information available would suggest that gp126 is active in vegetative and aggregation-competent cells and that in the latter cells it is joined by gp80. Later in development the two are joined by gp95 (and possibly gp150). However, since turnover is slow and they appear to decay only over an extended period, gp126, gp80, gp95 and possibly gp150 may all be employed during late morphogenesis. During development it seems likely that differential cell adhesion is involved in cell sorting.

A further consideration is that when the cells enter the grex stage they become surrounded by a mucopolysaccharide slime sheath which acts as the substratum for cell movement (Shaffer, 1965). A new contact site could play a role in adhesion to this. From a functional viewpoint it would be reasonable to expect that the grex cells can readily revert to their prior complement of contact sites. If a grex is crushed in the presence of bacteria its cells will return to the vegetative phase; however, if it is dissociated in the absence of food it will reaggregate (Raper, 1940). The latter reaggregation appears to be divided into two phases, the first is dependent on cell adhesion alone and the second which begins about one hour after dissociation is dependent on chemotaxis with the formation of aggregation streams (Garrod, 1974b).

6.7 The nature of adhesive mechanisms

At the beginning of this chapter we classifed cell adhesion as an example of cellular recognition thus implying that it is dependent on some type of specific complementary molecular interaction between the surface of adjacent cells. In conclusion we would like to re-examine this classification in the light of the evidence we have discussed.

Four cell-surface molecules, gp80, gp95, gp126 and gp150, have now been identified by means of specific Fab (or Fab') binding as candidates for cohesion molecules in *D. discoideum* alone. In no case, however, do we have any information as to the mechanism by which these molecules bring about intercellular binding. Moreover there are a few puzzling facts which should be considered. Firstly, in its purified form the CSA glycoprotein, GP80, absorbs all the cohesion-blocking activity of anti-CSA Fab. The Fab is therefore monospecific and does not reveal the existence of two complementary molecules as would be expected for a ligand–receptor type interaction. A way out of this is to suggest that both ligand and receptor activity reside in the same molecule and that intercellular bonding involves dimer formation (Muller & Gerisch, 1978).

Secondly, addition of purified gp80 to cells has no effect on their cohesion (Muller & Gerisch, 1978). This is somewhat surprising if cohesion is dependent on ligand–receptor interaction irrespective of whether or not the latter involves dimer formation. Of course it may be argued that denaturation of the molecule occurs when it is removed from the cell membrane and this causes loss of activity. Thus, with gp80 as with other molecules identified by specific cohesion-blocking Fabs, the evidence that they are really involved in cohesion remains indirect.

It was for this reason that we preferred our approach to vegetative cohesion based on our low molecular weight inhibitor and inhibition of cohesion by isolated membranes. Here was a direct approach not involving the use of an external agent, namely an antibody fragment. However, the inhibitor remains unidentified, cohesion-inhibiting membrane extracts proved toxic to cells and we would be the first to admit that our ligand–receptor model of vegetative cell cohesion (Jaffé *et al.* 1979) is highly speculative.

We may add to this the fact that although slime mould lectins exist, as well as receptors for them, the evidence that they perform any cohesive function remains poor. Could it be that cell–cell adhesion in slime moulds (and elsewhere) is not dependent upon ligand–receptor interaction at all?

It seems to us that the cohesion and adhesion of slime mould cells receive contributions from sources other than ligand–receptor interaction. After all the cells are capable of adhesion to a variety of substrata including agar, glass and plastic. Such adhesion probably involves non-specific molecular interactions. Some time ago Born & Garrod (1968) showed that cohesion of *D. discoideum* cells could be induced by raising the ionic strength of the surrounding medium, divalent cations being more effective than monovalent. The cells became more sensitive to this effect as they approached the aggregation stage; in modern parlance the effect was developmentally regulated. Subsequently, Garrod & Gingell (1970) demonstrated that the electrophoretic mobility and hence the surface-charge density of *D. discoideum* cells decreased as the cells approached the aggregation stage. This has subsequently been confirmed by Lee (1972) and Maeda (1980). We felt that these results could be consistent with the view that cohesion was of a colloidal nature being dependent on the interaction between London–van der Waals attractive forces and electrostatic repulsive forces (Curtis, 1962). However we also had some evidence against this (Gingell & Garrod, 1969). Recently, Gingell & Vince (1982b) have shown that for *D. discoideum* cells adhering to glass (1) increasing the ionic strength of the medium causes cell--substratum separation to decrease (2) divalent cations are more effective in this than monovalent and (3) there is an apparently finite separation of several tens of nanometres between cell and substratum which cannot be decreased further by increasing ionic strength or valency. They conclude that there is cell coat material between cell and substratum which acts as a cationic polyelectrolyte, and suggest that adhesion could result from the action of short-range forces between coat and substratum. They have evidence, however, that ionic binding with the substratum is not involved (Gingell & Vince, 1982a).

Another possibility which should be restated here is that Vogel *et al.* (1980) have suggested that binding of particles to cells may occur by hydrophobic interactions.

The crucial question still remains 'By what mechanism does adhesive bonding between cell surfaces occur?' Final proof that any of the molecules we have mentioned is really a cohesion molecule will depend upon showing how they bring about such bonding.

We thank Dr Adrian Simmonds for his comments on the manuscript and Mrs Lynette Adam for photography. Original work repeated on this paper was supported by the Science Research Council and facilities were provided by the Cancer Research Campaign.

6.8   References

Alexander, S., Shinnick, T. M. & Lerner, R. A. (1983). Mutants of Dictyostelium discoideum blocked in expression of all members of the developmentally regulated discoidin multigene family. *Cell.* **34**: 467–75.

Barondes, S. H., Springer, W. R. & Cooper, D. N. (1982). Cell adhesion. In *The Development of Dictyostelium discoideum*, ed. W. F. Loomis, pp. 195–231. New York: Academic Press.

Bartles, J. R. & Frazier, W. A. (1982). Discoidin I membrane interactions 1. Discoidin binds to two types of receptor on fixed *Dictyostelium discoideum* cells. *Biochim. Biophys. Acta* **687**: 121–8.

Bartles, J. R., Santaro, B. C. & Frazier, W. A. (1982). Discoidin I membrane interactions III. Interaction of Discoidin I with living *Dictyostelium discoideum* cells. *Biochim. Biophys. Acta* **687**: 137–46.

Berger, E. A. & Armant, D. R. (1981). Discoidins I and II: common and unique regions on two lectins involved in cell recognition in *Dictyostelium discoideum*. *J. Cell Biol.* **91**: 103a.

Beug, H., Gerisch, G., Kempff, S., Riedel, V. & Cremer, G. (1970). Specific inhibition of cell contact formations in *Dictystelium* by univalent antibodies. *Exp. Cell Res.* **63**: 147–58.

Beug, H., Gerisch, G. & Müller, E. (1971). Cell dissociation univalent antibodies as a possible alternative to proteolytic enzymes. *Science* **173**: 742–3.

Beug, H., Katz, F. E. & Gerisch, G. (1973). Dynamics of antigenic membrane sites relating to cell aggregation in *Dictyostelium discoideum*. *J. Cell Biol.* **56**: 647–58.

Beug, H., Katz, F. E., Stein, A. & Gerisch, G. (1973). Quantitation of membrane sites in aggregating *Dictyostelium* cells by use of tritiated univalent antibody. *Proc. Natl. Acad. Sci. U.S.A.* **70**: 3150–4.

Blumberg, D. D., Margolskee, J. P., Barkliss, E., Chung, S. N., Cohen, N. S. & Lodish, H. F. (1982). Specific cell–cell contacts are essential for induction of gene expression during differentiation of *Dictyostelium discoideum*. *Proc. natl. Acad. Sci. U.S.A.* **79**: 127–31.

Bonner, J. T. (1944). A descriptive study of the development of the slime mold *Dictyostelium discoideum*. *Am. J. Bot.* **31:** 175–82.

Bonner, J. T. (1947). Evidence for the formation of cell aggregates by chemotaxis in the development of the slime mold *Dictyostelium discoideum*. *J. Exp. Zool.* **106:** 1–26.

Bonner, J. T. (1967). *The Cellular Slime Molds*. Princeton, N.J.: Princeton University Press.

Bonner, J. T., Sieja, T. W. & Hall, E. M. (1971). Further evidence for the sorting out of cells in the differentiation of the cellular slime mould *Dictyostelium discoideum*. *J. Embriol Exp. Morph.* **25:** 457–65.

Born, G. V. R. & Garrod, D. (1968). Photometric demonstration of aggregation of slime mould cells showing effects of temperature and ionic strength. *Nature* **220:** 616–18.

Bozzaro, S. & Gerisch, G. (1978). Contact sites in aggregating cells of *Polysphondylium pallidum*. *J. Molec. Biol.* **120:** 265–79.

Bozzaro, S. & Gerisch, G. (1979). Developmentally regulated inhibitor of aggregation in cells of *Dictyostelium discoideum*. *Cell Diff.* **8:** 117–27.

Bozzaro, S., Perlo, C., Ceccarelli, A. & Mangiarolti, G. (1984). Regulation of gene expression in *Dictyostelium discoideum* cells exposed to immobilized carbohydrates. *EMBO J.* **3:** 193–200.

Bozzaro, S. & Roseman, S. (1982). Adhesion of *Dictyostelium discoideum* cells to sugar-derivatised polyacrylamide gels. In *Embryonic Development B: Cellular Aspects*, pp. 183–92. New York: Alan Liss Inc.

Bozzaro, S., Tsugita, A., Janku, M., Monok, G., Opatz, K. & Gerisch, G. (1981). Characterisation of a purified cell surface glycoprotein as a contact site in *Polysphondylium pallidum*. *Exp. Cell Res.* **134:** 181–91.

Brodie, C., Klein, C. & Swierkosz, J. (1983). Monoclonal antibodies: use to detect developmentally regulated antigens on *Dictyostelium discoideum* amoebae. *Cell* **32:** 1115–23.

Burger, E. A. & Clark, J. M. (1983). Specific cell-cell contact serves as the development signal to deactivate discoidin I gene expression in *Dictyostelium discoideum*. *Proc. natl. Acad. Sci. U.S.A.* **80:** 4983–7.

Cano, A. & Pestanna, A. (1984). The role of membrane lectins in *Dictyostelium discoideum* aggregation as ascertained by specific univalent antibodies against discoidin I. *J. Cell Biochem.* **25:** 31–43.

Chadwick, C. M., Ellison, J. E. & Garrod, D. R. (1984). Dual role for *Dictyostelium* contact site B in phagocytosis and developmental size regulation. *Nature* **307:** 646–7.

Chadwick, C. M. & Garrod, D. R. (1983). Identification of the cohesion molecule, contact sites B, of *Dictyostelium discoideum*. *J. Cell Sci.* **60:** 251–66.

Chang, C. M., Reitherman, R. W., Rosen, S. D. & Barondes, S. H. (1975). Cell surface location of discoidin, a developmentally regulated carbohydrate-binding protein from *Dictyostelium discoideum*. *Exp. Cell. Res.* **95:** 136–42.

Chang, C. M., Rosen, S. D. & Barondes, S. H. (1977). Cell surface location of an endogenous lectin and its receptor in *Polysphondylium pallidum*. *Exp. Cell Res.* **104:** 101–9.

Cowin, P., Mattey, D. L. & Garrod, D. R. (1984). Identification of desmosomal surface components (desmocollins) and inhibition of desmosome formation by specific Fab'. *J. Cell Sci.* **70:** 41–60.

Curtis, A. S. G. (1962). Cell contact and adhesion. *Biol. Rev.* **37:** 82–129.

Dehaan, R. L. (1959). The effects of the chelating agent ethylenediamine tetracetic acid on cell adhesion in the slime mould *Dictyostelium discoideum*. *J. Embryol. Exp. Morph.* **7:** 335–43.

Drake, D. K. & Rosen, S. D. (1982). Identification and purification of an endogenous receptor for the lectin pallidin from *Polysphondylium pallidum*. *J. Cell Biol.* **903:** 383–9.

Duffy, K. T. I. & Vogel, G. (1984). Linkage analysis of two phagocytosis receptor loci in *Dictyostelium discoideum. J. Gen. Microbiol.* **130:** 2071–7.

Durston, A. J. & Vork, F. (1979). A cinematographical study of vitally stained *Dictyostelium discoideum. J. Cell Sci.* **36:** 261–79.

Edelman, G. M. (1983). Cell adhesion molecules. *Science* **219:** 450–45.

Erdos, G. W. & Whitaker, D. (1983). Failure to detect immunocytochemically reactive endogenous lectin on the cells surface of *Dictyostelium dioscoideum. J. Cell. Biol.* **97:** 993–7.

Finney, R., Varnum, B. & Soll, D. R. (1979). 'Erasure' in *Dictyostelium:* a dedifferentiation involving the programmed loss of chemotactic functions. *Dev. Biol.* **73:** 290–303.

Forman, D. & Garrod, D. R. (1977a). Pattern formation in *Dictyostelium discoideum* I. Development of prespore cells and its relationship to the pattern of the fruiting body. *J. Embryol. Exp. Morph.* **40:** 215–28.

Forman, D. & Garrod, D. R. (1977b). Pattern formation in *Dictyostelium discoideum* II. Differentiation and pattern formation in non-polar aggregates. *J. Embryol. Exp. Morph* **40:** 229–43.

Frazier, W. A., Rosen, S. D., Reitherman, R. W. & Barondes, S. H. (1975). Purification and comparison of two developmentally regulated lectins from *Dictyostelium discoideum*. Discoidin I and II. *J. Biol. Chem.* **250:** 7714–21.

Garrod, D. R. (1969). The cellular basis of movement of the migrating grex of the slime mould *Dictyostelium discoideum J. Cell Sci.* **4:** 781–98.

Garrod, D. R. (1972). Acquisition of cohesiveness by slime mould cells prior to morphogenesis. *Exp. Cell. Res.* **72:** 588–91.

Garrod, D. R. (1973). *Cellular Development.* London: Chapman and Hall.

Garrod, D. R. (1974a). Cellular recognition and specific cell adhesion in cellular slime mould development. *Arch. Biol.* **85:** 7–31.

Garrod, D. R. (1974b). The cellular basis of movement of the migrating grex of the slime mould *Dictyostelium discoideum*: chemotactic and reaggregation behaviour of grex cells. *J. Embryol. Exp. Morph.* **32:** 57–68.

Garrod, D. R. (1981). Adhesive interactions of cells in development: specificity, selectivity and cellular adhesive potential. *Fortschritte der Zoologie* **26:** 183–95.

Garrod, D. R. (1984). Aggregation, cohesion, adhesion, phagocytosis and morphogenesis in *Dictyostelium* – mechanisms and implications. In *Dahlem Conference Workshop on 'Microbial adhesion and aggregation'* ed. K. C. Marshall, pp. 337–49. Berlin, Heidelberg, New York, Tokyo: Springer-Verlag.

Garrod, D. R. & Ashworth, J. M. (1973). Development of the cellular slime mould *Dictyostelium discoideum. Symp. Soc. Gen. Microbiol.* **23:** 407–35.

Garrod, D. R. & Forman, D. (1977) Pattern formation in the absence of polarity in *Dictyostelium discoideum. Nature* **265:** 144–6.

Garrod, D. R. & Gingell, D. (1970). A progressive change in the electrophoretic mobility of preaggregation cells of the slime mould, *Dictyostelium discoideum. J. Cell Sci.* **6:** 277–84.

Garrod, D. R. & Nicol, A. (1981). Cell Behaviour and molecular mechanisms of cell–cell adhesion. *Biol. Rev.* **56:** 199–242.

Garrod, D. R., Swan, A. P., Nicol, A. & Forman, D. (1978). Cellular recognition in slime mould development. *Society for Experimental Biology Symp.* **32:** *Cell-Cell Recognition*, pp. 173–202, Cambridge: Cambridge University Press.

Geltosky, J. E., Birdwell, C. R., Weseman, J. & Lerner, R. A. (1980). A glycoprotein, involved in aggregation of *D. discoideum*, is distributed on the cell surface in a nonrandom fashion favoring cell junctions. *Cell* **21:** 339–45.

Geltosky, J. E., Siu, C. H. & Lerner, R. A. (1976). Glycoproteins of the plasma membrane of *Dictyostelium discoideum* during development. *Cell* **8:** 391–6.

Geltosky, J. E., Weseman, J., Bakke, A. & Lerner, R. A. (1979). Identification of a cell surface glycoprotein involved in cell aggregation in *D. discoideum*. *Cell* 18: 391–8.

Gerisch, G. (1960). Zellfunktionen und Zellfunktionswechsel in der Entwicklung von *Dictyostelium discoideum*. V. Stadienspezifische Zellkontactbildung und ihre quantitative Erfassung. *Exp. Cell Res.* 25: 535–54.

Gerisch, G. (1968). Cell aggregation and differentiation in *Dictyostelium discoideum*. *Current Topics in Dev. Biol.* 3: 157–97.

Gerisch, G. (1977). Membrane sites implicated in cell adhesion. Their developmental control in *Dictyostelium discoideum*. In *International Cell Biology*, ed. Brinkley & Porter. New York: Rockefeller Press.

Gerisch, G., Beug, H., Malchow, D., Schwarz, H. & Stein, A. V. (1974). Receptors for intercellular signals in aggregating cells of the slime mold, *Dictyostelium discoideum*. *Miami Winter Symposium* 7, in *Biology and Biochemistry of Eukaryotic Cell Surfaces*, pp. 49–66. New York and London: Academic Press.

Gilkes, N. R., Laroy, K. & Weeks, G. (1979). An analysis of the protein, glycoprotein and monosaccharide composition of *Dictyostelium discoideum* plasma membranes during development. *Biochim. Biophys, Acta* 551: 349–62.

Gingell, D. & Garrod, D. R. (1969). Effect of EDTA on electrophoretic mobility of slime mould cells and its relationship to current theories of cell adhesion. *Nature* 221: 192–3.

Gingell, D. & Vince, S. (1982a). Substratum wetability and charge influence the spreading of *Dictyostelium* amoebae and the formation of ultrathin cytoplasmic lamellae. *J. Cell Sci.* 54: 255–85.

Gingell, D. & Vince, S. (1982b). Cell–glass separation depends on salt concentration and valency: measurements on *Dictyostelium* amoebae by finite aperture interferometry. *J. Cell Sci.* 54: 299–310.

Grabel, L. & Loomis, W. F. (1977). Cellular interactions regulating early biochemical differentiation in *Dictyostelium*. In *Development and Differentiation in the Cellular Slime Moulds*, ed. P. Cappuccinelli & J. M. Ashworth, pp. 189–99. Amsterdam: Elsevier/North Holland.

Gregg, J. H. (1965). Regulation in the cellular slime molds. *Dev. Biol.* 12: 377–93.

Gross, J. D., Town, C. D., Brookman, J. J., Jermyn, K. A., Peacey, M. J. & Kay, R. R. (1981). Cell patterning in *Dictyostelium*. *Phil. Trans. R. Soc. Lond. B* 295: 497–508.

Gustafson, T. & Wolpert, L. (1967). Cellular movement and contact in sea urchin morphogenesis. *Biol. Rev.* 42: 442–98.

Hayashi, M. & Takeuchi, I. (1976). Quantitative studies on cell differentiation during morphogenesis of the cellular slime mould *Dictyostelium discoideum*. *Dev. Biol.* 50: 302–9.

Heusgen, A. & Gerisch, G. (1975). Solubilised contact sites A from cell membranes of *Dictyostelium discoideum*. *FEBS Lett.* 56: 46–9.

Hintermann, R. & Parish, R. W. (1979). Synthesis of plasma membrane proteins and antigens during development of the cellular slime mold *Polysphondylium pallidum*. *FEBS Lett.* 108: 219–25.

Hirano, T., Yamada, H. & Miyazaki, T. (1984). Cross-reactivity of contact site A to antibody produced against stage-specific antigen from phenol/water extract of *Dictyostelium discoideum*. *J. Biochem.* 95: 1355–6.

Jaffé, A. R. (1980). Studies on the cohesion mechanism of axenic *Dictyostelium discoideum* cells. Ph.D. Thesis, University of Southampton.

Jaffé, A. R. & Garrod, D. R. (1979). Effect of isolated plasma membranes on cell-cell cohesion in the cellular slime mould. *J. Cell Sci.* 40: 245–56.

Jaffé, A. R., Swan, A. P. & Garrod, D. R. (1979). A ligand-receptor model for the cohesive behaviour of *Dictyostelium discoideum* axenic cells. *J. Cell Sci.* 37: 157–67.

Jermyn, K. A., Kilpatrick, D. C., Schmidt, J. A. & Stirling, J. L. (1977). Components of the plasma membrane of *Dictyostelium discoideum* during aggregation. In *Development and Differentiation in the Cellular Slime Moulds*, ed. P. Cappuccinelli & J. M. Ashworth, pp. 79–84. Amsterdam: Elsevier/North-Holland.

Jumblatt, J. E., Schlup, V. & Burger, M. M. (1980). Cell to cell recognition: specific binding of *Microciona* sponge aggregation factor to homotypic cells and the role of calcium ions, *Biochemistry* **19**: 1038–42.

Krefft, M., Voet, L., Gregg, J. H., Mairhofer, H. & Williams, K. L. (1984). Evidence that positional information is used to establish the prestalk-prespore pattern in *Dictyostelium discoideum* aggregates. *EMBO J.* **3**: 201–6.

Kunzli, M. & Parish, R. W. (1983). A developmentally regulated glycoprotein implicated in adhesion of *Dictyostelium* slugs is predominantly in prespore cells. *FEBS Lett.* **155**: 253–6.

Lam, T. Y., Pickering, G., Geltosky, J. E. & Sui, C. H. (1981). Differential cell cohesiveness expressed by prespore and prestalk cells of *Dictyostelium discoideum*. *Differentiation* **20**: 22–8.

Laroy, K. & Weeks, G. (1982). Extraction of membrane factors that inhibit aggregation in *Dictyostelium discoideum*. *J. Cell Sci.* **55**: 277–86.

Leach, C. K., Ashworth, J. M. & Garrod, D. R. (1973). Cell sorting out during the differentiation of mixtures of metabolically distinct populations of *Dictyostelium discoideum*. *J. Embryol. Exp. Morph.* **29**: 647–61.

Lee, K.-C. (1972). Cell electrophoresis of the cellular slime mould *Dictyostelium discoideum*. II. Relevance of the changes in cell surface charge density to cell aggregation and morphogenesis. *J. Cell Sci.* **10**: 249–65.

Lodish, H. F. & Alton, T. H. (1977). Translational and transcriptional control of protein synthesis in *Dictyostelium discoideum*. In *Development and Differentiation in the Cellular Slime Moulds*, ed. P. Cappuccinelli & J. M. Ashworth, pp. 253–72. Amsterdam: Elsevier/North Holland.

Loomis, W. F. Murray, B. A., Yee, L. & Jongens, T. (1983). Adhesion-blocking antibodies prepared against gp150 react with gp80 of *Dictyostelium*. *Exp. Cell Res.* **147**: 231–4.

Madely, I. C., Herries, D. G. & Hames, B. D. (1981). Common cell-surface receptors for discoidin I and discoidin II in *Dictyostelium discoideum*. *Differentiation* **20**: 278–90.

Maeda, Y. (1980). Changes in charged groups on the cell surface during development of the cellular slime mould *Dictyostelium discoideum*: an electron microscope study. *Devel. Growth and Diff.* **22**: 679–85.

Maeda, Y. & Maeda, M. (1974). Heterogeneity of the cell population of the cellular slime mold *Dictyostelium discoideum* before aggregation, and its relation to subsequent locations of the cells. *Exp. Cell Res.* **84**: 88–94.

Marin, F. T., Goyette-Boulay, M. & Rothman, F. G. (1980). Regulation of development in *Dictyostelium discoideum* III. Carbohydrate-specific intercellular interactions in early development. *Develop. Biol.* **80**: 301–12.

McConaghy, J. R., Saxe, C. L., Williams, G. B. & Sussman, M. (1984). Reversible inhibition of aggregation-mediated cohesivity in *Dictyostelium discoideum* by diffusible metabolites. *Develop. Biol.* (in press).

Müller, K. & Gerisch, G. (1978). A specific glycoprotein as the target site of adhesion blocking Fab in aggregating *Dictyostelium* cells. *Nature* **274**: 445–9.

Müller, K., Gerisch, G., Frommes I., Mayer, H. & Tsugita, A. (1979). A membrane glycoprotein of aggregating *Dictyostelium* cells with the properties of contact sites. *Europ. J. Biochem.* **99**: 419–26.

Müller, W. E. G. (1982). Cell membranes in sponges. *Int. Rev. Cytol.* **77**: 129–81.

Murray, B. A., Wheeler, S., Jongens, T. & Loomis, W. F. (1984). Mutation affecting a surface glycoprotein, gp80, of *Dictyostelium discoideum. Molec. and Cellular Biol.* **4**: 514–19.

Murray, B. A., Yee, L. D. & Loomis, W. F. (1981). Immunological analysis of a glycoprotein (Contact site A) involved in intercellular adhesion of *Dictyostelium discoideum. J. Supromolec. Struct. and Cellular Biochem.* **17**: 197–211.

Newell, P. C., Franke, J. & Sussman, M. (1972). Regulation of four functionally related enzymes during shifts in the developmental program of *Dictyostelium discoideum. J. Molec. Biol.* **63**: 373–82.

Newell, P. C., Longlands, M. & Sussman, M. (1972). Control of enzyme synthesis by cellular interaction during development of the cellular slime mould *Dictyostelium discoideum. J. Molec. Biol.* **56**: 541–54.

Nicol, A. & Garrod, D. R. (1978). Mutual cohesion and cell sorting-out among four species of cellular slime moulds. *J. Cell Sci.* **32**: 377–87.

Ochai, H., Schwarz, H., Merkl, R., Wagle, G. & Gerisch, G. (1982). Stage-specific antigens reacting with monoclonal antibodies against contact site A, a cell-surface glycoprotein of *Dictyostelium discoideum. Cell Diff.* **11**: 1–13.

Ochai, H., Stadler, J., Westphal, M., Wagle, G., Merkl, R. & Gerisch, G. (1982). Monoclonal antibodies against contact sites A of *Dictyostelium discoideum*: detection of modifications of the glycoprotein in tunicamycin-treated cells. *EMBO Journal* **1**: 1011–16.

Parish, R. W. & Schmidlin, S. (1979). Resynthesis of developmentally regulated plasma membrane proteins following disaggregation of *Dictyostelium pseudoplasmodia. FEBS Lett.* **99**: 270–4.

Parish, R. W., Schmidlin, S. & Parish, C. R. (1978). Detection of developmentally controlled plasma membrane antigens of *Dictyostelium discoideum* cells in SDS – polyacrylamide gels. *FEBS Lett.* **95**: 366–70.

Raper, K. B. (1940). Pseudoplasmodium formation and organisation in *Dictyostelium discoideum. J. Elisha Mitchell Sci. Soc.* **56**: 241–82.

Ray, J. & Lerner, R. A. (1982). A biologically active receptor for the carbohydrate-binding proteins of *Dictyostelium discoideum. Cell* **28**: 91–8.

Ray, J., Shinnick, T. & Lerner, R. (1979). A mutation altering the function of a carbohydrate binding protein blocks cell-cell cohesion in developing *Dictyostelium discoideum. Nature* **279**: 215–20.

Reitherman, R. W., Rosen, S. D., Frazier, W. A. & Barondes, S. H. (1975). Cell surface species-specific high affinity receptors for discoidin: developmental regulation in *Dictyostelium discoideum. Proc. Natl. Acad. Sci. U.S.A.* **72**: 3541–5.

Rosen, S. D. & Barondes, S. H. (1978). Cell adhesion in the cellular slime molds. In *Specificity of Embryological Interactions*, ed. D. R. Garrod, pp. 233–64. *Receptors and Recognition series B*, vol. 4. London: Chapman and Hall.

Rosen, S. D., Chang, C. M. & Barondes, S. H. (1977). Intercellular adhesion in the cellular slime mold *Polysphondylium pallidum* inhibited by interaction of asiolofetuin or specific univalent antibody with endogenous cell surface lectin. *Develop. Biol.* **61**: 202–13.

Rosen, S. D., Haywood, P. L. & Barondes, S. H. (1976). Inhibition of intercellular adhesion in the cellular slime mould by univalent antibody against a cell surface lectin. *Nature* **263**: 425–7.

Rosen, S. D., Kafka, J. A., Simpson, D. L. & Barondes, S. H. (1973). Developmentally regulated carbohydrate-binding protein in *Dictyostelium discoideum. Proc. Natl. Acad. Sci. U.S.A.* **70**: 2554–7.

Rosen, S. D., Simpson, D. L., Rose, J. E. & Barondes, S. H. (1974). Carbohydrate-binding protein from *Polysphondylium pallidum* implicated in intercellular adhesion. *Nature* **252**: 149–52.

Samuel, E. W. (1961). Orientation and rate of locomotion of individual amoebas in the life cycle of the cellular slime mold *Dictyostelium mucoroides*. *Dev. Biol.* **3**: 317–35.

Saxe, C. L. & Sussman, M. (1982). Induction of stage-specific cell cohesion in *D. discoideum* by a plasma-membrane-associated moiety reactive with wheat germ agglutinin. *Cell* **29**: 755–9.

Schaap, P., van der Molen, L. & Konijn, T. M. (1983). The organization of fruiting-body formation in *Dictyostelium minutum*. *Cell Diff.* **12**: 287–97.

Schmidt, J. A. & Loomis, W. F. (1982). Phosphorylation of contact site A glycoprotein (gp80) of *Dictyostelium discoideum*. *Dev. Biol.* **91**: 296–304.

Shaffer, B. M. (1962). The Acrasina. *Adv. Morphogen.* **2**: 109–82.

Shaffer, B. M. (1964). The Acrasina (continued). *Adv. Morphogen.* **4**: 301–22.

Shaffer, B. M. (1965). Cell movement within aggregates of the slime mould *Dictyostelium discoideum* revealed by surface markers. *J. Embryol. Exp. Morph.* **13**: 97–117.

Sharon, N. & Lis, H. (1979). Comparative biochemistry of plant glycoproteins. *Biochem. Soc. Trans.* **7**: 783–98.

Sharpe, P. T., Treffry, T. E. & Watts, D. J. (1982). Studies of early stages of differentiation of the cellular slime mould *Dictyostelium discoideum*. *J. Embryol. Exp. Morph.* **67**: 181–93.

Siu, C.-H., Lerner, R. A., Ma, G., Firtell, R. A. & Loomis, W. F. (1976). Developmentally regulated proteins of the plasma membrane of *Dictyostelium discoideum*. The carbohydrate-binding protein. *J. Molec. Biol.* **100**: 157–78.

Siu, C.-H., Des Roches, B. & Lam, T. Y. (1983). Involvement of a cell surface glycoprotein in the cell-sorting process of *Dictyostelium discoideum*, *Proc. Natl. Acad. Sci. U.S.A.* **80**: 659–600.

Springer, W. R. & Barondes, S. H. (1978). Direct measurement of species-specific cohesion in cellular slime moulds. *J. Cell Biol.* **78**: 937–42.

Springer, W. R. & Barondes, S. H. (1982). Externalisation of the endogenous intracellular lectin of a cellular slime mold. *Exp. Cell Res.* **138**: 231–30.

Springer, W. R. & Barondes, S. H. (1983). Monoclonal antibodies block cell-cell adhesion in *Dictyostelium discoideum*. *J. Biol. Chem.* **258**: 4698–701.

Springer, W. R., Cooper, D. W. N. & Barondes, S. H. (1984). *Cell* **39**: 557–64.

Springer, W. R., Haywood, P. L. & Barondes, S. H. (1980). Endogenous cell surface lectin in *Dictyostelium*: quantitation, elution by sugar, and elicitation by divalent immunoglobulin. *J. Cell Biol.* **87**: 682–90.

Stadler, J., Bauer, G. & Gerisch, G. (1984a). Acylation *in vivo* of the contact site A glycoprotein and of other membrane proteins in *Dictyostelium discoideum*. *FEBS Lett.* **172**: 326–30.

Stadler, J., Bauer, G., Westphal, M. & Gerisch, G. (1984b). Monoclonal antibody against cytoplasmic lectins of *Dictyostelium discoideum*: cross-reactivity with a membrane glycoprotein contact site A, and with *E. coli*-galactosidase and *lac* repressor. *Hoppe-Seyler's Z. Physiol. Chem.* **365**: 283–8.

Stadler, J., Gerisch, G., Bauer, G., Suchanek, C. & Huttner, W. B. (1983). *In vivo* sulfation of the contact site A glycoprotein of *Dictyostelium discoideum*. *EMBO Journal* **2**: 1137–43.

Steinberg, M. S. (1964). The problem of adhesive selectivity in cellular interaction. In *Cellular Membranes in Development*, ed. M. Locke, pp. 321–66. New York and London: Academic Press.

Steinemann, C., Hintermann, R. & Parish, R. W. (1979). Identification of a developmentally regulated plasma membrane glycoprotein involved in adhesion of *Polysphondylium pallidum* cells. *FEBS Lett.* **108**: 379–84.

Steinemann, C. & Parish, R. W. (1980). Evidence that a developmentally regulated

glycoprotein is target of adhesion-blocking Fab in reaggregating *Dictyostelium*. *Nature*
286: 621–3.

Sternfeld, J. (1979). Evidence for differential cellular adhesion as the mechanism of
sorting-out of various cellular slime mould species. *J. Embryol. Exp. Morph.* 53:
167–78.

Sternfeld, J. & Bonner, J. T. (1977). Cell differentiation in *Dictyostelium* under
submerged conditions. *Proc. Natl. Acad. Sci. U.S.A.* 74: 268–71.

Sternfeld, J. & David, C. N. (1981). Sorting out during pattern formation in
Dictyostelium. *Differentiation* 20: 61–4.

Sussman, M. (1976). The genesis of multicellular organization and the control of gene
expression *Dictyostelium discoideum*. *Progress in Molec. and Subcell. Biol.* 4: 103–31.

Swan, A. P. (1978). An inhibitor of cell cohesion from *Dictyostelium discoideum*.
Ph.D.Thesis. University of Southampton.

Swan, A. P. & Garrod, D. R. (1975). Cohesive properties of axenically grown cells of
the slime mould *Dictyostelium discoideum*. *Exp. Cell Res.* 93: 479–84.

Swan, A. P., Garrod, D. R. & Morris, D. (1977). An inhibitor of cell cohesion from
axenically grown cells of the slime mould, *Dictyostelium discoideum*. *J. Cell Sci.* 28:
107–16.

Takeuchi, I. (1963). Immunochemical and immunohistochemical studies on the
development of the cellular slime mold *Dictyostelium mucoroides*. *Dev. Biol.* 8: 1–26.

Takeuchi, I. (1969). Establishment of polar organisation during slime mould
development. In *Nucleic Acid Metabolism, Cell Differentiation and Cancer Growth*,
ed. E. V. Cowdry & S. Serio, pp. 298–304. Oxford: Pergamon Press.

Tasaka, M., Noce, T. & Takeuchi, I. (1983). Prestalk and prespore differentiation in
Dictyostelium as detected by cell type-specific monoclonal antibodies. *Proc. Natl. Acad.
Sci. U.S.A.* 80: 5340–4.

Toda, K., Bozzaro, S., Lottspeich, F., Merkl, R. & Gerisch, G. (1984). Monoclonal
anti-glycoprotein antibody that blocks cell adhesion in *Polysphondylium pallidum*. *Eur.
J. Biochem.* 140: 73–81.

Trinkaus, J. P. (1969). *Cell into organs. The forces that shape the embryo*. Englewood
Cliffs, New Jersey: Prentice Hall Inc.

Vogel, G., Thilo, L., Schwarz, H. & Steinhart, R. (1980). Mechanism of phagocytosis
in *Dictyostelium discoideum*: phagocytosis is mediated by different recognition sites
as disclosed by mutants with altered phagocytotic properties. *J. Cell Biol.* 86: 456–65.

Wilcox, D. K. & Sussman, M. (1978). Spore differentiation by isolated *Dictyostelium
discoideum* cells, triggered by prior cell contact. *Differentiation* 11: 125–31.

Wilcox, D. K. & Sussman, M. (1981a). Serologically distinguishable alterations in the
molecular specificity of cell cohesion during morphogenesis in *Dictyostelium
discoideum*. *Proc. Natl. Acad. Aci. U.S.A.* 78: 358–62.

Wilcox, D. K. & Sussman, M. (1981b). Defective cell cohesivity expressed late in
development of a *Dictyostelium* mutant. *Developmental Biol.* 82: 102–12.

7

The plasma membrane of higher plant protoplasts

L. C. Fowke

7.1 Introduction

The plant plasma membrane performs a number of vital functions including the regulation of materials moving into and out of cells, the synthesis and deposition of cell wall components and the reception of a variety of external stimuli. The presence of a complex cell wall surrounding plant cells has limited research on this very important cellular component.

During the 1960s, techniques were developed for the removal of cell walls yielding naked plant protoplasts (Cocking, 1979). The availability of protoplasts has greatly facilitated the study of the plant plasma membrane. Protoplasts have been isolated from a broad range of tissues and cells (e.g. see Vasil & Vasil, 1980; Fowke & Constabel, 1985); the most readily available and thus most thoroughly studied protoplasts are those isolated from leaf mesophyll and suspension-cultured cells. Research with protoplasts has provided considerable information regarding the structure, chemistry and function of the plant plasma membrane. This chapter will examine some of the approaches used to study the plasma membrane of higher plants and the information which has accumulated from these studies.

7.2 General structure and chemistry

The basic organization of the plasma membrane in higher organisms including both plants and animals is believed to be similar. The plasma membrane consists of two layers of lipids with associated proteins and carbohydrates. Details concerning the chemistry and arrangement of these components in the plant plasma membrane are very limited (Leonard & Hodges, 1980). The protein and lipid are believed to constitute approximately 80% of the membrane with carbohydrates mainly as glycolipids or glycoproteins making up the balance.

218   *L. C. Fowke*

Plant protoplasts which lack a bounding cell wall are particularly useful for studying the plasma membrane. Under proper isolation conditions it is possible to remove all detectable cell wall from the surface of plant cells. This can best be demonstrated by using electron-microscope techniques which permit visualization of large areas of the surface of the plasma membrane. The most sensitive techniques available are freeze-fracture, surface replication (Figs. 7.5–7.7, p. 228) and negative staining (see discussion in Fowke & Gamborg, 1980). The plasma membrane of higher plant protoplasts appears as a typical trilamellar unit membrane when viewed in thin sections by transmission electron microscopy (Fig. 7.8, p. 228).

7.2.1   Membrane staining

The absence of cell wall around protoplasts facilitates cytochemical investigation of the plasma membrane with the electron microscope. Most research has focussed on the carbohydrate components at the surface of protoplasts and very little attention has been paid to the nature of the lipids and proteins comprising the plasma membrane. Studies of surface carbohydrates are complicated by the possibility that some carbohydrates represent either remnants of the cell wall or wall precursors moving through the membrane. This problem has not been completely resolved; however, the results of Williamson (1979) support the generally held view that carbohydrates are an integral component of the plant plasma membrane.

A number of different stains have been used to demonstrate carbohydrates within the protoplast plasma membrane. Both the phosphotungstate-chromate and silicotungstate-chromate stains have been used successfully to identify carbohydrates in the plasma membrane of intact cells and tissues (see reviews by Roland, 1978; Quail, 1979). Silicotungstate-chromate has been applied to plant protoplasts with variable results (Taylor & Hall, 1978). The protoplast plasma membrane does react with the modified periodic acid-Schiff procedure of Thiery (Roland & Prat, 1973), which is used to stain carbohydrates. Pictures have also been published which suggest that the protoplast plasma membrane reacts with ruthenium red (Prat & Williamson, 1976; Taylor & Hall, 1978), a stain which is believed to bind to acid polysaccharides. Perhaps the most convincing demonstration of carbohydrates in the plasma membrane has come from studies utilizing lectins and artificial carbohydrate antigens (see sections 7.5.1, 7.5.2). These studies have demonstrated different sugar groups at the surface of protoplasts and have provided evidence for mobility of these sugars in the plane of the plasma membrane.

7.2.2 Surface charge

The nature of the surface charge on plant protoplasts has been investigated primarily to provide information regarding the mechanisms of protoplast fusion and endocytotic uptake by protoplasts. The results of electrophoretic studies (e.g. Grout, Willison & Cocking, 1972; Nagata & Melchers, 1978; Fischer, 1979; Griffing, Cutler, Shargool & Fowke, 1984) and research with cation binding (Taylor & Hall, 1978) indicate that most protoplasts exhibit a net negative surface charge. It has been suggested that the negative charge is due primarily to phosphate groups (Nagata & Melchers, 1978). The results of Gould, Ashmore & Gibbs (1981) indicate that the electrostatic charge on the surface of plant protoplasts varies with the phase of the cell cycle. Gould *et al.* observed a decrease in net negative charge on tobacco protoplasts during G1 and an increase in G2.

Unfortunately the presence of a negative charge on the plasma membrane of protoplasts does not necessarily reflect the situation in intact plant cells. The removal of the cell wall with enzymes undoubtedly results in some modification of the surface molecules. Some alterations may also be due to actual plasmolysis of the cell during wall removal. Plasmolysis may, for example, affect membrane components that determine cellular potential (Racusen, Kinnersley & Galston, 1977, but also see Briskin & Leonard, 1979; Goldsmith & Goldsmith, 1978) as well as alter the distribution of integral proteins within the plasma membrane (Wilkinson & Northcote, 1980).

Protoplasts have recently been shown to exhibit rather interesting behaviour in response to externally applied electrical fields. When the protoplast plasma membrane is exposed to field pulses of high intensity and short duration, local electrical breakdown occurs (Zimmermann & Scheurich, 1981; Zimmermann & Vienken, 1984). The breakdown, which is reversible, is associated with large permeability changes of the plasma membrane. This line of investigation with plant protoplasts should provide basic information concerning transport properties of the plasma membrane.

7.2.3 Freeze-fracture studies

The protoplast plasma membrane has also been examined using the freeze-fracture technique which permits visualization of the internal

structure of membranes. Intramembranous particles, generally believed
to be integral proteins, can be seen within the lipid portion of the membrane. The particles within the membrane are of particular interest since
they may be involved in the process of cellulose microfibril formation.
In a number of algal and higher plant cells, membrane particles have
been shown to be associated closely with the ends of microfibrils and
have been designated terminal complexes (e.g. Montezinos, 1982). Terminal complexes are believed to consist of enzymes responsible for the
synthesis and deposition of cellulose microfibrils.

Research with higher plant protoplasts has revealed intramembranous
particles distributed either randomly or in complex arrays (e.g. hexa-
gonal arrays, rows). It has been suggested that the arrays of particles may
function either in membrane-transport phenomena or in the process
of cellulose deposition (Robenek & Peveling, 1977; Schnabl, Vienken
& Zimmermann, 1980). However, it is more likely that the particle arrays
described are artifacts due either to glycerol treatment (Davey &
Mathias, 1979) or to plasmolysis during protoplast formation (Wilkinson
& Northcote, 1980). In any case such arrays do not seem to be necessary
for cell wall regeneration by protoplasts. Terminal complexes have not
been observed at the ends of cellulose microfibrils in higher plant protoplasts. The absence of terminal complexes may be due to a number of
factors (see review by Willison & Klein, 1982).

7.2.4   Plasma membrane isolation

In order to investigate the biochemistry of the plant plasma
membrane, it is necessary to be able to isolate pure plasma membrane
fractions. The isolation of the plasma membrane from cells has proved
very difficult. Conventional methods requiring high shear forces to disrupt the cell walls result in organelle breakage and damage to the plasma
membrane. The disruption of cell organelles reduces the characteristic
differences in size and density between membrane structures, making
separation of the plasma membrane and organelles difficult. The isolation
of the plasma membrane is further hampered by the lack of suitable
markers. The major plasma membrane markers used are phosphotung-
state–chromate staining, enzymes and surface labels. The advantages
and disadvantages of each of these markers have been reviewed previously (Quail, 1979; Hall & Taylor, 1979).

Plant protoplasts offer distinct advantages for isolation of the plant
plasma membrane. Protoplasts can be lysed gently thus preserving the
internal organelles intact and preventing plasma membrane damage.
Furthermore, the absence of the cell wall facilitates the application of

plasma membrane labels which can be used to identify plasma membrane fractions following isolation. A variety of labels have been utilized in attempts to isolate the plasma membrane from plant protoplasts e.g. radioactive diazotized sulphanilic acid (soybean – Galbraith & Northcote, 1977; corn – Perlin & Spanswick, 1980) radioactive Concanavalin A (carrot – Boss & Ruesink, 1979), radioactive carbohydrate-binding protein (ryegrass – Schibeci, Fincher, Stone & Wardrop, 1982), and lanthanum (tobacco – Taylor & Hall, 1979).

The fact that many protoplasts have a net negative charge and will bind tightly to polylysine-treated surfaces (see section 7.6, A) suggests an alternative technique for plasma membrane isolation. it may be possible to attach protoplasts to polylysine-treated beads, burst them, and collect the plasma membrane adhering to the beads. This technique has been successfully used for isolation of the plasma membrane from a number of animal cells (e.g. Jacobson & Branton, 1977).

7.3   Protoplast fusion

Techniques have been developed for the routine fusion of plant protoplasts. Fusions have been achieved between widely separated groups of plants as well as between plant protoplasts and animal cells (see reviews by Schieder & Vasil, 1980; Constabel & Cutler, 1985). These studies suggest a basic overall similarity of plasma membranes within living organisms. Fusion technology therefore provides a rather novel approach for investigating the interaction between plasma membranes of a broad spectrum of plant protoplasts and should provide interesting information regarding the organization of this important membrane system.

7.3.1   *Methodology*

Protoplast fusion involves the initial attachment or agglutination of protoplasts followed by induced fusion. A number of different techniques have been used to fuse plant protoplasts. These include the use of sodium nitrate, alkaline calcium solution, polyethylene glycol (PEG), polyvinyl alcohol, a synthetic phospholipid and electrical fields (see reviews by Schieder & Vasil, 1980; Constabel & Cutler, 1985). The most successful and reliable fusogen appears to be PEG. It is currently used widely for fusing plant protoplasts and seems to work equally well for animal cell fusions.

The technique involves the agglutination of protoplasts with high concentrations (*c.* 30% w/v) of high molecular weight PEG (e.g. 1540–6000) followed by the gradual elution of the PEG which results in fusion of the plasma membranes (Figs 7.1–7.4). Fusion products can be cultured and will often develop new cell walls, divide and regenerate whole plants.

7.3.2   Research potential

Most fusion research is directed towards the production of hybrid plants and very little attention has been paid to the actual mechanism of fusion. Even ultrastructural studies of the processes of agglutination and fusion have contributed very little to our understanding of membrane fusion (e.g. Fowke, 1978).

Plant protoplast fusion provides an excellent model system for investigating a number of properties of the plant plasma membrane. The following example illustrates the potential of protoplast fusion for membrane studies. First, it should be possible to attach labels to the surfaces of protoplasts, fuse them and follow the fate of the labels during and after membrane fusion. Such studies would provide information regarding

Fig. 7.1. Diagram illustrating induced fusion of plant protoplasts: (a) protoplast derived from cell suspension culture, (b) leaf protoplast, (c) agglutination of protoplasts using PEG and (d) fusion product formed after elution of PEG.

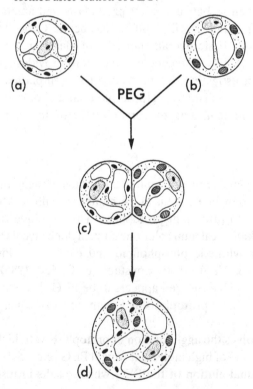

Fig. 7.2. Light micrograph showing agglutination of pea leaf and *Vicia* cell culture protoplasts. Note the chloroplasts (arrows) in the pea protoplast. Bar = 10 μm.

Fig. 7.3. Electron micrograph showing agglutination of pea (P) leaf and *Vicia* (V) cell culture protoplasts. Bar = 1 μm.

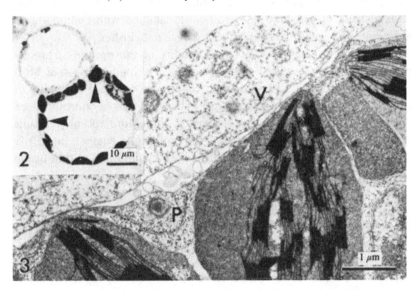

Fig. 7.4. Electron micrograph showing fusion product from *Vicia narbonensis* leaf protoplast and *Vicia hajastana* cell culture protoplast. The small leaf nucleus and chloroplasts (C) are easily distinguished from the larger cell culture nucleus and leucoplasts (L). (From Rennie, Weber, Constabel & Fowke, 1980.) Bar = 10 μm.

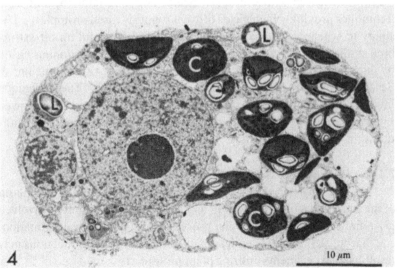

the fluidity of the plasma membrane as evidenced by the extent of membrane mixing following fusion. Labelling experiments of this type could be undertaken at the light-microscope level using fluorescent dyes such as fluorescein isothiocyanate or with the electron microscope using labels such as ferritin, hemocyanin or colloidal gold (see section 7.5.1). It has already been established that protoplasts labelled with colloidal gold are capable of agglutination (Burgess, Linstead & Fisher, 1977).

It is also possible to examine the fluidity of the protoplast plasma membrane with the electron-spin resonance technique. Boss & Mott (1980), using a fatty acid spin label, reported a decrease in fluidity of carrot plasma membrane in response to calcium ions at concentrations commonly used to promote protoplast fusion. More recently, results of spin-label studies with fusing carrot protoplasts suggest that PEG also directly affects the fluidity of the plasma membrane during fusion (Boss, 1983).

The freeze-fracture technique might also be used to investigate changes in protoplast plasma membranes during the fusion process. Freeze-fracture studies of plant and animal cells have demonstrated the movement of intramembranous particles, presumably integral proteins, away from the site of membrane fusion (e.g. Volkmann, 1981; Orci, Perrelet & Friend, 1977). Robenek & Peveling (1978) reported similar particle mobility during the spontaneous fusion of protoplasts of *Skimmia japonica*. Further freeze-fracture studies of fusing plant protoplasts would clarify the nature of intramembranous particle movement and contribute to our general understanding of plasma membrane fluidity.

Future research with protoplast fusion would certainly benefit from techniques providing high yields of synchronously fused protoplasts. The ability to synchronize fusion would be particularly useful for ultrastructural studies of the initial stages of membrane fusion. Synchronous fusion of higher plant protoplasts has been reported (Zimmermann & Scheurich, 1981; Zimmermann & Vienken, 1984). The technique involves agglutination of protoplasts by dielectrophoresis followed by application of a high external electrical field pulse of very short duration, which causes fusion.

7.4   Physiology of the plasma membrane

The plasma membrane, which separates the living cytoplasm of the plant cell from the cell wall and external environment, is involved in a number of important cellular processes (e.g. uptake, wall formation, cold hardiness). A better understanding of some of these functions may result from experiments utilizing plant protoplasts.

7.4.1 Transport properties

Attempts to study the transport properties of the plasma membrane using cells or tissues are hampered by the presence of the cell wall. Freshly isolated protoplasts provide an excellent system for investigating transport across the plasma membrane (Leonard & Rayder, 1985). Removal of the cell wall permits direct experimental access to the plasma membrane and eliminates absorption effects attributable to the wall. However, if protoplasts are to be used for membrane transport studies it is essential to establish that the isolation techniques involving enzyme treatment do not damage the plasma membrane and thus alter its transport characteristics.

Transport of ions, sugars and amino acids across the plasma membrane has been examined with a number of different protoplasts. Ion transport has been studied with protoplasts from tobacco cell suspensions (Mettler & Leonard, 1979) and protoplasts from corn roots (Lin, 1980). These studies indicate that ion transport by protoplasts is essentially the same as that observed in intact cells. Protoplasts also appear capable of regulating uptake of sugars and amino acids (e.g. Guy, Reinhold & Laties, 1978; Paszkowski, Lorz, Potrykus & Dierks-Ventling, 1980) as well as the efflux of photosynthetically produced sugars (Huber & Moreland, 1980).

The results of these studies suggest little or no damage to the plasma membrane due to isolation procedures and encourage further research on plasma membrane transport using protoplasts. Since techniques are now available for isolation of plant vacuoles from protoplasts (e.g. Leonard & Rayder, 1985) it should be possible to extend membrane transport studies to include a comparison of the plasma membrane and tonoplast.

7.4.2 Endocytosis

Uptake into animal cells commonly occurs by endocytosis, a process involving engulfment of material by an invagination of the plasma membrane. When the process involves uptake of large structures such as bacteria by protozoans it is generally referred to as phagocytosis. Endocytosis also occurs in plants, but rather infrequently. During infection of legumes by *Rhizobium*, for example, the bacteria penetrate the cell wall to the plasma membrane and are then taken into the cells by an endocytotic process (e.g. Newcomb, 1976; Bassett, Goodman & Novacky, 1977). It is perhaps not surprising that endocytosis is not commonly observed in plants. Plant cells are surrounded by a cell wall which filters out particles of a size suitable for endocytosis. In addition, plant cells are usually under high turgor pressure, a condition not conducive to endocytosis.

Nevertheless, the process does occur in some plant cells. The actual mechanism of endocytosis by the plant plasma membrane can be investigated using protoplasts. The uptake of polystyrene spheres by protoplasts provides one of the clearest examples of endocytosis (e.g. Willison, Grout & Cocking, 1971; Suzuki, Takebe, Kajita, Honda & Matsui, 1977). The spheres attach to the surface of protoplasts by electrostatic forces and are then rapidly taken into the protoplast by an energy dependent mechanism. The overall process bears a number of similarities to endocytosis in animal cells (Suzuki, Takebe, Kajita, Honda & Matsui, 1977).

Uptake into plant protoplasts of a variety of other particles, cell organelles and microorganisms has been reported. For example, considerable information is available concerning the infection of protoplasts by plant viruses (Takebe, 1983). While there is still some disagreement as to the mechanism of infection, the available evidence favours uptake by endocytosis. Studies of the uptake into plant protoplasts of organelles (nuclei, chloroplasts) and microorganisms have in general suffered from lack of information regarding their viability before uptake, and lack of critical ultrastructural information regarding their mode of uptake and fate within the protoplasts (see reviews by Fowke & Gamborg, 1980; Lorz, 1984).

Endocytosis of molecules or macromolecules by animal cells constitutes pinocytosis. Uptake of proteins often involves specific receptor-mediated endocytosis by coated pits and coated vesicles (see reviews by Pearse & Bretscher, 1981; Brown, Anderson & Goldstein, 1983). Recently, coated vesicles have also been implicated in the process of endocytosis by plant cells. Protoplasts of both bean leaves (Joachim & Robinson, 1984) and suspension-cultured soybean cells (Tanchak, Griffing, Mersey & Fowke, 1984) endocytosed cationized ferritin via coated vesicles. The process in soybean protoplasts involves binding of ferritin to the plasma membrane, rapid internalization of ferritin via coated pits to coated vesicles, uncoating of coated vesicles and eventual transport of the ferritin to dictyosomes and other membranous organelles. Further research is required to characterize this interesting pathway and to determine whether it operates in intact plant cells.

7.4.3   Frost hardiness

Many plants are able to undergo a process known as acclimation making them less susceptible to freezing injury. Such plants develop a level of frost hardiness which enables them to withstand severe winter conditions. It has been suggested that the process of acclimation is a direct function of the plasma membrane of plant cells (e.g. Steponkus, 1984). If so, protoplasts, which are free of any interference from the cell wall, should provide an excellent experimental system for studying the behaviour of the plasma membrane during the freeze-thaw cycle. In fact a number of recent studies have focussed on the osmotic behaviour of protoplasts and have provided considerable insight into the nature of freezing injury particularly as it relates to the plasma membrane. Results suggest that freezing injury involves an irreversible loss of plasma membrane material. Distinct differences have been noted between the responses of acclimated and non-acclimated protoplasts with the former being more resistant to freezing injury. Cold acclimation seems to decrease sensitivity of the plasma membrane to mechanical stresses during osmotic contraction and expansion (for a thorough discussion, see Steponkus, 1984).

7.5   Plasma membrane outer surface

The external surface of plant protoplasts has been examined using a number of different probes (e.g. lectins, artificial carbohydrate antigens, antibodies) in an attempt to clarify the chemical nature of the molecules exposed on the plasma membrane.

7.5.1   Lectins

Lectins are proteins or glycoproteins which are capable of binding to specific sugar groups. A variety of lectins with different sugar specificities have been isolated and purified (see reviews by Nicolson, 1974; Knox & Clarke, 1978). Lectins are providing a valuable tool for exploring the surface of the protoplast plasma membrane. The most widely used lectin for protoplast studies is Concanavalin A (Con A), a metalloprotein derived from the seeds of Jack Bean. It exists in either dimeric or tetrameric forms depending on the pH. Each monomeric unit has a single binding site specific for α-D-glucose and α-D-mannose.

Con A will bind to the surface of a wide variety of plant protoplasts causing agglutination. This ability of Con A to bind to protoplasts has been utilized in ultrastructural studies designed to explore the distribution of sugar groups on the plasma membrane. Various labels (e.g. ferritin, hemocyanin, colloidal gold) may be attached to lectins for electron-microscope studies (see reviews by Knox & Clarke, 1978; Fowke & Gamborg, 1980).

228   *L. C. Fowke*

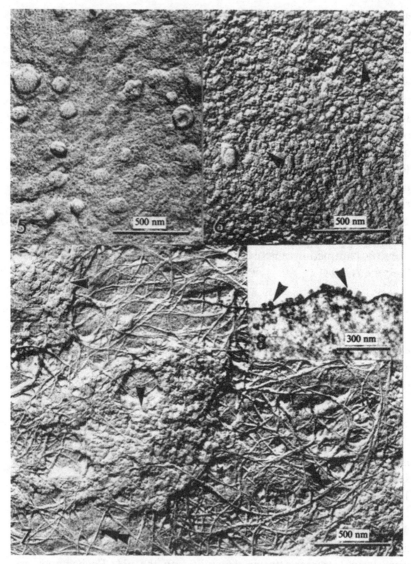

Figs. 7.5–7.8. From Williamson, Fowke, Constabel & Gamborg, 1976.
Figs. 7.5–7.7. Platinum-palladium replicas of the surface of soybean
protoplasts.

Fig. 7.5. Protoplast prefixed with glutaraldehyde and treated with
hemocyanin alone (no Con A). Hemocyanin is not bound.
Bar = 500 nm.

Fig. 7.6. Protoplast prefixed with glutaraldehyde and treated with Con
A/hemocyanin. Note the even distribution of hemocyanin molecules
(arrows). Bar = 500 nm.

One particularly useful technique involves the preparation of shadowed replicas of protoplast surfaces labelled with Con A-hemocyanin. This technique permits high-resolution analysis of large expanses of the protoplast surface (Figs. 7.5–7.8; and Williamson, Fowke, Constabel & Gamborg, 1976). When labelling is carried out using protoplasts prefixed with glutaraldehyde an even distribution of binding is observed (Fig. 7.6). Incubation of unfixed protoplasts at room temperature after labelling results in clustering of the Con A-binding sites (Fig. 7.7). These results suggest mobility of membrane components, thus indicating considerable fluidity of the plant plasma membrane. Similar results have been obtained using other labelling techniques and a variety of plant protoplasts (see review by Fowke & Gamborg, 1980).

Other lectins with different sugar specificities have been shown to bind to the surface of plant protoplasts (e.g. Larkin, 1978; Gruber, Glimelius, Eriksson & Frederick, 1984). These lectins might facilitate future studies of sugars on the surface of the plasma membrane.

7.5.2 *Artificial carbohydrate antigens*

The research with lectins clearly demonstrates the presence of surface carbohydrates on plant protoplasts. Agglutination studies using the β-glycosyl artificial carbohydrate antigens (Yariv antigens) indicate that a specific group of carbohydrates, the arabino 3,6-galactans are also located at the plasma membrane surface of a wide variety of plant protoplasts (Larkin, 1977). The Yariv antigens are phenolic glycosides which specifically interact with arabino 3,6-galactan proteins (Clarke, Anderson & Stone, 1979). The function of these arabinogalactan proteins at the surface of plant cells is not understood. Clarke *et al.* (1979) suggest that they may protect the plasma membrane from desiccation or they may interact with other molecules to provide some type of specific recognition mechanism at the cell surface. For example, they may serve as receptors for the glucan elicitors of the phytoalexin response during fungal infection. Such elicitors have been shown to bind to the surface

Captions for figs. 7.5–7.8. (*cont.*)

Fig. 7.7. Protoplast treated with Con A/hemocyanin and then incubated 16 h before fixation. The hemocyanin molecules are in distinct clusters (single arrows). Note the newly formed cellulose microfibrils (double arrows). Bar = 500 nm.

Fig. 7.8. Electron micrograph showing hemocyanin molecules (arrows) attached to the plasma membrane of a soybean protoplast. Bar = 300 nm.

of potato leaf protoplasts causing them to agglutinate (Peters, Cribbs & Stelzig, 1978).

7.5.3   Antibodies

Very little information is available concerning antigenic properties of the plasma membrane of plant protoplasts. Early immunological studies were undertaken in order to establish methods for agglutination and fusion of protoplasts. Hartmann, Kao, Gamborg & Miller (1973) reported agglutination with antibodies prepared against protoplasts from suspension cultures of *Vicia*, soybean and brome grass. Agglutination occurred with both homologous and heterologous antisera suggesting considerable cross-reactivity. However, it is uncertain whether the observed interactions were indeed due to antibodies or whether they were due to some other serum component interacting with the arabino-galactan proteins associated with the plasma membrane (Larkin, 1977).

Raff, McKenzie & Clarke (1980) clearly demonstrated the presence of specific antigenic determinants on the surface of *Prunus avium* protoplasts. Antibodies raised to crude tissue preparations bound to protoplast surfaces but not to intact cells. The location of specific antigens at the plasma membrane surface in plants raises the interesting possibility that they may be involved in cellular recognition phenomena.

7.6   Plasma membrane inner surface

A technique has been devised which permits the visualization of the inner surface of large fragments of protoplast plasma membrane. Two cytoplasmic organelles, microtubules and coated vesicles, are closely associated with the plasma membrane and can be examined using this technique. Study of protoplast plasma membrane fragments should help to clarify the role of these organelles in plant cells as well as provide basic information regarding their structure, chemistry and distribution.

7.6.1   Methodology

The technique for examining the inner surface of the plasma membrane was initially developed for studies of the plasma membrane of algal protoplasts and has been applied to studies of higher plants (e.g. Fowke, Griffing, Mersey & Van der Valk, 1983). The technique involves the attachment of protoplasts to either glass coverslips or electron-microscope grids followed by gentle lysis of the protoplasts. The membrane fragments remaining attached to the substrate are washed, fixed and either treated with fluorescent antibodies for immunofluorescence or stained and examined in the electron microscope (Figs.

7.9–7.16). The method produces large, relatively clean fragments of the protoplast plasma membrane.

7.6.2 Microtubules

Microtubules on plasma membrane fragments from plant protoplasts have been successfully stained with fluorescent anti-tubulin for examination in the fluorescence microscope (Fig. 7.10; Lloyd, Slabas, Powell & Lowe, 1980; Van der Valk, Rennie, Connolly & Fowke, 1980). Such preparations have proved useful for studying the general distribution of microtubules associated with the plasma membrane, particularly

Fig. 7.9. Diagram illustrating the preparation of protoplast plasma membrane fragments: (a) freshly isolated protoplast, (b) attachment of protoplast to substratum (grid or coverslip) with polylysine, (c) osmotic bursting of protoplast. The resulting attached membrane fragments are washed, fixed and stained for electron microscopy (on grids) or immunofluorescence (on coverslips).

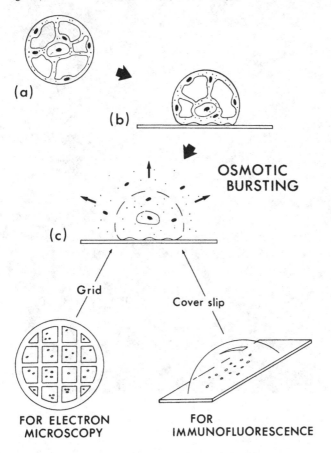

(a)

(b)

OSMOTIC
BURSTING

(c)

Grid

Cover slip

FOR ELECTRON
MICROSCOPY

FOR
IMMUNOFLUORESCENCE

Fig. 7.10. Light micrograph showing fluorescent microtubules on inner surface of plasma membrane fragment from tobacco cell culture protoplast. The membrane was treated with antibodies to porcine brain tubulin followed by a fluorescent second antibody. Bar = 20 μm.

Fig. 7.11. Electron micrograph of plasma membrane fragment similar to that in Fig. 7.10 showing numerous microtubules (arrows). (From Van der Valk, Rennie, Connolly & Fowke, 1980.) Bar = 3 μm.

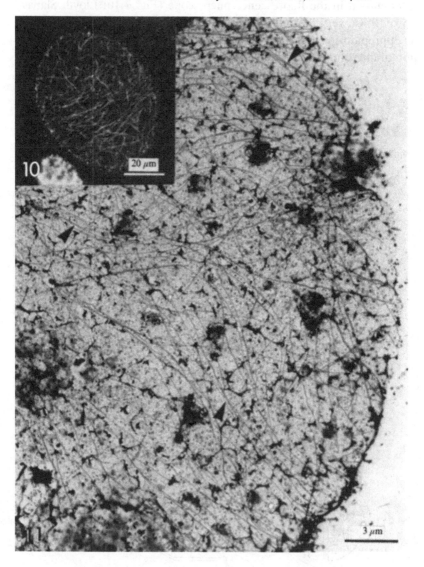

in relation to cell wall regeneration. Fluorescence studies coupled with electron-microscope observations show a direct correlation between wall-regenerating capacity of protoplasts and the number of microtubules attached to the plasma membrane (Marchant & Hines, 1979; Van der Valk *et al.*, 1980). This is not surprising since microtubules are believed to play an important role in the orientation of cellulose microfibrils during cell wall formation (reviewed by Heath & Seagull, 1982; Robinson & Quader, 1982). Ultrastructural studies of plasma membrane fragments have also provided information concerning the frequency and lengths of microtubules as well as the nature of the interaction of microtubules in bundles (Doohan & Palevitz 1980; Van der Valk *et al.*, 1980). The microtubules in the bundles appear to be linked by bridges and preliminary results of a microdensitometer analysis are providing information regarding the spacing of the bridges (C. Jensen, personal communication).

7.6.3   *Coated vesicles*

Coated vesicles are widespread in plants and available evidence suggests that they may be involved in both exocytosis and endocytosis in plant cells (e.g. Newcomb, 1980; Fowke *et al.*, 1985).

Protoplasts contain an abundance of coated vesicles primarily associated with the plasma membrane as coated pits and thus are particularly well suited for investigating this organelle. Recent studies of protoplasts involving conventional thin sectioning techniques and the examination of plasma membrane fragments have made an important contribution to our understanding of the structure and distribution of coated vesicles and the nature of their association with the plasma membrane (Figs. 7.12–7.16; Doohan & Palevitz, 1980; Van der Valk & Fowke, 1981; Fowke, Rennie & Constabel, 1983). Experiments which have demonstrated the internalization of cationized ferritin into protoplasts by coated vesicles (Joachim & Robinson, 1984; Tanchak *et al.*, 1984) have provided the first direct evidence which supports a role for coated vesicles in the process of endocytosis (see also section 7.4.2 above).

7.7   **Summary**

The study of higher plant protoplasts is contributing significantly to our knowledge of the structure, chemistry and function of the plant plasma membrane. Carbohydrates are present on the plasma membrane as indicated by staining reactions and binding studies using lectins and artificial carbohydrate antigens. Immunological work has demonstrated specific antigens on the surface of the plasma membrane as well. It

234 *L. C. Fowke*

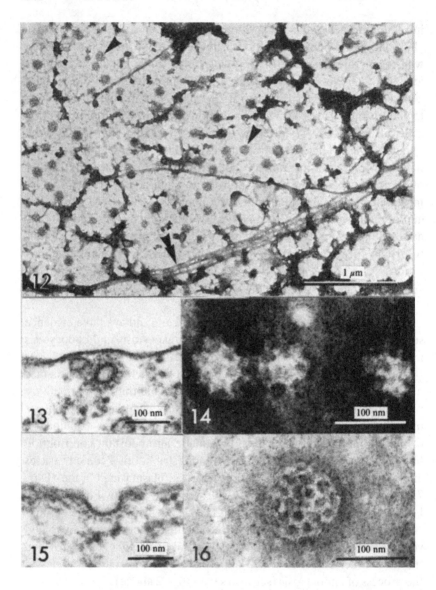

Fig. 7.12. Electron micrograph of plasma membrane fragment from tobacco cell culture protoplast showing numerous coated vesicles or coated pits (single arrows). Note the bundle of microtubules (double arrow). Bar = 1 μm.

Figs. 7.13–7.16. Tobacco protoplasts, from Van der Valk & Fowke, 1981. Bar = 100 nm.

Fig. 7.13. Electron micrograph showing coated vesicle associated with plasma membrane.

will be interesting to study the distribution and chemical nature of these surface molecules further and investigate the possibility that they are involved in cellular recognition reactions.

Results of studies using lectin binding, electron-spin resonance and freeze-fracture illustrate the general fluidity of the plant plasma membrane. Future research of this nature coupled with studies of the mechanism of protoplast fusion should contribute further to our understanding of the mobility of membrane components.

Plant protoplasts are also providing interesting information regarding transport across the plasma membrane. Protoplasts appear capable of regulating the uptake of ions, sugars and amino acids in a manner similar to that of intact cells. The mechanism of endocytosis by the plant plasma membrane can also be investigated using plant protoplasts.

Finally, protoplasts are proving useful for studying cytoplasmic organelles which are closely associated with the inner surface of the plant plasma membrane. Recent research has furnished information concerning the structure, chemistry, distribution and function of both microtubules and coated vesicles.

The research contributions of Fred Constabel, Pieter Van der Valk, Francis Williamson and Joe Connolly as well as the excellent technical assistance of Pat Rennie, Jim Kirkpatrick and Jerry Shyluk are gratefully acknowledged. I wish to thank my colleagues who kindly supplied reprints and preprints of their research work. Thanks also to Bev Garnett and Dennis Dyck for their help in preparing the manuscript. Financial assistance was provided by the Natural Sciences and Engineering Research Council of Canada.

7.8   References

Bassett, B., Goodman, R. N. & Novacky, A. (1977). Ultrastructure of soybean nodules 1: release of rhizobia from the infection thread. *Can. J. Microbiol.* **23:** 573–82.

Boss, W. F. (1983). Poly(ethylene glycol) – induced fusion of plant protoplasts. A spin-label study. *Biochimica et Biophysica Acta* **730:** 111–18.

Boss, W. F. & Mott, R. L. (1980). Effects of divalant cations and polyethylene glycol on the membrane fluidity of protoplasts. *Plant Physiol.* **66:** 835–7.

Boss, W. F. & Ruesink, A. W. (1979). Isolation and characterization of concanavalin A-labelled plasma membrane of carrot protoplasts. *Plant Physiol.* **64:** 1005–11.

Captions for figs. 7.12–7.16 (*cont.*)

Fig. 7.14. Electron micrograph showing negatively stained coated vesicles on the inner surface of a plasma membrane fragment.

Fig. 7.15. Electron micrograph showing coated pit on plasma membrane.

Fig. 7.16. Electron micrograph of plasma membrane fragment showing negatively stained coated pit on plasma membrane.

236   L. C. Fowke

Briskin, D. P. & Leonard, R. T. (1979). Ion transport in isolated protoplasts from tobacco suspension cells. *Plant Physiol.* **64:** 959–62.

Brown, M. S., Anderson, R. G. W. & Goldstein, J. L. (1983). Recycling receptors: the round-trip itinerary of migrant membrane proteins. *Cell* **32:** 663–7.

Burgess, J., Linstead, P. J. & Fisher, V. E. L. (1977). Studies on higher plant protoplasts by scanning electron microscopy. *Micron* **8:** 113–22.

Clarke, A. E., Anderson, R. L. & Stone, B. A. (1979). Form and function of arabinogalactans and arabinogalactan-proteins. *Phytochemistry* **18:** 521–40.

Cocking, E. C. (1979). Protoplasts: past and present. In *Advances in Protoplast Research*, Proceedings, 5th Int. Protoplast Symposium, Hungarian Academy Sciences, Szeged.

Constabel, F. & Cutler, A. J. (1985). Protoplast fusion. In *Plant Protoplasts*, ed. L. C. Fowke & F. Constabel. Boca Raton: CRC Press Inc.

Davey, M. R. & Mathias, R. J. (1979). Close-packing of plasma membrane particles during wall regeneration by isolated higher plant protoplasts – fact or artifact. *Protoplasma* **100:** 85–99.

Doohan, M. E. & Palevitz, B. A. (1980). Microtubules and coated vesicles in guard-cell protoplasts of *Allium cepa* L. *Planta* **149:** 389–401.

Fischer, D. J. (1979). Studies of plant membrane components using protoplasts. *Plant Sci. Lett.* **15:** 127–33.

Fowke, L. C. (1978). Ultrastructure of isolated and cultured protoplasts. In *Frontiers of Plant Tissue Culture 1978*, ed. T. A. Thorpe. Calgary: Int. Assoc. Plant Tissue Cult.

Fowke, L. C. & Constabel, F. (eds.) (1985). *Plant Protoplasts*. Boca Raton: CRC Press Inc. (In press).

Fowke, L. C. & Gamborg, O. L. (1980). Applications of protoplasts to the study of plant cells. *Int. Rev. Cytol.* **68:** 9–51.

Fowke, L. C., Griffing, L. R., Mersey, B. G. & Tanchak, M. A. (1985). Protoplasts for studies of cell organelles. In *Plant Protoplasts*, ed. L. C. Fowke & F. Constabel. Boca Raton: CRC Press.

Fowke, L. C., Griffing, L. R., Mersey, B. G. & Van der Valk, P. (1983). Protoplasts for studies of the plasma membrane and associated cell organelles. In *Protoplasts 1983 (Lecture Proceedings)*, ed. I. Potrykus, C. T. Harms, A. Hinnen, R. Hutter, P. J. King & R. D. Schillito. Basel: Birkhauser Verlag.

Fowke, L. C., Rennie, P. J. & Constabel, F. (1983). Organelles associated with the plasma membrane of tobacco leaf protoplasts. *Plant Cell Reports* **2,** 292–5.

Galbraith, D. W. & Northcote, D. H. (1977). The isolation of plasma membrane from protoplasts of soybean suspension cultures. *J. Cell Sci.* **24,** 295–310.

Goldsmith, T. H. & Goldsmith, M. H. M. (1978). The interpretation of intracellular measurements of membrane potential, resistance and coupling in cells of higher plants. *Planta* **143,** 267–74.

Gould, A. R., Ashmore, S. E. & Gibbs, A. J. (1981). Cell cycle related changes in the quantity of TMV virions bound to protoplasts of *Nicotiana sylvestris*. *Protoplasma* **108:** 211–23.

Griffing, L. R., Cutler, A. J., Shargool, P. D. & Fowke, L. C. (1985). Isoelectric focussing of plant cell protoplasts. Separation of different protoplast types. *Plant Physiol.* **77:** 765–9.

Grout. B. W. W., Willison, J. H. M. & Cocking, E. C. (1972). Interactions at the surface of plant cell protoplasts: an electrophoretic and freeze-etch study. *Bioenergetics* **4:** 585–602.

Gruber, P. J., Glimelius, K., Eriksson, T. & Frederick, S. E. (1984). Interactions of galactose-binding lectins with plant protoplasts. *Protoplasma* **121:** 34–41.

Guy, M., Reinhold, L. & Laties, G. G. (1978). Membrane transport of sugars and amino acids in isolated protoplasts. *Plant Physiol.* **61:** 593–6.

Hall, J. L. & Taylor, A. R. D. (1979). Isolation of the plasma membrane from higher plant cells. In *Plant Cell Organelles – Methodological Surveys (b) Biochemistry*, vol. 9, ed. E. Reid, Chichester, U.K.: Horwood.

Hartmann, J. X., Kao, K. N., Gamborg, O. L. & Miller, R. A. (1973). Immunological methods for the agglutination of protoplasts from cell suspension cultures of different genera. *Planta* 112: 45–56.

Heath, I. B. & Seagull, R. W. (1982). Oriented cellulose fibrils and the cytoskeleton: a critical comparison of models. In *The Cytoskeleton in Plant Growth and Development*, ed. C. W. Lloyd. New York: Academic Press.

Huber, S. C. & Moreland, D. E. (1980). Translocation, *Plant Physiol.* 65: 560–2.

Jacobson, B. S. & Branton, D. (1977). Plasma membrane: rapid isolation and exposure of the cytoplasmic surface by use of positively charged beads. *Science* 195: 302–4.

Joachim, S. & Robinson, D. G. (1984). Endocytosis of cationic ferritin by bean leaf protoplasts. *Eur. J. Cell Biol.* 34: 212–16.

Knox, R. B. & Clarke, A. E. (1978). Localization of proteins and glycoproteins by binding to labelled antibodies and lectins. In *Electron Microscopy and Cytochemistry of Plant Cells*, ed. J. L. Hall. Amsterdam: Elsevier/North Holland.

Larkin, P. J. (1977). Plant protoplast agglutination and membrane-bound β-lectins. *J. Cell Sci.* 26, 31–46.

Larkin, P. J. (1978). Plant protoplast agglutination by lectins. *Plant Physiol.* 61, 626–9.

Leonard, R. T. & Hodges, T. K. (1980). The plasma membrane. In *The Biochemistry of Plants*, vol. 1, *The Plant Cell*, ed. N. E. Tolbert. New York: Academic Press.

Leonard, R. T. & Rayder, L. (1985). The use of protoplasts for studies of membrane transport in plants. In *Plant Protoplasts*, ed. L. C. Fowke & F. Constabel. Boca Raton: CRC Press.

Lin, W. (1980). Corn root protoplasts. *Plant Physiol.* 66, 550–4.

Lloyd, C. W., Slabas, A. R., Powell, A. J. & Lowe, S. B. (1980). Microtubules, protoplasts and plant cell shape. *Planta* 147: 500–6.

Lorz, H. (1985). Isolated cell organelles and subprotoplasts – their role in somatic cell genetics. In *Plant Genetic Engineering*, ed. J. H. Dodds. Cambridge: Cambridge University Press.

Marchant, H. J. & Hines, E. R. (1979). The role of microtubules and cell-wall deposition in elongation of regenerating protoplasts of *Mougeotia*. *Planta* 146: 41–8.

Mettler, I. J. & Leonard, R. T. (1979). Ion transport in isolated protoplasts from tobacco suspension cells. *Plant Physiol.* 63: 191–4.

Montezinos, D. (1982). The role of the plasma membrane in cellulose microfibril assembly. In *The Cytoskeleton in Plant Growth and Development*, ed. C. W. Lloyd. New York: Academic Press.

Nagata, T. & Melchers, G. (1978). Surface charge of protoplasts and their significance in cell-cell interaction. *Planta* 142: 235–8.

Newcomb, E. H. (1980). Coated vesicles: their occurrence in different plant cell types. In *Coated Vesicles*, ed. C. J. Ockleford & A. Whyte. Cambridge, New York: Cambridge University Press.

Newcomb, W. (1976). A correlated light and electron microscope study of symbiotic growth and differentiation in *Pisum sativum* root nodules. *Can. J. Bot.* 54: 2163–86.

Nicolson, G. L. (1974). The interactions of lectins with animal cell surfaces. *Int. Rev. Cytol.* 39: 89–190.

Orci, L., Perrelet, A. & Friend, D. S. (1977). Freeze-fracture of membrane fusions during exocytosis in pancreatic B-cells. *J. Cell Biol.* 75: 23–30.

Paszkowski, J., Lorz, H., Potrykus, I. & Dierks-Ventling, C. (1980). Amino acid uptake and protein synthesis in cultured cells and protoplasts of *Zea mays* L. *Z. Pflanzenphysiol.* 99: 251–9.

Pearse, M. F. & Bretscher, M. S. (1981). Membrane recycling by coated vesicles. *Ann. Rev. Biochem.* 50: 85–101.

Perlin, D. S. & Spanswick, R. M. (1980). Labeling and isolation of plasma membranes from corn leaf protoplasts. *Plant Physiol.* **65:** 1053–7.

Peters, B. M., Cribbs, D. H. & Stelzig, D. A. (1978). Agglutination of plant protoplasts by fungal cell wall glucans. *Science* **201:** 364–5.

Prat, R. & Williamson, F. A. (1976). Chronologie de la sécrétion de parois par les protoplastes végétaux. *Soc. Bot. Fr. Coll. Sécrét. Végét.* **123:** 33–45.

Quail, P. H. (1979). Plant cell fractionation. *Ann. Rev. Plant Physiol.* **30:** 425–84.

Racusen, R. H., Kinnersley, A. M. & Galston, A. W. (1977). Osmotically induced changes in electrical properties of plant protoplast membranes. *Science* **198:** 405–8.

Raff, J., McKenzie, I. F. C. & Clarke, A. E. (1980). Antigenic determinants of *Prunus avium* are associated with the protoplast surface. *Z. Pflanzenphysiol.* **98:** 225–34.

Rennie, P. J., Weber, G., Constabel, F. & Fowke, L. C. (1980). Dedifferentiation of chloroplasts in interspecific and homospecific protoplast fusion products. *Protoplasma* **103,** 253–62.

Robenek, H. & Peveling, E. (1977). Ultrastructure of the cell wall regeneration of isolated protoplasts of *Skimmia japonica* Thunb. *Planta* **136:** 135–45.

Robenek, H. & Peveling, E, (1978). Beobachtungen am Plasmalemma wahrend der Fusion isolierter Protoplasten von *Skimmia japonica* Thunb. *Ber. Deutsch. Bot. Ges.* **91:** 351–9.

Robinson, D. G. & Quader, H. (1982). The microtubule–microfibril syndrome. In *The Cytoskeleton in Plant Growth and Development*, ed. C. W. Lloyd. New York: Academic Press.

Roland, J.-C. (1978). General preparation and staining of thin sections. In *Electron Microscopy and Cytochemistry of Plant Cells*, ed. J. L. Hall. Amsterdam: Elsevier/North Holland.

Roland, J.-C. & Prat, R. (1973). Les protoplastes et quelques problèmes concernant le rôle et l'élaboration des parois. In *Protoplastes et Fusion de Cellules Somatiques Végétales*. Paris: C.N.R.S.

Schibeci, A., Fincher, G. B., Stone, B. A. & Wardrop, A. B. (1982). Isolation of plasma membrane from protoplasts of *Lolium multiflorum* (ryegrass) endosperm cells. *Biochem. J.* **205:** 511–19.

Schieder, O. & Vasil, I. K. (1980). Protoplast fusion and somatic hybridization. In *Perspectives in Plant Cell and Tissue Culture, Int. Rev. Cytol.*, 11B, ed. I. K. Vasil. New York: Academic Press.

Schnabl, H., Vienken, J. & Zimmermann, U. (1980). Regular arrays of intramembranous particles in the plasmalemma of guard cell and mesophyll cell protoplasts of *Vicia faba*. *Planta* **148:** 231–7.

Steponkus, P. L. (1984). Role of the plasma membrane in freezing injury and cold acclimation. *Ann. Rev. Plant Physiol.* **35:** 543–84.

Suzuki, M., Takebe, S., Kajita, S., Honda, Y. & Matsui, C. (1977). Endocytosis of polystyrene spheres by tobacco leaf protoplasts. *Exptl. Cell Res.* **105:** 127–35.

Takebe, I. (1983). Protoplasts in plant virus research. In *Plant Protoplasts*, ed. K. L. Giles, *Int. Rev. Cytol. Suppl.* 16. New York: Academic Press.

Tanchak, M. A., Griffing, L. R., Mersey, B. G. & Fowke, L. C. (1984). Endocytosis of cationized ferritin by coated vesicles of soybean protoplasts. *Planta* **162:** 481–6.

Taylor, A. R. D. & Hall, J. L. (1978). Fine structure and cytochemical properties of tobacco leaf protoplasts and comparison with the source tissue. *Protoplasma* **96:** 113–26.

Taylor, A. R. D. & Hall, J. L. (1979). An ultrastructural comparison of lanthanum and silicotungstic acid/chromic acid as plasma membrane stains of isolated protoplasts. *Plant Sci. Lett.* **14:** 139–44.

Van der Valk, P. & Fowke, L. C. (1981). Ultrastructural aspects of coated vesicles in tobacco protoplasts. *Can. J. Bot.* **59:** 1307–13.

Van der Valk, P., Rennie, P. J., Connolly, J. A. & Fowke, L. C. (1980). Distribution of cortical microtubules in tobacco protoplasts. An immunofluorescence microscopic and ultrastructural study. *Protoplasma* 105: 27–43.

Vasil, I. K. & Vasil, V. (1980). Isolation and culture of protoplasts. In *Perspectives in Plant Cell and Tissue Culture, Int. Rev. Cytol.*, 11B, ed. I. K. Vasil. New York: Academic Press.

Volkmann, D. (1981). Structural differentiation of membranes involved in the secretion of polysaccharide slime by root cap cells of cress (*Lepidium sativum* L.). *Planta* 151: 180–88.

Wilkinson, M. J. & Northcote, D. H. (1980). Plasma membrane ultrastructure during plant protoplast plasmolysis, isolation and wall regeneration: a freeze-fracture study. *J. Cell Sci.* 42: 401–15.

Williamson, F. A. (1979). Concanavalin A binding sites on the plasma membrane of leek stem protoplasts. *Planta* 144: 209–15.

Williamson, F. A., Fowke, L. C., Constabel, F. C. & Gamborg, O. L. (1976). Labelling of concanavalin A sites on the plasma membrane of soybean protoplasts. *Protoplasma* 89: 305–16.

Willison, J. H. M., Grout, B. W. W. & Cocking, E. C. (1971). A mechanism for the pinocytosis of latex spheres by tomato fruit protoplasts. *Bioenergetics* 2: 371–82.

Willison, J. H. M. & Klein, A. S. (1982). Cell wall regeneration by protoplasts isolated from higher plants. In *Cellulose and Other Natural Polymer Systems: Biogenesis, Structure and Degradation*, ed. R. M. Brown. New York: Plenum Press.

Zimmermann, U. & Scheurlich, P. (1981). High frequency fusion of plant protoplasts by electric fields. *Planta* 151: 26–32.

Zimmermann, U. & Vienken, J. (1984). Electric field-mediated cell-to-cell fusion. In *Cell Fusion: Gene Transfer and Transformation*, ed. R. F. Beers & E. G. Bassett. New York: Raven Press.

8

Cell recognition and adhesion in yeast mating

Clinton E. Ballou and Michael Pierce

8.1 Overview of the yeast mating reaction

Heterothallic yeasts (those with a stable haplophase) may grow vegetatively by budding or fission in both the haploid and diploid forms, and the heterozygous diploid cell may convert to the haplophase by meiosis followed by sporulation whereas the haploid cells of opposite mating type may mate to form a zygote and re-enter the diplophase (Fowell, 1969). The mating reaction is a highly regulated process that occurs only between closely related strains, that is initiated by the exchange of species-specific diffusible pheromones, and that is facilitated by specific recognition-adhesion factors (Crandall & Brock, 1968; Crandall, 1977; Manney & Meade, 1977; Thorner, 1980; Yanagishima & Yoshida, 1981). The object of this article is to review our knowledge regarding the molecular biology of the recognition and adhesion processes. In doing this, we describe briefly the organization of the yeast cell wall, the nature of the wall components, the changes that occur in the wall in response to the sex pheromones and, finally, what is known about the structures and properties of the recognition–adhesion factors.

8.2 Organization of the yeast cell wall

The yeast cell wall has the appearance of a matrix of mannoprotein in which a fibrous glucan network is embedded to provide strength and continuity (Bartnicki-Garcia & McMurrough, 1971). These two components, mannoprotein and glucan, predominate in the wall; and electron micrographs of thin sections provide little evidence that they are intermixed in other than a random order (Matile, Moor & Robinow, 1969), although activity staining for acid phosphatase suggests that this mannoprotein is confined to the inner and outer surfaces of the wall (Linnemans, Boer & Elbers, 1977). Although the yeast cell wall lacks the obviously layered organization of some bacterial envelopes (Salton, 1964), recognizable structures are formed in the septum region during budding, with the mother cell acquiring a highly organized bud scar, whereas the daughter cell is released with only a vaguely defined and apparently impermanent birth scar (Cabib & Shematek, 1981). The primary bud scar is composed of chitin, but mannoprotein and glucan are incorporated during maturation of this structure (Cabib & Shematek, 1981). The importance of chitin to the budding process is revealed by the fact that chitin synthesis is enhanced during the cell separation stage of the cycle (Cabib, Ulane & Bowers, 1974) and by the dramatic effects on septum organization caused by inhibitors of chitin synthesis (Cabib *et al.*, 1974) or glucosamine auxotrophy (Ballou, Maitra, Walker & Whelan, 1977). Lipid is not normally a significant component of the wall (Phaff, 1971), although some yeasts, when grown on a medium rich in long-chain fatty acids, may incorporate covalently linked fatty acid into the wall (Kappeli, Muller & Fiecther, 1978).

The structural integrity of the yeast cell wall is altered by treatment with disulphide reducing agents (Lipke, Taylor & Ballou, 1976) or by incorporation of certain mannoprotein mutations (Ballou, Cohen & Ballou, 1980), and both factors enhance digestion of the wall with β-glucanase. These results suggest that the wall mannoproteins cover the glucan, that they are cross-linked by disulphide bonds, and that the carbohydrate part of the mannoproteins has a general structural role in addition to more specialized functions. Cell wall mannoproteins with specialized functions include glucanases, which probably are involved in wall morphogenesis during budding, mating and spore germination, and the mating-specific adhesion factors (Ballou, 1982). Several inducible mannoprotein enzymes are found in the periplasm, but not all appear to be incorporated into the cell wall proper as is the acid phosphatase (Linnemans *et al.*, 1977). There is no evidence for a physical association of the plasmalemma with the wall (Farkaš, 1979), such as the adhesion bridges that connect the inner and outer membranes of the Gram-negative bacterial cell envelope, and during glucanase digestion of the yeast cell the protoplast can exit freely through the first opening in the wall if the reaction is done under proper osmotic conditions (Ballou, 1981).

8.3   Structures of yeast cell wall components

Saccharomyces cerevisiae cell walls contain three main components, chitin, glucan and mannoprotein, in order of increasing amount (Phaff, 1971). The chitin is a typical $\beta1\rightarrow4$-N-acetyl-D-glucosamine polymer that, in the mature bud scar where most of it is found, is difficult to dissolve and relatively resistant to chitinase attack. Indirect evidence suggests that chitin may also be distributed at certain times in small amounts throughout the *S. cerevisiae* cell wall (Cabib & Shematek, 1981; Phaff, 1971; Schekman & Brawley, 1979).

Yeast cell wall glucan is a mixture of $\beta1 \rightarrow 3$ and $\beta1 \rightarrow 6$-linked polymers, and some molecular forms possess both linkages to yield a branched polysaccharide (Phaff, 1971). The manner in which the glucan chains interact to form fibres is not well understood although $\beta1 \rightarrow 3$-glucans, such as laminarin, readily associate to give insoluble material. Some studies indicate that the chains can form double or triple helices (Atkins, Parker & Preston, 1969), but a physiological role for such ordered structures has not been demonstrated in yeast (Ballou, 1982). The evidence suggesting there may be a differential distribution of the glucans in the cell wall has been reviewed (Cabib & Shematek, 1981).

The bulk mannoprotein obtained by extracting whole cells of *S. cerevisiae* with hot citrate buffer followed by precipitation as the cetyltrimethyl-ammonium-borate complex usually contains about 90% mannose (Ballou, 1976). About 10% of this mannose is released by β-elimination and it can be separated into a mixture of mannose, mannobiose, mannotriose and mannotetraose, which indicates that these units are linked to the hydroxyls of serine and threonine in the mannoprotein. The remainder of the carbohydrate is linked as polymannose chains to asparagine by way of the typical di-*N*-acetylchitobiose (Nakajima & Ballou, 1974). These chains are composed of a core unit of about 10 mannoses adjacent to the polypeptide and an outer chain of 150 or more mannoses attached to the core.

A different pattern is observed when one analyses individual mannoproteins. Thus, the vacuolar carboxypeptidase has four carbohydrate chains the size of core units (Hashimoto, Cohen, Zhang & Ballou, 1981), whereas the periplasmic invertase has about six core units and three polymannose chains per protein subunit of 60 000 daltons (Lehle, Cohen & Ballou, 1979). Neither of these glycoproteins contains serine or threonine-linked carbohydrate, in contrast to the sexual agglutinin from *Hansenula wingei* 5-cells that is a glycoprotein in which nearly all of the mannose (85% by weight) is linked in this fashion (Yen & Ballou, 1974).

The yeast mannoprotein core oligosaccharides are identical to some mammalian high-mannose glycoprotein core units (Kornfeld & Kornfeld, 1976), and they are similar to plant oligosaccharides such as those from pineapple stem bromelain (*Ananas comosus*) (Ishihara, Takahashi, Oguri & Tejima, 1979), although the latter contain xylose and fucose not found in *S. cerevisiae* core units. This similarity of core structures reflects the involvement of similar biosynthetic precursors (Lehle, 1980), the structures of which have been conserved during evolution. Divergence has occurred, however, in the processing of the core precursor, and yeasts appear unique in the steps that lead to addition of the large polymannose outer chain structure found on the extracellular mannoproteins (Ballou, 1980).

The mannose chains of the cell wall mannoproteins are the immuno-dominant structures on the surfaces of most yeasts (Ballou, 1974). Antisera made in rabbits against whole, heat-killed *S. cerevisiae* cells will aggluti-nate the cells, and this agglutination is completely reversed by adding pure mannooligosaccharides that represent the hapten determinants of the particular yeast strain. This indicates that neither the protein nor glucan components of the intact wall is immunogenic. Because immuni-zation with soluble mannoprotein raises antibodies against the protein component, the above result probably means that the protein is covered by polysaccharide in the intact cell. In contrast, isolated cell wall glucan is not immunogenic, so one can infer nothing from these results regarding its exposure on the cell surface.

Yeast strains differ from each other immunogenically owing to differ-ences in mannoprotein structure. Some strains, such as *S. cerevisiae* X2180, express a single immunodominant structure, in this instance ter-minal $\alpha 1 \rightarrow 3$-linked D-mannopyranose; whereas other strains may show two or more codominant structures, such as *Kluyveromyces lactis* with terminal $\alpha 1 \rightarrow 3$-linked D-mannopyranose and terminal $\alpha 1 \rightarrow 2$-linked 2-acetamido-2-deoxy-D-glucopyranose (Ballou & Raschke, 1974). The immunogenic determinants of yeasts are readily altered by mutagenesis with ethylmethane sulphonate, and it is easy to enrich for mannoprotein mutants by agglutination with specific antiserum (Ballou & Raschke, 1974), by cell sorting after labelling the cells with fluorescent antibody or lectin (Douglas & Ballou, 1980), and by binding to a basic ion-exchange resin (Sikkema, Ballou, Letters & Ballou, 1982). The latter selection is based on the fact that many mannoproteins contain phosphate and that mutation can eliminate the sites of phosphorylation or the enzymes that catalyse the addition of phosphate. The mannoprotein mutants obtained in *S. cerevisiae* X2180 have been assigned to 10 comple-mentation groups and each can be rationalized on the basis of a defective glycosyltransferase activity in the affected strain (Ballou, 1980).

8.4   Yeast sex pheromones

Haploid cells of *S. cerevisiae* secrete into the growth medium polypeptide pheromones that have the ability to induce a mating response in the cell of opposite mating type (for a review see Thorner, 1981). α-Cells secrete α-pheromone, a tridecapeptide, whereas a-cells secrete an undecapeptide. In mixed culture, the pheromones act reciprocally to alter cell–cell interaction, arrest cell growth in G1 phase of the cycle, and initiate the reactions that lead to cell fusion and zygote formation. There is evidence that these events occur in the α-haploid as a consequence of the inhibition of membrane adenylate cyclase (Liao & Thorner, 1980), which may in turn alter the state of cellular protein phosphorylation. For the purpose of this review, we will consider only the changes that affect the first of these properties. In this discussion it is important to avoid confusing the pheromones, usually called 'factors', with the sexual agglutinins, often called 'factors'. Thus, *S. cerevisiae* α-factor (pheromone) is not analogous to *Pichia amethionina* α-factor (sexual agglutinin).

S. cerevisiae a-cells respond to α-pheromone by showing an enhanced agglutinability with α-cells, a change that occurs within 30 min or less (Fehrenbacher, Perry & Thorner, 1978). As a somewhat slower response, the cells begin to elongate (Levi, 1956) and a fuzzy surface layer appears at the point of the cell extension (Lipke *et al.*, 1976). This fuzzy layer is characterized by an enhanced ability to bind conconavalin A (Tkacz & MacKay, 1979) or antibodies directed against the unstimulated parent a-cell (Lipke & Ballou, 1980). It is probable that this property is a result of a general loosening of the cell wall such that the mannoprotein side chains become more accessible, although there is also some suggestion that an a-cell specific macromolecule may be increased in amount or accessibility during this morphogenesis (Lipke & Ballou, 1980).

When a-cells are exposed to α-pheromone, they continue to synthesize cell wall materials. The newly made mannoprotein is slightly different in structure from that pre-existing in the wall, and the ratio of glucan to mannoprotein that is made increases slightly (Lipke *et al.*, 1976). Although 2-deoxyglucose, an inhibitor of polysaccharide synthesis, prevents the morphogenesis, it does not prevent the increased agglutinability induced by α-pheromone (Thorner, 1981).

Saccharomyces kluyveri exhibits a strong constitutive sexual agglutination but it also produces sexual pheromones, and *S. kluyveri* 16-cells respond to *S. cerevisiae* α-pheromone even though these two strains are not interfertile. Immunochemical cross-reactivity has been demonstrated between the pheromones from the α-mating types of these two strains, which have similar although not identical amino acid compositions (Thorner, 1981). Diffusible pheromones have not been demonstrated for *H. wingei*, another yeast with constitutive agglutionation factors, but inspection in the electron microscope of thin sections of two haploid cells in the process of mating reveals that the cells have undergone an obvious morphogenetic change while the walls remain intact (Conti & Brock, 1965). It is probable, therefore, that the cells exchange chemical signals as a prelude to mating and, because they are in intimate contact as a result of prior agglutination, there is no need for the massive secretion of pheromone that is observed in *S. cerevisiae*.

8.5   Yeast cell recognition and adhesion

8.5.1   Saccharomyces cerevisiae – *an inducible system*

Laboratory strains of *S. cerevisiae*, such as X2180, show only a very weak agglutinative reaction when diploid cells of opposite mating type are mixed. To assay their interaction, advantage is usually taken of the 'preconditioning' that occurs when each cell type is exposed to the pheromone secreted by the other. If the two strains contain complementing auxotrophic markers, the increased frequency of prototrophic diploids that result with preconditioning can be used to follow the induction of the agglutinative state (Fehrenbacher *et al.*, 1978), whereas the change in turbidity that occurs on formation of cell clumps has also been measured (Terrance & Lipke, 1981).

Only limited progress has been made in characterizing the agglutination factors from uninduced *S. cerevisiae* cells, although it is reported (Terrance & Lipke, 1981) that the properties of the factors on uninduced cells are similar to those of the factors on induced cells. In this instance, the agglutinative activity of *a*-cells was found to be heat-stable and sensitive to reducing agents, whereas that on α-cells was heat labile. Solubilized extracts of uninduced α-cells did not agglutinate *a*-cells but did inactivate them, suggesting that this factor is monovalent. The purified factor is a glycoprotein with $M_r = 10^6$ (Lipke & Terrance, 1982).

Some strains of *S. cerevisiae* show a constitutive agglutination reaction, and the factors released by brief autoclaving have been purified, the *a*-cell yielding a 23 000 dalton glycoprotein and the α-cell a 130 000 dalton glycoprotein (Yanagishima & Yoshida, 1981). Genetic analysis of the control of the constitutive and inducible agglutination factors of the two mating types suggests that they are controlled by specific genes (MAT) that are closely related but not identical (Thorner, 1981). In spite of the clear evidence of genetic control, the agglutinative property can be influenced by growth conditions, as has also been observed for *H. wingei* (Crandall & Caulton, 1975).

8.5.2 Yeasts with constitutive agglutination factors

Hansenula wingei – This yeast is found in symbiotic association with bark beetles that colonize coniferous trees. When the adult female insect bores into the tree and lays its egg, an inoculum of yeast is left that grows on the tree sap that seeps into the tunnel. The insect larva eventually utilizes the yeast as a food source and, in the process, becomes inoculated with yeast cells that remain with it on metamorphosis to the adult stage (Crandall & Brock, 1968).

H. wingei has two mating types, called 5- and 21-cells, and when mixed together the cells undergo an immediate and massive agglutination (Crandall & Broxk, 1968). The reaction occurs optimally between pH 4.5 and 5.5 and is stimulated by Mg^{2+}. At pH 2, the agglutination is completely reversed. The activity of the 21-cells is destroyed by heating whereas the 5-cell activity is not affected, although the latter cell type loses it activity on treatment with reducing agents such as mercaptoethanol.

Active agglutination factors have been isolated from *H. wingei* cells after their release by proteolysis. Subtilisin digestion of 5-cells releases a glycoprotein with the ability to agglutinate 21-cells (5-agglutinin), an indication that it is multivalent, whereas trypsin digestion of 21-cells releases a monovalent molecule with the ability to inhibit the agglutination of 21-cells by 5-agglutinin. The properties of these agglutination factors are summarized below, but it should be remembered that the factors were obtained by limited proteolysis so that some part of the molecule presumed to be anchored in the cell wall is lost in the isolation process.

The 21-cell agglutination factor ($M_r = 27\,000$) has the properties of a 'recognizer molecule' in that it is a protein with a secondary structure that is essential to its activity, as demonstrated by its sensitivity to heat and other denaturants. The high content of acidic amino acids (25% aspartate plus glutamate) is consistent with its low pI and may be related to the role of Mg^{2+} as an activator (Burke, Mendonça-Previato & Ballou, 1980).

The isolated 5-agglutinin has an M_r of 960 000, it contains 85% mannose, 10% protein (55% serine and 8% threonine), and 5% phosphate (Yen & Ballou, 1974). In contrast to 21-factor, the 5-agglutinin activity is stable to prolonged boiling or treatment with acid at 25°C, but it is inactivated by digestion with pronase, α-mannosidase and by mercaptoethanol. The latter effect is known to result from the reduction of disulphide bonds that leads to the release of small glycopeptides ($M_r = 12\,500$) that retain the ability to bind to 21-cells. The effects of enzymatic digestion suggest that both the peptide and the carbohydrate parts are important for activity. The amino acid composition of 5-agglutinin is very unusual, with an extremely high content of serine and threonine. All of the carbohydrate is attached to these amino acids and it can be released by treatment with 0.1 N NaOH (β-elimination) to yield an homologous series of mannose-containing oligosaccharides with 1 to 15 sugar units. These oligosaccharides do not inhibit the agglutination reaction, so it is unlikely that the recognition involves a simple protein –carbohydrate interaction. Although 5-agglutinin from wild-type *H. wingei* contains phosphate, a mutant that lacks phosphate in its mannoproteins still retains full agglutination activity (Sing, Yeh & Ballou, 1976).

As a model for the active portion of 5-agglutinin, one can visualize an extended core polypeptide chain of about 1000 amino acids, most of which are serine or threonine substituted by carbohydrate. To approximately six cysteine residues distributed along this core polypeptide are attached by disulphide bond six small glycopeptide recognition sites. These recognition sites contain a single cysteine along with about 27 other amino acids and 60 mannose units. The intact 5-agglutinin binds to 21-cells with an apparent free energy of association of $-14\,\text{kcal/mol}$, which is assumed to be the result of multiple interactions each of which has an association free energy of about $-5\,\text{kcal/mol}$ (Taylor & Orton, 1971).

Pichia amethionina – This is a heterothallic yeast that was isolated from exudates of necrotic cactus tissue (Starmer, Phaff, Miranda & Miller, 1978). The two mating types are designated *a* and *α*, and the haploids agglutinate strongly when mixed. The agglutination properties of *P. amethionina a*-cells parallel those of *H. wingei* 21-cells, whereas those of *P. amethionina α*-cells are similar to those of *H. wingei* 5-cells.

The agglutination factors of this yeast have been isolated after solubilization of the cell walls by digestion with Zymolyase, a β-glucanase (Mendonça-Previato, Burke & Ballou, 1982). This procedure yields what probably represent the intact macromolecules that retain the portion that holds the factors in the cell wall. As its name implies, the α-agglutinin from α-cells is a multivalent molecule, and it contains about 80% mannose, 12% protein and 3% phosphate. From its gel-filtration properties, it appears to have a molecular weight of about 10^6, and the compositional analysis shows 50% hydroxyamino acids. Treatment with 0.1 N NaOH releases 75% of the carbohydrate, which suggests that a significant portion (25%) may be linked to asparagine and could represent that part of the agglutinin molecule normally embedded in the cell wall.

P. amethionina a-cell agglutination factor, released by Zymolyase digestion, is a large mannoprotein that is excluded during filtration on Bio-Gel P-60, but after subtilisin digestion of this mannoprotein an active component is released that is included by Bio-Gel P-60 and has a molecular weight of about 27 000 based on gel electrophoresis (Mendonça-Previato *et al.*, 1982). Like *H. wingei* 21-factor, this *a*-factor is very labile to heat, is unaffected by reducing agents, and has a high content of acidic amino acids.

Although there is a striking similarity between the agglutination factors from *H. wingei* and *P. amethionina*, the isolated molecules show no cross-reactivity. The agglutinins from both yeast strains probably have extended shapes with little secondary protein structure in the region that carries the binding sites, since almost every other amino acid is substituted by carbohydrate. That portion of the *P. amethionina* α-agglutinin held in the cell wall may, however, represent a more typical mannoprotein. The evidence suggests that the *P. amethionina* *a*-factor is structurally differentiated, with the recognition function being carried by a non-glycosylated polypeptide while the intact molecule is anchored to the cell wall by a section of mannoprotein.

Saccharomyces kluyveri – This yeast, isolated from *Drosophila pinicola* in the Yosemite region of California, is heterothallic with haploid mating types designated as 16- and 17-cells (Barker & Miller, 1969). The sexual agglutination activity of 16-cells is stable to heat and destroyed by reducing agents, whereas that of 17-cells is inactivated by heat (Burke *et al.*, 1980).

Digestion of 16-cells with Zymolyase releases a monovalent recognition factor that contains 95% carbohydrate and has an apparent M_r of 500 000 (Pierce & Ballou, 1982). An active binding fragment can be released from this 16-factor by reduction, but a more convenient isolation involves direct treatment of the cells with dithiothreitol. The released fragment, treated with iodoacetamide and purified 120-fold by affinity chromatography, has an M_r of 17 000 and contains 30% carbohydrate.

This 16-cell recognition fragment can be coupled to Affi-gel and used for the affinity purification of 17-factor that is solubilized from 17-cells by Zymolyase digestion. Purified intact 17-factor has an apparent M_r of 47 000 and contains 20% carbohydrate. Its activity is not affected by reducing agents or by mannosidase digestion.

For quantitative analysis of binding, the isolated 16-factor fragment has been labelled with ^{125}I by the chloramine-T method and its interaction with different yeast cell types determined. A strict specificity is observed in that binding occurs only with *S. kluyveri* 17-cells or the isolated 17-

factor. Scatchard analysis of the data for binding of 16-factor to 17-cells reveals an apparent K_d of 3×10^{-8} M and an average of 6×10^5 sites per cell. The activity of the [^{125}I]16-factor is not affected by heat or by 0.1 N HCl at 25°C for 90 min, but it is destroyed by digestion with pronase or jack bean α-mannosidase, and by treatment with 0.3 mM sodium periodate at 25°C for 6 h or by 0.1 N sodium hydroxide at 25°C for 2 h. The conclusion from such experiments is that both the peptide and carbohydrate structures are involved in the recognition activity.

8.5.3   *Comparative aspects of yeast sexual agglutination*

The parallel properties shown by the sexual agglutination factors from the two mating types of each yeast strain so far analysed have interesting taxonomic implications with regard to the assignment of mating types to incompatible strains. *H. wingei* 5-agglutinin, *P. amethionina* α-agglutinin and *S. kluyveri* 16-factor are heat-stable macromolecules from which the recognition sites can be released by disulphide reducing agents. Since the sexual agglutination property is under control of the mating type locus (Thorner, 1981), we suggest that these three haploids have the same mating type. These properties are also shown by *S. cerevisiae* a-cells (Terrance & Lipke, 1981), and we conclude that all of these strains have the mating type that is analogous to the a-mating type of *S. cerevisiae*. This conclusion is supported by the observation that *S. kluyveri* 16-cells respond to the α-pheromone produced by *S. cerevisiae* α-cells even though these strains are not interfertile (McCullough & Herskowitz, 1979).

Parallel properties are also shown by *H. wingei* 21-factor, *P. amethionina* a-factor and *S. kluyveri* 17-factor. All are heat-labile glycoproteins from which a proteinaceous recognizer molecule that retains full biological activity in the presence of reducing agents can be released by controlled proteolysis. *S. cerevisiae* α-cells have similar properties (Terrance & Lipke, 1981) and, on this basis, we propose that all of these haploids have the mating type of *S. cerevisiae* α-cells.

Although the generalization may lack validity, it appears that the kind of unidirectional recognition in which one cell type carries the recognizer molecule and the other the site of recognition may be characteristic of many fertilization reactions. During fertilization in sea urchins, a heat-labile binding protein in the acrosome of the sperm head recognizes and attaches to heat-stable glycoprotein recognition sites on the egg coat (Vacquier & Moy, 1977; Kinsey & Lennarz, 1981). Thus, in recognition property, the sperm parallels the yeast α-mating type and the egg the a-mating type.

8.6 References

Atkins, E. D. T., Parker, K. D. & Preston, R. D. (1969). The helical structure of β-1,3-linked xylan in some siphoneous green algae. *Proc. Roy. Soc. B* **173**: 209–21.

Ballou, C. E. (1974). Some aspects of the structure, immunochemistry and genetic control of yeast mannan. *Adv. Enzymol.* **40**: 239–70.

Ballou, C. E. (1976). Structure and biosynthesis of the mannan component of the yeast cell envelope. *Adv. Microbial. Physiol.* **14**: 93–158.

Ballou, C. E. (1980). Genetics of yeast mannoprotein synthesis. In *Fungal Polysaccharides*, ed. P. A. Sandford & K. Matsuda, ACS Symposium Series 126, pp. 1–14. Washington D.C.: American Chemical Society.

Ballou, C. E. (1982). Structure, organization and function of the yeast cell envelope. In *The Molecular Biology of the Yeast Saccharomyces cerevisiae*, ed. J. N. Strathern, E. W. Jones & J. R. Broach, pp. 335–60. New York: Cold Spring Harbor Laboratory, Cold Spring Harbor.

Ballou, C. E., Maitra, S. K., Walker, J. W. & Whelan, W. L. (1977). Developmental defects associated with glucosamine auxotrophy in *Saccharomyces cerevisiae*. *Proc. Natl. Acad. Sci. USA* **74**: 4351–5.

Ballou, C. E. & Raschke, W. C. (1974). Polymorphism of the somatic antigen of yeast. *Science* **184**: 127–34.

Ballou, D. L., Cohen, R. E. & Ballou, C. E. (1980). *Saccharomyces cerevisiae* mutants that make mannoproteins with a truncated carbohydrate outer chain. *J. Biol. Chem.* **225**: 5986–91.

Barker, E. R. & Miller, M. W. (1969). Some properties of *Saccharomyces kluyveri*. *Antonie van Leeuwenhoek* **35**: 159–71.

Bartnicki-Garcia, S. & McMurrough, I. (1971). Biochemistry of morphogenesis in yeasts. In *The Yeasts*, vol. 2, ed. A. H. Rose & J. S. Harrison, pp. 441–91. New York: Academic Press.

Burke, D., Mendonça-Previato, L. & Ballou, C. E. (1980). Cell–cell recognition in yeast: purification of *Hansenula wingei* 21-cell sexual agglutination factor and comparison of the factors from three genera. *Proc. Natl. Acad. Sci. USA* **77**: 318–22.

Cabib, E. & Shematek, E. M. (1981). Structural polysaccharides of plants and fungi: comparative and morphogenetic aspects. In *Biology of Carbohydrates*, vol. 1, ed. V. Ginsburg & P. Robbins, pp. 51–90. New York: John Wiley and Sons.

Cabib, E., Ulane R. & Bowers, B. (1974). A molecular model for morphogenesis: the primary septum of yeast. *Curr. Top. Cell Regul.* **8**: 1–32.

Conti, S. F. & Brock, T. D. (1965). Electron microscopy of cell fusion in conjugating *Hansenula wingei*. *J. Bacteriol.* **90**: 524–33.

Crandall, M. & Caulton, J. H. (1975). Induction of haploid glycoprotein mating factors in diploid yeasts. *Methods Cell Biol.* **12**: 185–207.

Crandall, M. J. (1977). Mating-type interaction in micro-organisms. In *Receptors and Recognition*, Series A, vol. 3, ed. P. Cuatrecasas & M. F. Greaves, pp. 45–100. London: Chapman and Hall.

Crandall, M. J. & Brock, T. D. (1968). Molecular basis of mating in the yeast *Hansenula wingei*. *Bacteriol. Rev.* **32**, 139–63.

Douglas, R. H. & Ballou, C. E. (1980). Isolation of *Kluyveromyces lactis* mannoprotein mutants by fluorescence-activated cell sorting. *J. Biol. Chem.* **255**: 5975–85.

Farkaš, V. (1979). Biosynthesis of cell walls of fungi. *Microbiol. Rev.* **43**: 117–44.

Fehrenbacher, G., Perry, K. & Thorner, J. (1978). Cell–cell recognition in *Saccharomyces cerevisiae*: regulation of mating-specific adhesion. *J. Bacteriol.* **134**: 893–901.

Fowell, R. R. (1969). Life cycles in yeasts. In *The Yeasts*, vol. 1, ed. A. H. Rose & J. S. Harrison, pp. 461–71. New York: Academic Press.

Hashimoto, C., Cohen, R. E., Zhang, W.-J. & Ballou, C. E. (1981). The carbohydrate chains of yeast carboxypeptidase Y are phosphorylated. *Proc. Natl. Acad. Sci. USA* **78**: 2244–8.

Ishihara, H., Takahashi, N., Oguri, S. & Tejima, S. (1979). Complete structure of the carbohydrate moiety of stem bromelain. *J. Biol. Chem.* **254:** 10715–19.

Kappeli, O., Muller, M. & Fiecther, A. (1978). Chemical and structural alterations at the cell surface of *Candida tropicalis* induced by hydrocarbon substrate. *J. Bacteriol.* **133:** 952–8.

Kinsey, W. H. & Lennarz, W. J. (1981). Isolation of a glycopeptide fraction from the surface of the sea urchin egg that inhibits sperm-egg binding and fertilization. *J. Cell Biol. 91:* 325–31.

Kornfeld, R. & Kornfeld, S. (1976). Comparative aspects of glycoprotein structure. *Ann. Rev. Biochem.* **45:** 217–37.

Lehle, L. (1980). Biosynthesis of the core region of yeast mannoproteins. Formation of a glucosylated dolichol-bound oligosaccharide precursor, its transfer to protein and subsequent modification. *Eur. J. Biochem.* **109:** 589–601.

Lehle, L., Cohen, R. E. & Ballou, C. E. (1979). Carbohydrate structure of yeast invertase. Demonstration of a form with only core oligosaccharides and a form with completed polysaccharide chains. *J. Biol. Chem.* **254:** 12209–18.

Levi, J. D. (1956). Mating reaction in yeast. *Nature* **177:** 753–4.

Liao, H. & Thorner, J. (1980). Yeast mating pheromone α factor inhibits adenylate cyclase. *Proc. Natl. Acad. Sci. USA* **77:** 1898–1902.

Linnemans, W. A. M., Boer, P. & Elbers, P. F. (1977). Localization of acid phosphatase in *Saccharomyces cerevisiae*: a clue to cell wall formation. *J. Bacteriol.* **131:** 638–44.

Lipke, P. N. & Ballou, C. E. (1980). Altered immunochemical reactivity of *Saccharomyces cerevisiae* a-cells after α-factor induced morphogenesis. *J. Bacteriol.* **141:** 1170–7.

Lipke, P. N., Taylor, A. & Ballou, C. E. (1976). Morphorgenic effect of α-factor on *Saccharomyces cerevisiae* a cells. *J. Bacteriol.* **127:** 610–18.

Lipke, P. N. & Terrance, K. (1982). Structure and properties of a cell adhesion molecule from *Saccharomyces cerevisiae*. *Fed. Proc.* **41:** 753.

Manney, T. R. & Meade, J. H. (1977). Cell–cell interactions during mating in *Saccharomyces cerevisiae*. In *Receptors and Recognition – Microbial Interactions*, Series B, vol. 3, ed. J. R. Reissig, pp. 281–321. London: Chapman and Hall.

Matile, P. H., Moor, H. & Robinow, C. F. (1969). Yeast cytology. In *The Yeasts*, vol. 1, ed. A. H. Rose & J. S. Harrison, pp. 219–302. London: Academic Press.

McCullough, J. & Herskowitz, I. (1979). Mating pheromones of *Saccharomyes kluyveri*: pheromone interactions between *Saccharomyces kluyveri* and *Saccharomyces cerevisiae*. *J. Bacteriol.* **138:** 146–54.

Mendonça-Previato, L., Burke, D. & Ballou, C. E. (1982). Sexual agglutination factors from the yeast *Pichia amethionina*. *J. Cell. Biochem.* **19:** 171–8.

Nakajima, T. & Ballou, C. E. (1974). Structure of the linkage region between the polysaccharide and protein parts of *Saccharomyces cerevisiae* mannan. *J. Biol. Chem.* **249:** 7685–94.

Phaff, H. J. (1971). Structure and biosynthesis of the yeast cell envelope. In *The Yeasts*, vol. 2, ed. A. H. Rose & J. S. Harrison, pp. 135–210. New York: Academic Press.

Pierce, M. & Ballou, C. E. (1982). Sexual agglutination factors from the yeast *Saccharomyces kluyveri*. *J. Biol. Chem.* **258:** 3576–82.

Salton, M. R. J. (1964). *The Bacterial Cell Wall*. New York: Elsevier Publishing.

Schekman, R. & Brawley, V. (1979). Localized deposition of chitin on the yeast cell surface in response to mating pheromone. *Proc. Natl. Acad. Sci. USA* **76:** 645–9.

Sikkema, W. D., Ballou, L., Letters, J. & Ballou, C. E. (1982). An enrichment technique for the isolation of yeast mutants with altered cell surface mannoproteins. *Fed. Proc.* **41:** 755.

Sing, V., Yeh, Y.-F. & Ballou, C. E. (1976). Isolation of a *Hansenula wingei* mutant with an altered sexual agglutinin. In *Surface Membrane Receptors*, ed. R. A. Bradshaw,

W. A. Frazier, R. C. Merrell, D. I. Gottlieb & R. A. Hogue-Angeletti, pp. 87–97. New York: Plenum Publishing Co.

Starmer, W. T., Phaff, H. J., Miranda, M. & Miller, M. W. (1978). *Pichia amethionina*, a new heterothallic yeast associated with the decaying stems of ceroid cacti. *Int. J. System. Bacteriol.* **28:** 433–41.

Taylor, N. W. & Orton, W, L. (1971). Cooperation among the active binding sites in the sex-specific agglutinin from the yeast, *Hansenula wingei. Biochemistry* **10:** 2043–8.

Terrance, K. & Lipke, P. N. (1981). Sexual agglutination in *Saccharomyces cerevisiae. J. Bacteriol.* **148:** 889–96.

Thorner, J. (1980). Intercellular interactions of the yeast *Saccharomyces cerevisiae.* In *The Molecular Genetics of Development*, ed. T. Leighton & W. F. Loomis, pp. 119–78. New York: Academic Press.

Thorner, J. (1981). Pheromonal regulation of development in *Saccharomyces cerevisiae.* In *The Molecular Biology of the Yeast Saccharomyces. Life Cycle and Inheritance*, ed. J. N. Strathern, E. W. Jones & J. R. Broach. New York: Cold Spring Harbor Laboratory.

Tkacz, J. S. & MacKay, V. (1979). Sexual conjugation in yeast. Cell surface changes in response to the action of mating hormones. *J. Cell Biol.* **80:** 326–33.

Vacquier, V. D. & Moy, G. W. (1977). Isolation of bindin. The protein responsible for adhesion of sperm to sea urchin eggs. *Proc. Natl. Acad. Sci. U.S.A.* **74:** 2456–60.

Yanagishima, N. & Yoshida, K. (1981). Sexual interactions in *Saccharomyces cerevisiae* with special reference to the regulation of sexual agglutinability. In *Sexual Interactions in Eukaryotic Microbes*, ed. D. H. O'Day & P. A. Horgen, pp. 261–95. New York: Academic Press.

Yen, P. H. & Ballou, C. E. (1974). Partial characterization of the sexual agglutination factor from *Hansenula wingei* Y-2340 type 5 cells. *Biochemistry* **13:** 2428–37.

9

Cell-surface receptors in the pollen–stigma interaction of Brassica oleracea

H. G. Dickinson and I. N. Roberts

9.1 Introduction

Interactions between the alighting pollen grain and the receptive tissues of the pistil precede fertilization in angiosperms, and in many families mechanisms have evolved which result in the preselection of gametes during this stage of the reproductive cycle. In *Brassica* such a mechanism functions to recognize and reject 'self'-pollen, preventing self-fertilization and therefore providing the plant with a valuable device to promote outcrossing. This phenomenon of intraspecific or self-incompatibility (S.I.) is widespread in flowering plants (East, 1940; Brewbaker, 1957) and, apart from its obvious practical significance to plant breeding, is of considerable academic interest. The acquisition of S.I. systems has been held to be responsible for the rapid ascendancy of the land plants (Whitehouse, 1950), for by promoting outbreeding and hence heterozygosity, they may well increase evolutionary potential. Equally, at a cellular level, S.I. mechanisms provide one of the best known examples of intercellular communication in higher plants (Heslop-Harrison, 1978*b*).

Brassica oleracea has proved a particularly suitable subject for the study of molecules involved in communication between pollen and pistil for a number of reasons. Firstly, the genetics have been studied in some detail; the system is completely homomorphic and the compatability of the pollen with respect to the stigma has been demonstrated to be sporophytically controlled, with multiple self-incompatability (S-) alleles acting in pairs at a single S-locus (Thompson, 1957). Secondly, the site of recognition and rejection of self-pollen has been localized precisely at the stigma surface (Ockenden, 1972). This 'superficial' interaction is also found in many other species with sporophytic control and is further correlated with the production of trinucleate pollen (Brewbaker, 1957) and the possession of conspicuously papillate stigmas with little, if any, surface exudate (Heslop-Harrison, Heslop-Harrison & Barber, 1975). Such stigmas have been found to possess a thin superficial proteinaceous pellicle (Mattsson, Knox, Heslop-Harrison & Helsop-Harrison, 1974) and, since the reactions resulting in rejection of self-pollen take place between largely ungerminated pollen and the stigma surface, this pellicle is a logical site to search for the receptor molecules involved in the recognition of self-pollen.

A further valuable aid to our understanding of the S.I. reaction is provided by the fact that self-pollination of immature stigmas of *Brassica* in the bud results in successful fertilization (el Murabaa, 1957); this 'bud compatibility' diminishes as the bud matures, suggesting that a new component conferring the ability to recognize and reject self-pollen is added during stigma maturation (Shivanna, Heslop-Harrison & Heslop-Harrison, 1978). Finally, work by Dickinson & Lewis (1973*a,b*) has indicated that the pollen grains carry a sporophytically derived coating which is the first part of the pollen to come into contact with the pellicle and which, in view of the sporophytic control of S.I. and the superficial nature of the reaction, must be considered likely to carry the S-determinants of the pollen. Indeed, extracts of the coat material can induce some typical symptoms of the incompatibility response (Heslop-Harrison, Knox & Heslop-Harrison, 1974; Dickinson & Lewis, 1975).

From the above it is clear that any molecule postulated to be a receptor or recognition molecule in the strictest sense must conform to a rigidly defined set of requirements including localization at the pollen or stigma surface, ability to interact with other molecules in an S-specific manner leading to rejection of self-pollen, and, in the case of the stigma receptor, absence or inaction in the bud. Before discussing the nature and mode of action of molecules that have been identified as potential receptors in pollen–stigma interactions, it is helpful to consider briefly the different stages of interaction that can be distinguished and the sequence by which they take place.

9.2   Sequence of events following pollination
9.2.1   *Acceptance of compatible pollen*
The proposal that pollen–pistil interactions in general proceed via a sequence of phases analogous to a dialogue between the interacting partners (Heslop-Harrison *et al.* 1975) has proved of considerable value in interpreting events in *Brassica*. Phases of adhesion, hydration, germination and tube growth may readily be identified and, for convenience, may be discussed separately.

Shortly after alighting on the stigma the pollen grain becomes adhered to the stigma surface such that subsequent removal by the pollinating agent is unlikely. The components responsible for the adhesion appear to be the pollen coat and the pellicle; three stigmatic proteins have been implicated in the adhesion event (Stead, Roberts & Dickinson, 1980) but no molecule of the pollen coat, a lipoprotein complex, has yet been identified as being specifically involved in adhesion. Most recently the use of new electron-microscopic techniques has revealed the presence of a superficial 'membrane' investing the pollen coat and exine (Elleman & Dickinson, 1985). While this layer has yet to be characterized fully, it is capable of binding several specific probes and seems likely to play an important part in any binding to the stigmatic pellicle. Nevertheless differences in the adhesion of cross- and self-pollen have been detected (Stead, Roberts & Dickinson, 1979) suggesting that this phase represents an important initial interaction between pollen and stigma.

Adhesion is followed by a second phase, hydration, during which the pollen grain acquires water from the stigma resulting in the rehydration and activation of the grain, which is shed in a dormant condition and in different degrees of dehydration. Rehydration is clearly an essential prerequisite for pollen germination and, in the case of trinucleate pollen, evidence exists indicating that this must be a controlled process. The rate of water supply has proved critical in attempts to germinate trinucleate pollens *in vitro* (Bar Shalom & Mattsson, 1977) and it has been suggested that such pollens lack normal membranes in their dehydrated state and that these need to be reconstituted prior to growth (Heslop-Harrison, 1979a). Since stigmatic water destined for the pollen protoplast must of necessity pass through the pollen coating, it is logical to consider the possibility that the coating may in some way control the flow of water to the hydrating grain. Using anhydrous fixatives it has recently proved possible to monitor structural changes as water passes through the coating (Elleman & Dickinson, 1985). These structural changes involve a conversion from a largely electron-lucent matrix containing small spherical inclusions to an electron-opaque, highly structured material featuring packed membrane-like profiles. These techniques reveal that as stigmatic water reaches the protoplast itself, spectacular changes also take place in the cytoplasm. In *Brassica* full rehydration on the stigma, as evidenced by swelling of the pollen grain, is achieved 1–4 h after pollination thus providing further evidence that a slow, regulated rehydration is required for optimal growth.

Hydration is followed by a phase during which the pollen grain germinates to produce a pollen tube. The onset of germination is marked by the appearance of a small swelling of cytoplasm through the germinal aperture, known as a tube initial. These may form *in vitro* in a humid atmosphere (Roberts, Stead, Ockendon & Dickinson, 1979*a*) and have been shown to be produced consistently at 98% relative humidity (R.H.), but not at 100% R.H. (Ferrari, Lee & Wallace, 1981), strongly suggesting that a critical turgor pressure must be reached for grain activation. The tube initial then develops to produce the pollen tube by a mechanism which appears to require some form of participation by the stigma (continued tube growth is seldom observed in a humid atmosphere) and involves a change in pollen metabolism (Ferrari & Wallace, 1977).

The phase of tube entry into the style in species such as *Brassica oleracea* which possess 'dry' stigmas with entire cuticles (Heslop-Harrison & Shivanna, 1977) necessitates the production of a 'cutinase' (Linskens & Heinen, 1962). Some evidence that this enzyme has also to be activated by a signal from the stigma obtains from experiments involving digestion of the stigma surface with proteolytic enzymes. Although quite large percentage germination is achieved from stigmas treated in this way, the tubes produced are unable to penetrate the stigma surface (Shivanna, *et al.* 1978). Whether this 'cutinase activation' is part of the same process which results in activation of the tube initial is not known. Once beneath the cuticle, the pollen tube grows rapidly down the papilla, travelling between the cellulosic cell wall and the cuticle, and enters the style. In sporophytic S.I. systems incompatibility is not expressed in the style and pollen achieving adhesion, rehydration, germination, tube growth and entry into the style will grow down to the ovary and effect fertilization. This is the sequence of events which leads to the acceptance of compatible pollen; as we shall see below, incompatible pollen may be rejected at a number of developmental stages.

9.2.2   *Rejection of self-incompatible pollen*
    In *Brassica* self-pollen is inhibited by constraints which act at all stages of pollen–stigma interaction. Initially, adhesion is weak relative to cross-pollen and, although self-pollen eventually becomes adhered to an extent comparable to cross (Stead *et al.* 1979), the percentage of grains that appear fully turgid is small compared to that obtained on a compatible stigma (Roberts, Stead, Ockendon & Dickinson, 1980). This is not to say that incompatible grains do not take up water, indeed recent experiments suggest that they do (Zuberi & Dickinson 1985), but rather that their development is inhibited before full turgidity is observed. The number of pollen grains which germinate on a self-stigma, although very variable, is also usually much smaller than that found

on a cross-stigma (Dickinson & Lewis, 1973*a*). When self-pollen tubes are produced they are frequently malformed and seldom penetrate the stigmatic cuticle; if they do they elicit the formation of a 'callosic lenticule' in the papillar cells of the stigma (Dickinson & Lewis, 1973*a*).

It seems likely therefore that a superficial localization of receptor molecules in the pollen coat, or its investing layer, and stigma pellicle results in an early recognition event which adversely affects the development of self-pollen during all phases of pollen–stigma interaction and effectively prevents self-fertilization. The behaviour of cross- and self-pollen during the successive phases of the pollen–stigma interaction is summarized in Fig. 9.1.

Fig. 9.1. A summary of the sequence of events that follows pollination in *Brassica*.

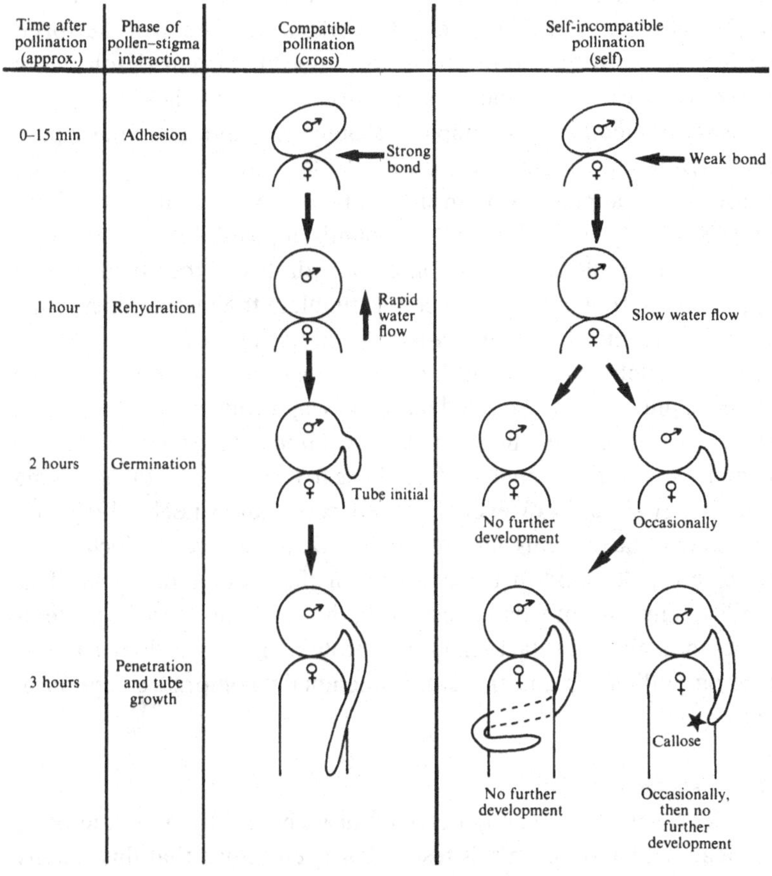

9.3   Receptors on the stigma surface
Although complex carbohydrate molecules of the cell wall have
been shown to play a major role in host–pathogen recognition reactions
(for review see Albersheim *et al.* 1981; Ralton, Howlett & Clarke, this
volume) the primary focus of attention in incompatibility research has
been the proteinaceous pellicle secreted by the stigma. Since its discovery
(Mattsson, Knox, Heslop-Harrison & Heslop-Harrison, 1974) it has
become evident that this layer is particularly well adapted not only for
a role in pollen capture and rehydration but also for recognizing the
compatibility of the alighting grain. Potential glycoprotein receptor sites
have been identified by binding experiments with Con A and pollen
proteins (Clarke *et al.* 1979), and proteins of the pellicle have been
shown to be involved in pollen adhesion, binding cross-pollen more
strongly than self-pollen (Stead *et al.* 1979, 1980). Further, the fissured
aspect presented by the pellicle of the turgid stigmatic papillae may
indicate a structural adaptation capable of regulating water flow to the
grain (Mattsson *et al.* 1974). These data taken together with the speed
with which recognition and rejection are deduced to occur (Kroh, 1966;
Ferrari & Wallace, 1977) would seem to argue against a 'host–pathogen'
type response mediated by complex carbohydrate interaction and phytoalexin synthesis. In fact, the time available between recognition and
response would appear to be insufficient for any kind of synthesis *de
novo*. It is not, however, beyond the bounds of possibility that enzymes
secreted by the pollen could act on carbohydrates of the stigmatic cell
wall, producing fragments which could stimulate the plasma membrane
to release preformed 'phytoalexin-like' biostatic molecules.

Thus, although cell wall and indeed cytoplasmic components of the
stigmatic papillae cannot be ruled out as playing a role in S.I., the pellicle
must be considered to be the most likely location for the stigmatic receptors, at least for those involved in the early stages of pollen–stigma
interaction. It should however be noted here that Ameele (1982) has
suggested that the proteins detected on the stigma surface by biochemical
tests are normally held in the cell wall of the untreated stigma. This
possibility is important when considering the secretion of these proteins
but does not alter the fact that such proteins must be drawn to the
surface on pollination and therefore functionally comprise a superficial
pellicle.

9.3.1   Pellicle secretion
The pellicle is readily detected histochemically by virtue of its
esterase activity and, using this test, it has been shown that this activity

increases during maturation of the stigma in the bud (Shivanna *et al.* 1978). The significance of the esterase activity is not known but it is present on even the youngest stigmas and may be associated with a pellicle component essential for pollen activation and germination. Stigmas have to have reached a certain stage before pollination is effective and removal of the pellicle using protease seriously inhibits pollen growth past the tube initial stage. Stigmas in the bud remain capable of accepting both cross- and self-pollen until shortly before flowering opening (Fig. 9.2) when it has been proposed that a new component is added to the pellicle enabling the recognition and rejection of self-pollen (Shivanna *et al.* 1978).

Measurements of pollen adhesion following protease and cycloheximide treatments of the mature stigma surface have revealed that the pellicle proteins have rapid turnover rates (Stead *et al.* 1980). Whether this is due to cycling of the proteins in the pellicle or continuous secretion and denaturation at the stigma surface has yet to be established. It is difficult to see how cycling could be achieved bearing in mind the extracuticular localization of the pellicle. On the other hand if the pellicle proteins are normally held in the cell wall prior to pollination (Ameele, 1982) then cycling of proteins between cell wall and cytoplasm becomes more feasible. Most recently the concept of cycling proteins at the stigma surface has been reinforced by data from autoradiography and inhibitor studies (Roberts, Harrod & Dickinson, 1984*a,b*) which, as does the earlier work of Stead *et al.* (1980), points to a half-life of some molecules of about 120 min.

Fig. 9.2. The changes in the percentage of seed-set following self- or cross-pollinations which follow the development of the pistil. (Reproduced from Hinata, 1981.)

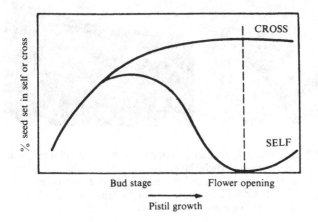

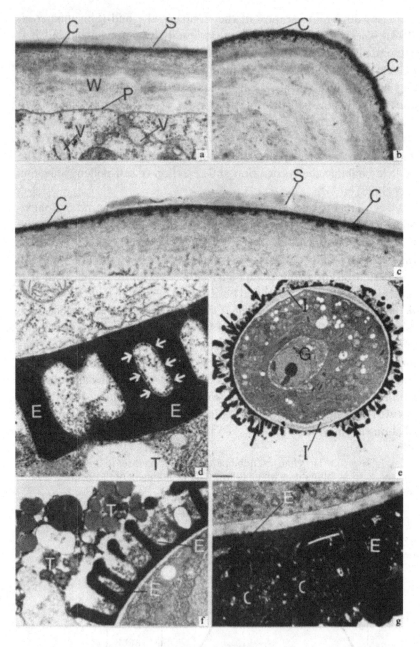

Fig. 9.3. (a) Transmission electron micrograph of a papillar cell wall revealing the waxy superficial layer(S) of cuticle (C), fibrillar cell wall (W), plasma membrane (P) and the subjacent cytoplasm containing vesicles (V) derived from rough endoplasmic reticulum. The 'pellicle' is not often visible in electron micrographs. ×27 000.

Our knowledge of the mechanism of secretion of the pellicle is also incomplete. Ultrastructural analyses of the papillar cells reveal an apparently active cytoplasm containing many dictyosomes budding off numerous small vesicles containing protein which appear to fuse with the plasma membrane (Fig. 9.3*a*). Whether the proteins are then expelled into the cell wall and reach the surface by simple diffusion through the dissected cuticle (Fig. 9.3*a,b,c*) or whether definite channels exist in the cell wall and proteins can be cycled between the cytoplasm and the surface, is not known. Some evidence that such channels may exist has been obtained from freeze-fracture studies; replicas of fracture planes through the cell wall indicate channels running through the fibrillar wall material (Roberts *et al.* 1984*a*). Similar structures have also been observed in the cell wall of the stigmatic papillae of *Gladiolus* (Clarke, Abbot, Mandel & Pettitt, 1980) but they may be artifactual, and their function is purely a matter of speculation at present. Studies by Carpita, Sabularse, Montezinos & Delmer (1979) have indicated that although the bulk of pores in the plant cell wall are small, there may be a number of large pores allowing the passage of large molecular weight proteins.

9.3.2 Stigmatic proteins

Analysis of pellicle proteins in isolation is limited by the small quantities present. Heslop-Harrison (1978*b*) using microgel electrofocus-

Caption for fig. 9.3 (*cont.*)

(b) Material as shown in Fig. 9.3a, but sectioned tangentially to reveal the particulate nature of the dark-staining cuticle (C). ×7510.

(c) Material as shown in Fig. 9.3a, but showing the particulate nature of the cuticle (C) in transverse section. The superficial waxy layer (S) is also evident. ×22960.

(d) Transmission electron micrograph showing the exine (E) of a young microspore just released from the tetrad. Material from the tapetum (T) has begun to accumulate onto the surface of this wall. Exact transverse sections reveal that a thin electron-lucent space (arrows) separates this material from the exine. ×27680.

(e) Low-power transmission electron micrograph depicting a young pollen grain immediately prior to its coating by the main tapetal mass. The exine of this grain is coated in a grano-fibrillar matrix (arrows). Within the grain the generative cell (G) is conspicuous, as is the development of the protein-rich intine (I). *c.* × 850.

(f) Material at a stage immediately subsequent to that shown in Fig. 9.3c. The lipoidal tapetal mass (T) is now moving into the exine (E) which is itself coated with a grano-fibrillar matrix. ×8840.

(g) Transmission electron micrograph of a pollen grain just prior to release from the anther. The pollen grain coating (C) has become compacted into the exine (E), assuming a very electron-opaque aspect. ×9000.

ing found some 24 protein components in stigma eluates of rye, but most analyses have been performed on extracts obtained by homogenizing entire stigmas. Nasrallah, Barber & Wallace (1970) and Sedgely (1974) were able to detect S-specific proteins using immunological techniques and Nishio & Hinata (1977) identified a range of S-specific glycoproteins using isoelectric focusing. These authors have recently been able to propose a classification of S-genotypes based on the isoelectric points of stigma glycoproteins (Hinata, 1981). Circumstantial evidence for the involvement of such molecules in S.I. was obtained when a glycoprotein was found to be acquired by maturing stigmas coincident with the ability to recognize and reject self-pollen (Roberts *et al.* 1979a). Significantly, this molecule was one of three proteins absent from isoelectric focusing profiles following protease digestion of the stigma surface (Stead *et al.* 1980).

Recently progress has been made in the purification and analysis of such glycoproteins. Ferrari, Bruns & Wallace (1981) have isolated the S_2S_2 glycoprotein and found it to have inhibitory effects on the germination of self-, but not cross-pollen. Similarly Nishio & Hinata (1979) have isolated the S_7S_7 molecule and analysed the amino acid and carbohydrate composition. The S_2S_2 and S_7S_7 glycoproteins are strikingly similar in gross composition but no sequencing data have yet been published. Neither has a mechanistic account of how these glycoproteins bring about the observed S.I. reaction been given. However we have recently proposed a model to account for many of the phenomena characteristics of S.I. based on experiments demonstrating the affinity of a stigma surface glycoprotein for pollen extracts *in vitro* (Roberts & Dickinson, 1983). Admixture of partially purified stigma and pollen extracts results in a 'molecular hybridization' between the stigma glycoprotein and an as yet unidentified pollen component. Clearly such an event occuring *in vivo* could form the basis of an S.I. system and we must conclude that overwhelming evidence for the glycoprotein nature of the stigma receptor and its localization in the pellicle now exists.

Immunodiffusion experiments performed by Nasrallah (1979) showed no cross-reactivity between S-specific antigens from 21 different incompatibility genotypes of *Brassica oleracea*. This lack of immunological relatedness suggests that the S-allele products are not polypeptide variants of one protein species but, perhaps, molecules responsible for linking specific markers onto a protein core (Nasrallah, 1979). It may well be, therefore, that the S-specificity of the pollen receptor resides in the carbohydrate moeity of the glycoprotein. Different carbohydrate chains linked to a protein core might also account for the large differences in pI observed between the different S-specific glycoproteins (Nishio & Hinata, 1977). In more recent work, these authors now consider that differences in the protein moieties of these molecules may have some significance (Nishio & Hinata, 1982).

9.4 Receptors in the pollen

In the same way that cytoplasmic and wall-held molecules cannot be ruled out as stigma receptors, components of the pollen intine, plasma membrane, and cytoplasm cannot be eliminated from any discussion of pollen receptors. Certainly a bewildering array of proteins diffuse from pollen on wetting and many of these have been shown to originate from the intine (Knox & Heslop-Harrison, 1971). Such molecules are not involved in germination *per se* and may well be involved in pollen- –stigma interactions (Gilissen, 1977; Stevens & Murray, 1982). Some authors have deduced the activation of S-alleles in sporophytic S.I. systems to occur in the grain before anaphase I (Brewbaker, 1957) or during meiotic divisions (Pandey, 1960). However no evidence exists supporting S-product synthesis by the pollen grain and, following the suggestion that S-gene products could be synthesized by the sporophyte and transferred to the grain (Heslop-Harrison, 1967), attention has focused on the pollen coating.

The pollen coat was found by histochemical and electron-microscopic techniques to be transferred from the tapetum of the sporophyte to the exine of the developing pollen grain some 60–70 h prior to anthesis (Dickinson & Lewis, 1973b). Extracts of this material have been shown to stimulate the production of callose in stigmas of the same S-genotype in a reaction comparable to that elicited by self-pollen (Heslop-Harrison et al. 1974; Keohas, Knox & Dumas, 1983). Extracts from compatible coating do not induce callose synthesis and thus the pollen coat may well carry the sporophytic determinants of S.I. Further indications of the role of the pollen coat in S.I. derive from the experiments of Roggen (1974); rinsing the pollen coat with organic solvents was found to modify the S.I. system resulting in the production of self-seed, and indicating a possible involvement of lipid or lipid-soluble materials.

9.4.1 *Pollen coat formation*

The deposition of an organized tryphine (Echlin, 1971) or pollen coat onto the exine of a member of the Cruciferae was first investigated by Dickinson & Lewis (1973b) on *Raphanus sativus*. Following the completion of the sporopollenin ectexine two phases of tapetal activity result in the deposition of first proteinaceous fibrils (Fig. 9.3e) and then a disorganized lipoidal mass onto the grain surface (Fig. 9.3*f*). The protein originates from cisternae of the tapetal endoplasmic reticulum, whilst the major part of the lipoid is derived from large elaioplasts also present in the tapetum. Particularly striking is the manner by which the tapetal products are almost 'attracted' to the exine surface and also the way in which the protein and lipoid apparently fuse to form a fairly homogenous electron-opaque mass, investing the entire grain, containing only a few angular electron-lucent spaces.

More recent examination of these events in *Brassica oleracea* (A. D. Stead & H. G. Dickinson, unpublished) has revealed the process basically to be the same in that a major portion of the tapetal cytoplasm is transferred to the surface of the young pollen grain. In *Brassica* the sequence of events appears to be more truncated than in *Raphanus* with small islets of cytoplasm containing presumably both the protein and lipoidal components moving from the locular periphery to the exine. Again these protoplasmic masses appear physically to be drawn into the cavities generated by the baculae and muri of the ectexine. Close examination with the electron microscope shows that the tryphine does not make intimate contact with the electron-opaque sporopollenin, but in all cases a narrow electron-lucent layer, some 8 nm in width separates the ectexine and the tryphine (Fig. 9.3*d*). This layer has recently been named the 'exinic outer layer' by Gaude & Dumas (1984). During the final stages of pollen development the tryphine of *Brassica* is reduced to a thick electron-opaque matrix, indistinguishable from that of *Raphanus* (Fig. 9.3*g*). Since individual vesicles may be observed in the tryphine of grains hydrated on the stigma surface, this homogeneous electron-opaque aspect clearly cannot signify that this coating is reduced to a few simple compounds, as is the case in species carrying a pollenkitt on their pollen grains (Heslop-Harrison, 1968). Further evidence that the coating may be of a somewhat more sophisticated structure comes from anhydrous fixation techniques (Elleman & Dickinson, 1985), which reveal the presence of a membrane-like layer enveloping the entire grain and lying closely appressed to the coating (Fig. 9.4).

9.4.2   Pollen coat components

Although it is clear from cytophysiological investigations that the pollen coat is essentially a lipoprotein complex, detailed information on the precise chemical nature of the constituent molecules is lacking. Analyses of organic extracts of the pollen coat reveal the presence of simple sugars in addition to lipid and protein but their role is unknown (Roggen, 1974). Fractionation of protein eluates of the coating indicates that proteins in the molecular weight range 10–25 kD are active in eliciting the callose rejection response (Heslop-Harrison *et al.* 1974). Also proteins of isoelectric points pH 8–9 are released from the lipidic matrix of the pollen coat by Triton X-100 (Roberts *et al.* 1980) but these do not appear to hybridize with stigma molecules in the same way that the stigma glycoprotein hybridizes with pollen extracts (Roberts & Dickinson, 1983). However, S-specific molecules have not been detected in pollen extracts by the techniques which revealed the presence of S-specific antigens and glycoproteins in stigma extracts (Section 9.3.2) and direct evidence of the involvement of pollen proteins in S.I. is limited to the experiments on callose elicitation performed by Heslop-Harrison *et al.* (1974). Although the immunodiffusion experiments of Nasrallah (1979) indicate that pollen receptors may be very different from stigmatic molecules, recent evidence suggests that the pollen coating does contain glycoproteins of the same charge and size as those found in the stigma (H. G. Dickinson, unpublished). Obtaining conclusive proof that these molecules are involved in the S.I. response is very difficult, particularly because there is no male parallel to 'bud pollination'.

Involvement of lipidic coat materials is apparent from the effectiveness of organic washes in overcoming the incompatibility barrier (Roggen, 1974). Although fundamental changes in seven fatty acid components of the coat lipids were not detected by GLC following cross- and self-pollination (Roberts, Stead & Dickinson, 1979*b*) the possibility remains that the lipid fraction is more than a simple seal or adhesive substance.

Fig. 9.4. Pollen wall of *Brassica oleracea* after 'dry' fixation (Elleman & Dickinson, 1985). The coating (C) appears electron lucent and the coating superficial layer (arrows) is clearly visible. ×21 300.

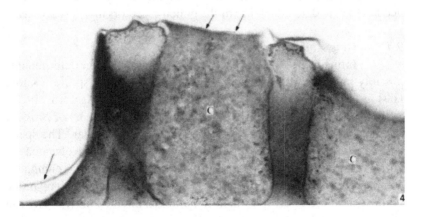

Preliminary studies using nuclear magnetic resonance techniques suggest that the lipidic matrix of the coating undergoes some change following pollination of either compatibility. So far it has, however, proved impossible to detect any qualitative or quantitative difference between cross- and self-matings (I. N. Roberts & H. G. Dickinson, unpublished). Further, taking into consideration the presence of carboyhydrate in the pollen coat (Roggen, 1974), it is tempting to suggest an involvement of glycolipid molecules in pollen–stigma interactions. The role of such molecules in cognitive interactions in bacterial, plant and animal cells has been discussed recently (Elbein, 1980) and it is conceivable that glycopolipids could well have been associated with the low molecular weight protein fraction found to be active in eliciting the callose response by Helsop-Harrison *et al.* (1974).

An involvement of glycolipids as described above would explain our inability to detect S-specific proteins in pollen extracts. An alternative explanation is that S-specific proteins are indeed present in the pollen coat but are not detected due to masking effects of coat lipids (Roggen, 1975). Equally such proteins might be present in minute amounts which are not detected in isoelectric focusing experiments, or may be masked by major pollen proteins focusing at identical pH levels.

Immunological techniques might reasonably be supposed to hold considerable promise in the investigation of pollen coat components. Certainly it is possible to identify any number of antigens in the pollen grain coating, but proving the involvement of any in the S.I. response is far more difficult. Since the S-genotypes available are far from isogenic in other respects, simple adsorption experiments are not possible (C. J. Ellemann & H. G. Dickinson, unpublished). The use of antibodies is, however, proving useful in the identification of new antigens arising following self- and cross-pollinations. Clearly much more research is needed to elucidate the nature of the pollen receptors in S.I. reactions.

9.4.3   Other components of the pollen

Intine proteins have been shown to carry antigenic determinants possibly involved in incompatibility responses in some species (Knox, Heslop-Harrison & Reed, 1970; Knox, Willing & Ashford, 1972). Although similar proteins are found in the intine (Fig. 9.3c) of *Brassica* pollen their role in pollen–stigma interactions is not clear. The speed and nature of the S.I. response argue against a direct involvement in initial recognition reactions. This critical factor of time also eliminates complex carbohydrates of the intine functioning in a manner comparable to those found to be involved in host–pathogen interactions (Albersheim *et al.* 1981). However, components of the intine must certainly be involved in later pollen–stigma interactions.

The role of the pollen plasma membrane must also be taken into account. Recent evidence suggests that the pollen plasma membrane does not exist as a functional entity in some dehydrated grains and must re-form prior to pollen germination (Heslop-Harrison, 1979*a*). Although *Brassica* possesses a plasma membrane at all times (Elleman & Dickinson, 1985) the manner of rehydration is critical for all trinucleate pollens (Bar Shalom & Mattsson, 1977). We have observed that initial tube growth may occur *in vitro* in a humid environment (Roberts *et al.* 1979*a*) and Ferrari *et al.* (1981) have reported that tube initials are consistently produced *in vitro* at R.H. 98%. They interpret this as an indication that tube emergence is triggered by a precise level of cell turgor which is achieved at this R.H. Clearly any mechanism preventing this turgor level from being reached could constitute a simple, effective barrier to self-fertilization. Thus the different rates of turgor development observed between cross- and self-pollen (Roberts *et al.* 1980; Roberts & Dickinson, 1983) may well indicate the fundamental mode of operation of the S.I. system. It should be emphasized that the achievement of turgor requires two components, one the presence of sufficient water and two, the maintenance of an effective semipermeable membrane. The S.I. system could thus have its principal action on the pollen plasma membrane, rather than on the passage of water into the grain.

9.5 Towards a molecular model

9.5.1 Lectins in plant sexual interactions

That lectins have the potential to act as 'identity tags' during short-range intercellular communication has been amply demonstrated in numerous biological systems (Bolwell, Callow, Callow & Evans, 1979; Heslop-Harrison, 1978*a*). The basis of this specificity resides in the affinity of the lectin-like molecule for a particular complementary sequence of sugars in a carbohydrate moiety which, of course, may form part of a glycoprotein. Since these interactions are adhesive in nature a simple incompatibility mechanism may be envisaged where 'illegitimate' unions are prevented by a lack of complementation inhibiting cell–cell adhesion. Indeed, as noted by Pandey (1969) 'the evolution of a primitive specificity gene involved in sexuality can be traced to the simplest of plants (including prokaryotic bacteria and algae) where mechanisms exist by which two apparently similar cells are conjugationally attracted or repelled by each other'. Pandey (1980) further considers that incompatibility systems in plants are 'extremely ancient developments derived by parallel evolution from a single specificity element which has grown in complexity and become intregrated with a wide variety of genes controlling different aspects of sexual physiology'.

It is therefore interesting that where a molecular basis for these genetic relationships has been sought glycoprotein molecules have frequently been implicated in the primary adhesive interactions. For example, mating in yeasts (Burke, Medonça-Previato & Ballou, 1980; Ballou & Pierce, this volume), flagellar adhesion in algae (Snell, 1976) and sexual development in *Volvox* (Kurn, 1981) all involve glycoprotein–lectin-like binding reactions. Similar molecules are also found to play significant roles in fertilization events in *Fucus serratus* (Bolwell *et al.* 1979), in gymnosperms (Hagman, 1975) and in angiosperm gametophytic S.I. systems (Herrero & Dickinson, 1979, 1980; Dickinson, Moriarty & Lawson, 1982).

Thus the involvement of a glycoprotein in the sporophytic S.I. system of *Brassica* is not particularly surprising. However its mode of action must be somewhat more complex than a simple failure to adhere by self-pollen, as discussed below.

9.5.2   The significance of adhesion and hydration

Ferrari *et al.* (1981) have proposed that the stigma glycoprotein blocks development of self-pollen by specifically interfering with attachment of pollen grains and tubes to female tissues. Likewise, Gaude & Dumas (1982) have suggested a model in which adhesion is regarded as directly controlling the availability of water to the grain. However, we have found that self-pollen does eventually adhere to an extent comparable to cross-pollen (Stead *et al.* 1979) but still fails to effect fertilization. Also self-compatible varieties show a similar pattern of adhesion to self-incompatible varieties on self-pollination but, despite the lack of adhesion, self-fertilization is not prevented.

For these and other reasons (Roberts *et al.* 1980) we prefer to interpret differences in adhesion as a consequence of the recognition reaction rather than the actual rejection reaction. We hypothesize that the single most important factor governing grain development is the interaction of stigmatic water with the region of the pollen coating between the papilla and the grain. The consequences of this interaction may be the denial of stigmatic water to the pollen or, more likely, the production of low levels of inhibitor within the pollen. Certainly the profound effects of raised atmospheric humidity, and protein-synthesis inhibitors (Roberts *et al.* 1984*a,b*), on pollen germination and S.I. support this conclusion (Section 9.5.5). The adhesive force described by Ferrari *et* *al.* (1981) is no doubt essential for the penetration but this must surely occur at a stage later than the initial responses.

One, albeit very simplistic, explanation for these events would be if both self-incompatible and self-compatible species of *Brassica* possessed a simple glycoprotein-based binding system which resulted in the full binding of pollen to the stigma in some three hours. This system would be essential for establishing the intimate contact necessary for pollen hydration, and the stigmatic element would be manufactured very early – such that it was present in bud stigmas. Indeed, it would be the equivalent of this system in the Caryophyllaceae that can be removed by protease (Heslop-Harrison & Heslop-Harrison, 1975) rendering the pollen incapable of entry. We would propose that this system would operate under all circumstances.

In self-incompatible varieties we suggest that another very similar glycoprotein-based binding system has evolved, but in which a part of each molecule involved contained an area which was S-specific. In cross-pollinations this system would operate in a manner identical to the nonspecific binding system described above, and would bind the pollen to the stigma. Since there is little doubt from the electrofocusing work (Roberts *et al.* 1979*a*) that at least the stigmatic part of this system is present in large quantities, it is reasonable to propose that this binding might be very rapid and effective, producing the intimacy necessary for pollen hydration.

The events following an incompatible cross remain far from clear. From both light- and electron-microscope observations (Zuberi & Dickinson, 1985; Elleman & Dickinson, 1985) it is clear that some stigmatic water moves through the pollen coat into the incompatible grain. This is evidenced not only by a change in shape of the pollen grain (Zuberi & Dickinson, 1985) but also cytoplasmic reorganization, characteristic of hydration, within the coating and the protoplast (Elleman & Dickinson, 1985). It is during this rehydration, the extent of which is closely linked both to genotype and environment, that the self-pollen becomes inhibited. The source of the inhibitor is unclear, but it may be swept through the coat into the grain, following a recognition event at the coating surface involving the second family of glycoproteins referred to earlier; alternatively it may be synthesized within the pollen protoplast itself, stimulated by an activator molecule produced as a result of recognition. The surprising observation that cycloheximide applied to the stigma can 'free' self-pollen from the S.I. system (Roberts *et al.* 1984*a*) could thus be explained were the synthesis of the stigmatic recognition glycoprotein inhibited, or the continued synthesis of inhibitor within the pollen prevented. Either alternative demands that the level of inhibitor within the grain is finely balanced between synthetic and degenerative processes. Thus prevention of synthesis will result in release of the pollen from the 'hold' of the S.I., as will dilution of the inhibitor by the supply of atmospheric water (Carter & McNeilly, 1975). We know little of the nature of the inhibitor at present, but new evidence from Hodgkin & Lyon (1984) suggests that it is possible to extract comparatively large quantities of inhibitor from selfed stigmas. The analysis of these molecules is currently under way.

9.5.3   Active versus passive responses

The majority of known biological recognition systems act to discriminate between 'self' and 'non-self' and actively to reject 'non-self'. The rejection of self-pollen by the S.I. systems is therefore an unusual phenomenon. In attempting to understand this rejection it is first necessary to distinguish between two alternative mechanisms that could be operating. For either the development of self-pollen could be actively inhibited or that of cross-pollen could be stimulated. These alternative hypotheses have been designated 'oppositional' and 'complementary' respectively (Bateman, 1952).

Extracts from the pistils of several species have now been shown to inhibit the development of self-pollen *in vitro*; in *Primula* extracts from pin styles will retard the growth of pin but not thrum pollen tubes and thrum stylar extracts have the converse effect (Golynskaya, Bashnikova & Tomchuk, 1976; Shivanna, Heslop-Harrison & Heslop-Harrison, 1981) whereas in *Lilium longiflorum*, fluid taken from the stylar canal will, when added to a normal culture medium, inhibit self-pollen tube elongation (Dickinson *et al.* 1982). These types of experiment are not easily carried out on *Brassica*, for, in common with many species with trinucleate pollen, it is particularly difficult to germinate *in vitro*. However Ferrari & Wallace (1975) have reported a series of experiments in which the development of self-pollen *in vitro* is inhibited by stigmatic extracts, work which has in many ways been amplified in later studies (Ferrari *et al.* 1981). A medium has now been developed in which pollen of *Brassica* will germinate freely (Roberts, Gaude, Harrod & Dickinson, 1983). At the time of writing a clear inhibition of pollen development by extracts of self-stigmas has yet to be obtained.

Nevertheless the very specific nature of the S.I. response, the inhibition of the response by irradiation, the fact that *Brassica* pollen will grow in our medium for considerable distances (i.e. well in excess of the distances from stigma to ovary), and indeed the discovery of inhibitors (Hodgkin & Lyon, 1984) would indicate that sporophytic systems of S.I., in common with the gametophytic, involve a mechanism by which development of self-pollen is positively inhibited.

9.5.4 *The stimulation of callose synthesis*

The callose rejection response cannot be assigned a central role in any hypothesis proposed to explain the mechanisms of S.I. It occurs at a stage following adhesion and hydration events and, although frequently present, is by no means a regular feature of the rejection of self-pollen. Grains are frequently observed to fail on a self-stigma without eliciting any apparent response in the papillar cytoplasm.

It is possible that callose deposition may represent a secondary response to self-pollination acting as a 'back-up' system where the S.I. system has failed to block pollen germination and a self-pollen tube has formed. Callose lenticules are almost without exception found where a self-pollen tube has attempted penetration of the stigmatic cuticle. Such a response is closely reminiscent of events in the Gramineae where, as part of the gametophytically controlled S.I. system, stigmatic callose is formed in response to challenge by an incompatible pollen-tube (Heslop-Harrison, 1979*b*). It is thus not impossible, particularly in view of recent findings concerning the inheritance of S.I. in *Brassica* (Zuberi, Zuberi & Lewis, 1981) that a secondary, gametophytically controlled system is activated at this point, and that the full S.I. mechanism in *Brassica* contains both sporophytically and gametophytically controlled elements. Callose has, however, also been reported to be formed when 'self'-pollen coat or tapetal extracts are applied to stigmatic papillae (Heslop-Harrison *et al.* 1974). For this reason it is also possible that callose is formed in response to 'recognition' between the large pool of S-specific glycoprotein in the papillae and sporophytic coat molecules that may find their way to the stigmatic cytoplasm, either through discontinuities in the cuticle or as a consequence of penetration by the pollen tube. An alternative, and perhaps more likely, explanation is that the callose-rejection response is merely a consequence of gametophytic proteins leaching from the inefficient self-pollen tubes (Zuberi & Dickinson, 1985). Such molecules would not be immobilized at the pellicle by adhesion events, nor would they be coordinated by normal pollen metabolism, and they might therefore be expected to elicit a typical wound response. Callose is frequently formed in plant cells as a consequence of mechanical or chemical injury and there is little doubt that the stigmatic papillae of *Brassica* are particularly well adapted for the production of this glucan (Dickinson & Lewis, 1973*a*).

9.5.5   *Environmental effects and pollen transfer experiments*

Perhaps the most significant clues to the manner by which the S.I. system of *Brassica* operates come from experiments involving alteration of the microenvironment experienced by the germinating pollen grain. Raising the atmophseric humidity has proved effective in overcoming the S.I. system (Carter & McNeilly, 1975, 1976) and pollen transferred from a self-stigma to cross-stigma will behave as in a normal cross-pollination (Kroh, 1966). These data give strong indications that in the short term the pollen cytoplasm is not universally inhibited as a consequence of self-recognition and that the main limiting factor is the inability of self-pollen to achieve the turgor pressure necessary for optimal germination, probably as a result of the action of an inhibitor. Stimulation for the synthesis of this inhibitor presumably comes from a recognition event in the coating. Indeed transfer of the pollen coat from self-pollen to cross-pollen which has had its own coating removed prevents the cross-pollen from hydrating; conversely cross-pollen coat applied to coat-less self-pollen enables this pollen to hydrate (Roberts & Dickinson, 1983). However, in all cases so far examined, and we must emphasize that these experiments are at a very early stage, coat transfer is not effective in enabling self-pollen tubes to penetrate the stigmatic papillae, whereas transfer of pollen from a 'cross' to a 'self'-stigma results in self-fertilization (Kroh, 1966). Thus the original interpretation of S.I. in terms of specific cutinase activation (Christ, 1959) retains some validity, but this must be seen as secondary to the primary rehydration responses, occurring as it does at a much later stage.

At first sight it is tempting to propose, as we have done (Roberts *et al.* 1980), that the observed rehydration responses result from a simple 'switching-off' of the water supply to self-pollen by a specifically induced change in pellicle permeability, or alternatively from the synthesis of a germination inhibitor (Ferrari & Wallace, 1977). However, we must also consider any other possible effects on self-pollen metabolism, especially since raised atmospheric $[CO_2]$ is a well known treatment which also overcomes S.I. (Nakanishii & Hinata, 1973). The stimulatory effect of high $[CO_2]$ on pollen germination and tube growth (Dhaliwal, Malik & Singh, 1981; I. N. Roberts & H. G. Dickinson, unpublished) indicates that elevated phosphoenolopyruvate (PEP) carboxylase activity resulting in the synthesis of organic acids might have pronounced effects on turgor pressure, as suggested by Dhaliwal *et al.* (1981). Thus effects of stigma molecules on self-pollen metabolism or even on the re-formation of the pollen plasma membrane could also result in the observed rehydration responses. But, since the pollen is not seriously disrupted on self-recognition and recovers on transfer to a compatible stigma, such effects on self-pollen would have to be biostatic in nature and not toxic.

9.6   Conclusions

From that information currently available we may conclude that a glycoprotein receptor situated either permanently or transitorily in the stigmatic pellicle is capable of recognizing a component of the coating superficial layer investing the self-pollen grain. This recognition event would seem to have two major direct or indirect effects: firstly, it prevents the complete binding of the grain to the stigmatic surface; and secondly, the pollen is unable to develop sufficient turgor to develop further. Data are lacking with regard to the component in the pollen coating that is recognized by the stigmatic glycoprotein, indeed it is not known whether it is a protein, glycoprotein or glycolipid. While full information on the character of the stigmatic glycoprotein has yet to become available, the first indications are that its specificity lies in the protein moiety (Nishio & Hanata, 1982).

How this recognition at the molecular level on the stigmatic surface manages to prevent pollen development is not clear. A first step in the elucidation of this problem would be to determine whether all chemical interactions remain restricted to the pellicle/pollen coat interface, or whether molecular hybrids formed from pollen and stigma receptors (or their immediate products) find their way into the main body of the pollen coat – or indeed into the pollen grain itself. Such experiments are, surprisingly, very difficult to carry out with any accuracy, but results from work involving light- and electron-microscope autoradiography point to some of the molecular hybrids remaining immobile (Roberts *et al.* 1984*b*). This type of result, once fully confirmed, would point to the recognition event stimulating the production of a mobile secondary message which may either act directly as an inhibitor, or stimulate the production of one in the pollen protolast. The third possibility also remains that, as is the case for some host–pathogen interactions, recognition produces a message which travels to the stigma plasma membrane and stimulates the production of a phytoalexin-like inhibitor.

9.7   References

Albersheim, P., Darvill, A. G., McNeil, M., Valent, B. S., Hahn, M. G., Lyon, G., Sharp, J. K., Desjardins, A. E., Spellman, M. W., Ross, L. M., Robertsen, B. K., Aman, P. & Franzen, L-E. (1981). Structure and function of complex carbohydrates active in regulating plant–microbe interactions. *Pure and Appl. Chem.* **53**: 79–88.
Ameele, R. J. (1982). The transmitting tract in *Gladiolus*. I. The stigma and the pollen–stigma interaction. *Am. J. Bot.* **69**: 389–401.

Bar Shalom, D. & Mattsson, O. (1977). Mode of hydration as an important factor in the germination of trinucleate pollen grains. *Bot. Tiddskr.* **71:** 245–51.

Bateman, A. J. (1952). Self-incompatibility systems in angiosperms. I. Theory. *Heredity* **6:** 285–310.

Bolwell, G. P., Callow, J. A., Callow, M. E. & Evans, L. V. (1979). Fertilisation in brown algae. II. Evidence for lectin-sensitive complementary receptors involved in gamete recognition in *Fucus serratus. J. Cell Sci.* **36:** 19–30.

Brewbaker, J. L. (1957). Pollen cytology and incompatibility systems in plants. *J. Hered.* **48:** 271–7.

Burke, D., Mendonça-Previato, L. & Ballou, C. E. (1980). Cell–cell recognition in yeast: purification of *Hansenula wingei* 21-cell sexual agglutination factor and comparison of the factors from three genera. *Proc. Natl. Acad. Sci. U.S.A.* **77:** 312–22.

Carpita, N., Sabularse, D., Montezinos, D. & Delmer, D. P. (1979). Determination of the pore size of cell walls of living plant cells. *Science* **205:** 1144–7.

Carter, A. L. & McNeilly, T. (1975). Effects of increased humidity on pollen growth and seed set following self-pollination in Brussels sprout (*Brassica oleracea* var. *gemmifera*). *Euphytica* **24:** 805–13.

Carter, A. L. & McNeilly, T. (1976). Increased atmospheric humidity post-pollination: a possible aid to the production of inbred line seed from mature flowers in the Brussels sprout (*Brassica olerace* var. *gemmifera*). *Euphytica* **25:** 531–6.

Christ, B. (1959). Entwicklungsgeschichtliche und physiologische Untersuchungen uber die Selsterilität von *Cardamine pratensis* L. *Z. Bot.* **47:** 88–112.

Clarke, A. E., Abbot, A., Mandel, T. E. & Pettitt, J. M. (1980). Organisation of the wall layers of the stigmatic papillae of *Gladiolus gandavensis:* a freeze-fracture study. *J. Ultrastruct. Res.* **73:** 269–81.

Clarke, A. E., Gleeson, P., Harrison, S. & Knox, R. B. (1979). Pollen–stigma interactions: identification and characterisation of surface components with recognition potential. *Proc. Natl. Acad. Sci. U.S.A.* **76:** 3358–62.

Dhaliwal, A. S., Malik, C. P. & Singh, M. B. (1981). Overcoming incompatibility in *Brassica compestris* L. by carbon dioxide, and dark fixation of the gas by self- and cross-pollinated pistils. *Ann. Bot.* **48:** 227–33.

Dickinson, H. G. & Lewis, D. (1973a). Cytochemical and ultrastructural differences between intraspecific compatible and incompatible pollinations in *Raphanus. Proc. R. Soc. Lond. B.* **183:** 21–38.

Dickinson, H. G. & Lewis, D. (1973b). The formation of the Tryphine coating the pollen grains of *Raphanus* and its properties relating to the self-incompatibility system. *Proc. R. Soc. Lond. B* **184:** 149–65.

Dickinson, H.G. & Lewis, D. (1975). Interaction between the pollen grain coating and the stigmatic surface during compatible and incompatible intraspecific pollinations in *Raphanus*. In *The Biology of the Male Gamete*, ed. J. G. Duckett & P. A. Racey, pp. 165–75. (Suppl. 1. Biol. J. Linn. Soc.)

Dickinson, H. G., Moriarty, J. F. & Lawson, J. R. (1982). Pollen–pistil interaction in *Lilium longiflorum:* the role of the pistil in controlling pollen tube growth following cross- and self-pollinations. *Proc. R. Soc. Lond. B* **215:** 45–62.

East, E. M. (1940). The distribution of self-sterility in the flowering plants. *Proc. Amer. Phil. Soc.* **82:** 449–518.

Echlin, P. (1971). The role of the tapetum during microsporogenesis of Angiosperms. In *Pollen Development and Physiology*, ed. J. Heslop-Harrison, pp. 41–62. London: Butterworth.

Elbein, A. D. (1980). Glycolipids pp. 571–587 In *The Biochemistry of Plants*, vol. 3, ed. P. K. Stumpf & E. E. Conn, pp. 571–87. New York: Academic Press.

Elleman, C. J. & Dickinson, H. G. (1985). Pollen–stigma interactions in *Brassica* IV: first events at the cell surface following compatible and incompatible matings *J. Cell Sci.* (in the Press).

Ferrari, T. E., Bruns, D. & Wallace, D. H. (1981). Isolation of a plant glycoprotein involved with control of intercellular recognition. *Plant Physiol.* **67**: 270–7.

Ferrari, T. E., Lee, S. S. & Wallace, D. H. (1981). Biochemistry and physiology of recognition in pollen–stigma interactions. *Phytopathology* **71**: 752–5.

Ferrari, T. E. & Wallace, D. H. (1975). Germination of *Brassica* pollen and expression of incompatibility *in vitro*. *Euphytica* **24**: 757–65.

Ferrari, T. E. & Wallace, D. H. (1977). A model for self-recognition and regulation of the incompatibility response of pollen. *Theor. Appl. Genet.* **50**: 211–25.

Gaude, T. C. & Dumas, C. (1982). Stigma–pollen recognition: importance of adhesion and hydration of the pollen grain in the incompatibility response. *Incompatibility Newsletter Assoc. EURATOM-ITAL, Wageningen* **14**: 7–16.

Gaude, T. C. & Dumas, C. (1984). A membrane-like structure on the pollen wall surface in *Brassica*. *Annals of Botany* **54**: 821–5.

Gilissen, L. J. W. (1977). The influence of relative humidity on the swelling of pollen grains *in vitro*. *Planta* **137**: 299–301.

Golynskaya, E. L., Bashnikova, N. V. & Tomchuk, N. N. (1976). Phytohaemagglutinins from the pistil of *Primula* as possible proteins of generative incompatibility. *Soviet Pl. Physiol.* **23(1)**: 69–76.

Hagman, M. (1975). Incompatibility in forest trees. *Proc. R. Soc. London.* B **188**: 313–26.

Herrero, M. & Dickinson, H. G. (1979). Pollen–pistil incompatibility in *Petunia hybrida*. Changes in the pistil following compatible and incompatible intraspecific crosses. *J. Cell Sci.* **36**: 1–18.

Herrero, M. & Dickinson, H. G. (1980). Pollen tube growth following compatible and incompatible intraspecific pollinations in *Petunia hybrida*. *Planta* **148**: 217–21.

Heslop-Harrison, J. (1967). Ribosomes sites and S gene action. *Nature* **218**: 90–1.

Heslop-Harrison, J. (1968). Tapetal origin of pollen coat substances in *Lilium*. *New Phytol.* **67**: 779–86.

Heslop-Harrison, J. (1978*a*). Recognition and response in the pollen stigma interaction. In *Cell–Cell Recognition*, pp. 121–38. (Symposia of the Society for Experimental Biology No. XXXII, Cambridge University Press).

Heslop-Harrison, J. (1978*b*). *Cellular recognition systems in plants. Studies in Biology, 100.* London: Edward Arnold.

Heslop-Harrison, J. (1979*a*). An interpretation of the hydrodynamics of pollen. *Amer. J. Botan.* **66**: 737–43.

Heslop-Harrison, J. (1979*b*). Aspects of the structure, cytochemistry and germination of the pollen of rye (*Secale cereale* L.). *Ann. Bot.* **44**: SI, 1–47.

Heslop-Harrison, J. & Heslop-Harrison, Y. (1975). Enzymic removal of the proteinaceous pellicle of the stigma papilla prevents pollen tube entry in the Caryophyllaceae. *Ann. Bot.* **39**: 163–5.

Heslop-Harrison, J., Hesop-Harrison, Y. & Barber, J. (1975). The stigma surface in incompatibility responses. *Proc. Roy. Soc. Lond.* B **188,** 287–97.

Heslop-Harrison, J., Knox, R. B. & Heslop-Harrison, Y. (1974). Pollen wall proteins: exine-held fractions associated with the incompatibility response in Cruciferae. *Theor. Appl. Genet.* **44**: 133–7.

Heslop-Harrison, Y. & Shivanna, K. R. (1977). The receptive surface of the angiosperm stigma. *Ann. Bot.* **41**: 1233–58.

Hinata, K. (1981). Self-incompatibility: physiological aspects for the breeding of cruciferous vegetables. In *Chinese cabbage (Proc. First. Int. Symp.)*, ed. N. S. Talekov & T. D. Griggs, pp. 321–34. Taiwan, China: AVRDC.

Hinata, K., Nishio, T., Kimura, J., (1982). Comparative studies on S-glycoproteins purified from different S-genotypes in self-incompatible *Brassica* species. II. Immunological specificities. *Genetics* **100**: 649–57.

Hodgkin, T. & Lyon, G. D. (1984). Pollen germination inhibitors in extracts of *Brassica oleracea* (L.) stigmas. *New Phytol.* **96**: 293–8.

Kehoas, C., Knox, R. B. & Dumas, C. (1983). Specificity of the callose response in stigmas of *Brassica*. *Annals of Botany* 52: 597–602.

Klein, J. (1971). Private and public antigens of the mouse H-2 system. *Nature* 229: 635–7.

Knox, R. B. & Heslop–Harrison, J. (1971). Pollen wall proteins: localisation of antigenic and allergenic proteins in the pollen grain walls of *Ambrosia* spp. (ragweeds). *Cytobios* 4: 49–54.

Knox, R. B., Heslop-Harrison, J. & Reed, J. (1970). Localisation of antigens associated with the pollen grain wall by immuno-fluorescence. *Nature* 225: 1066–8.

Knox, R. B., Willing, R. R. & Ashford, A. E. (1972). Role of pollen wall proteins as recognition substances in interspecific incompatibility in poplars. *Nature* 237: 381–3.

Kroh, M. (1966). Reaction of pollen after transfer from one stigma to another. *Der Zuchter* 36: 185–9.

Kurn, N. (1981). Altered development of the multicellular alga *Volvox carteri* caused by lectin binding. *Cell Biol. Int. Reports* 5: 867–75.

Lewis, D. (1952). Serological reactions of pollen incompatibility substances. *Proc. Roy. Soc. Lond. Series B* 140: 127–35.

Linskens, H. F. & Heinen, W. (1962). Cutinase-Nachweis in pollen. *S. Botan.* 50: 338–47.

Mann, S. L., Raff, J. & Clarke, A. E. (1982). Isolation and partial characterisation of components of *Prunus avium* (L.) styles, including an antigenic protein associated with a self-incompatibility genotype. *Planta* 156: 505–16.

Mattsson, O., Knox, R. B., Heslop-Harrison, J. & Heslop-Harrison, Y. (1974). Protein pellicle of stigmatic papillae as a probable recognition site in incompatibility reactions. *Nature* 213: 703–4.

Murabaa, A. I. M. el (1957). Factors affecting seed set in Brussels sprouts, radish and cyclamen. *Meded. Ladbouwhogesch. Wageningen* 57: 1–33

Nakanishii, T. & Hinata, K. (1973). An effect time for CO_2 gas treatment in overcoming self-incompatibility in *Brassica*. *Plant Cell Physiol.* 14: 873–9.

Nasrallah, M. E. (1979). Self-incompatibility antigens and S-gene expression in *Brassica*. *Heredity* 43: 259–63.

Nasrallah, M. E., Barber, J. T. & Wallace, D. H. (1970). Self-incompatibility proteins in plants: detection, genetics and possible mode of action. *Heredity* 25: 23–7.

Nishio, T. & Hinata, K. (1977). Analysis of S-specific proteins in stigma of *Brassica oleracea* L. by isoelectric focussing. *Heredity* 38: 391–6.

Nishio, T. & Hinata, K. (1979). Purification of an S-specific glycoprotein in self-incompatible *Brassica campestris* L. *Japan J. Genetics* 54: 307–11.

Nishio, T. & Hinata, K. (1980). Rapid detection of S-glycoproteins of self-incompatible crucifers using Con A reaction. *Euphytica* 29.

Nishio, T. & Hinata, K. (1982). Comparative studies on S-glycoproteins purified from different S-genotypes in self-incompatible *Brassica* species. I. Purification and chemical properties. *Genetics* 100: 641–7.

Ockendon, D. J. (1972). Pollen tube growth and site of incompatibility in *Brassica*. *New Phytol.* 71: 519–22.

Pandey, K. K. (1960). Evolution of gametophytic and sporophytic systems of self-incompatibility in angiosperms. *Evolution* 14: 98–115.

Pandey, K. K. (1969). Elements of the S-gene complex. V. Interspecific cross-compatibility relationships and the theory of evolution of the S-complex. *Genetica* 40: 447–74.

Pandey, K. K. (1980). Evolution of incompatibility systems, in plants: Origin of 'independent' and 'complementary' control of incompatibility in angiosperms. *New Phytol.* 84: 381–400.

Roberts, I. N. (1981). Pollen-stigma interactions in *Brassica oleracea*. Ph.D. Thesis, University of Reading.

Roberts, I. N. & Dickinson, H. G. (1983). Intraspecific incompatibility on the stigma of *Brassica* sp. *Phytomorphology* **31:** 165–74.

Roberts, I. N., Gaude, T. C., Harrod, G. & Dickinson, H. G. (1983). Pollen–stigma interactions in *Brassica oleracea*. A new pollen germination medium and its use in elucidating the mechanism of self-incompatibility. *Theor. appl. Genet.* **65:** 231–8.

Roberts, I. N., Harrod, G. & Dickinson, H. G. (1984a). I. The ultrastructure and physiology of the stigmatic papillar cells. *J. Cell Sci.* **66:** 241–53.

Roberts, I. N., Harrod, G. & Dickinson, H. G. (1984b). II. Pollen–stigma interaction in *Brassica*. The effect of stigma/surface proteins following pollination and their role in the self-incompatibility response. *J. Cell Sci.* **66,** 255–64.

Roberts, I. N., Stead, A. D. & Dickinson, H. G. (1979b). No fundamental changes in lipids of the pollen grain coating of *Brassica oleracea* following either self- or cross-pollinations. *Incompatibility Newslett.* **11:** 77–83.

Roberts, I. N., Stead, A. D., Ockendon, D. J. & Dickinson, H. G. (1979a). A glycoprotein associated with the acquisition of the self-incompatibility system by maturing stigmas of *Brassica oleracea*. *Planta* **146:** 179–83.

Roberts, I. N., Stead, A. D., Ockendon, D. J. & Dickinson, H. G. (1980). Pollen–stigma interactions in *Brassica oleracea*. *Theor. Appl. Genet.* **58:** 241–6.

Roggen, H. P. (1974). Pollen washing influences (in) compatibility in *Brassica oleracea* varieties. In *Fertilisation in Higher Plants*, ed. H. F. Linskens, pp. 273–8. Amsterdam: North-Holland Publishing Company.

Roggen, H. P. (1975). Stigma application of an extract from rape pollen (*Brassica napus* L.) effects on self-incompatibility in Brussels sprout (*Brassica oleracea* var. *gemmifera*). *Incomp. Newslett. Assoc.EURATOM-ITAL, Wageningen* **6:** 80–4.

Sedgely, M. (1974). Assessment of serological techniques for S-allele identification in *Brassica oleracea*. *Euphytica* **23:** 543–51.

Shivanna, K. R., Heslop-Harrison, Y. & Heslop-Harrison, J. (1978). The pollen–stigma interaction: bud pollination in the Cruciferae. *Acta Bot. Neerl.* **27:** 107–19.

Shivanna, K. R., Heslop-Harrison, J. & Heslop-Harrison, Y. (1981). Heterostyly in *Primula* 2. Sites of pollen inhibition, and effects of pistil constituents on compatible and incompatible pollen tube growth. *Protoplasma* **107:** 319–37.

Snell, W. J. (1976). Mating in *Chlamydomonas:* a system for the study of specific cell adhesion. I. Ultrastructural and electrophoretic analyses of flagellar surface components involved in adhesion. *J. Cell Biol.* **68:** 48–69.

Stead, A. D., Roberts, I. N. & Dickinson, H. G. (1979). Pollen–pistil interactions in *Brassica oleracea:* events prior to pollen germination. *Planta* **146:** 211–16.

Stead, A. D., Roberts, I. N. & Dickinson, H. G. (1980). Pollen–stigma interactions in *Brassica oleracea:* the role of stigmatic proteins in pollen grain adhesion. *J. Cell Sci.* **42:** 417–23.

Stevens, V. A. M. & Murray, B. G. (1982). Studies on heteromorphic incompatibility systems: physiological aspects of the incompatibility system of *Primula obconica*. *Theor. Appl. Genet.* **61:** 245–56.

Thompson, K. F. (1957). Self-incompatibility in marrow-stem kale, *Brassica oleracea* var. *acephala*. I. Demonstration of sporophytic system. *J. Genet.* **55:** 45–60.

Whitehouse, H. L. K. (1950). Multiple-allelomorph incompatibility of pollen and style in the evolution of the angiosperms. *Ann. Bot., New Series* **14:** 198–216.

Zuberi, M. I. & Dickinson, H. G. (1985). Pollen–stigma interaction in *Brassica III*, hydration of the pollen grains. *J. Cell Sci.* (In the Press).

Zuberi, M. I., Zuberi, S., Lewis, D. (1981). The genetics of incompatibility in *Brassica*. I. Inheritance of self-incompatibility in *Brassica campestris* L. var *Toria*. *Heredity* **46:** 175–90.

10

Receptors in host–pathogen interactions

Julie E. Ralton, Barbara J. Howlett and Adrienne E. Clarke

10.1 Introduction

Most microorganisms do not cause plant disease as most plants have evolved powerful mechanisms for preventing colonization. The relationships between pathogenic microorganisms and their host plants are diverse – ranging from pathogens which can attack a wide range of plants, to those which have a limited host range, or are confined to a single plant species. Pathogen specificity in host–pathogen interactions may be described at two levels (Heath, 1981):

(i) Plant species specificity, which determines the host species range of a pathogen. The basic distinction between host and non-host species is thought to be determined by whether a pathogen can overcome a range of pre-formed barriers and nonspecific defence reactions of the plant.

(ii) Host cultivar specificity, which determines the cultivar range within a given host species. Once a plant species has become host to a pathogen, there is continuing evolution of the interacting organisms so that detrimental effects on the host are offset by effects which limit the success of the pathogen. The host is under selection pressure to evolve resistance to the pathogen and this leads to the emergence of resistant cultivars. A subsequent mutation of the pathogen which enables it to colonize the host cultivar leads to establishment of a new pathogen race. These continuing genetically based interactions are superimposed on the basic susceptibility of a plant species to a pathogen, and may be described by the gene-for-gene concept, which states that for every gene for resistance in the host there is a corresponding gene for avirulence in the pathogen. Many specific interactions between specialized pathogens (including viruses, bacteria and fungi) and host plants can be described in terms of this hypothesis (Table 10.1), which was first proposed by Flor in 1955.

The resistance genes in plants (and avirulence genes in pathogens) are, with a few exceptions, inherited as dominant characters. However while explaining many specialized plant–pathogen relationships, the gene-for-gene theory may not be representative of all host–pathogen interactions (Day, Barrett & Wolfe, 1983). Generally, the systems it describes have been developed by classical plant breeding aimed at developing disease-resistant cultivars of agriculturally important plants by including resistance genes originating from different, but related, species of the crop plant. The hypothesis has not yet been shown to apply to any natural host–pathogen interaction involving a non-cultivated host plant (Day *et al.* 1983). The data are often limited to genetic analyses of the host and do not include direct analysis of the pathogen. Another limitation is that the relationships are usually determined by assessing symptom expression in the host plant, and in categorizing host plants as either 'susceptible' or 'resistant' many subtle interactions between host and pathogen may be overlooked (Bailey, 1983).

The genetic control in gene-for-gene relationships (for review see Keen, 1982) is used as a basis for consideration of the recognition events involved (for reviews see Bushnell & Rowell, 1981; Heath, 1981; Ellingboe, 1981; Vanderplank, 1982; Keen, 1982). Specific interactions between gene products of the resistance and avirulence genes are implied by the hypothesis, and these are presumed to be executed at the level of protein, being the direct gene products, or secondary gene products such as the saccharide components of glycoconjugates (Ellingboe, 1981). It is also possible, theoretically, that the exchange of information could take place at DNA level, with direct movement of DNA between the host and pathogen. The classic example of DNA transfer is the transfer of a segment of DNA (the T region) of the Ti plasmid of the pathogenic bacterium, *Agrobacterium tumefaciens*, to the host nuclear genome (Caplan *et al.* 1983). There is however, no evidence that these types of interaction occur in fungal pathogen–host interactions (Gilchrist & Yoder, 1984). The interactions, leading to the expression of specificity, presumably involve gene products of both the interacting partners in one form or another, either as receptors or signals, or both. These putative receptors and signals have not been identified for fungal diseases (except in a few diseases involving specific toxins), and we have no understanding of their nature that in any way approaches that of receptors in animal systems such as the asialoglycoprotein (Tanabe, Pricer

Table 10.1. *Examples of host–pathogen systems which can be described in terms of the gene-for-gene hypothesis*

Pathogen type	Host–Pathogen system
Viruses	Tobacco mosaic virus – tomato
Bacteria	*Pseudomonas syringae* – soybean
Oomycetes	*Phytophthora infestans* – potato
Ascomycetes	*Erysiphe graminis hordei* – barley
Deuteromycetes	*Cladosporium fulvum* – tomato
Basidiomycetes	*Melampsora lini* – flax
	Puccinia graminis avenae – oats
	Ustilago avenae – oats
	Tilletia caries – wheat

Adapted from Vanderplank (1982).

& Ashwell, 1979), insulin (Jacobs, Hazum, Schechter & Cuatrecasas, 1979) and epidermal growth factor receptors (Hunter, 1984). A great deal of effort is currently being made in an attempt to understand the molecular basis of specificity in plant pathology. If the genes involved can be identified, it is quite possible that they could be moved from one plant to another to create desirable cultivars having specific disease resistance. Genes for storage proteins of corn, *Zea mays*, have been transferred into sunflower cells via the Ti plasmid (Matzke *et al.* 1984), so that genetic engineering of plants using recombinant DNA technology is now a reality. The availability of this technology and the need to understand the disease process in detail (as background for rational design of ecologically acceptable disease control measures) are powerful incentives for studies in this field.

In this chapter we will primarily discuss fungal pathogenesis, focusing on the current ideas for the basis of specificity in the gene-for-gene interactions. Bacterial disease of plants is not as economically important as fungal disease; however, bacteria are easier to manipulate in the laboratory and we also refer to work with bacterial pathogens. We will not discuss viruses or viroids; the subject has recently been reviewed comprehensively by Gould & Symons (1983).

10.2 Fungal pathogens

Most economically important fungal pathogens come from four taxonomic classes: Oomycetes, Ascomycetes, Deuteromycetes and Basidiomycetes. They vary widely in the type of plant tissue they infect, their mode of penetration and their nutritional requirements. This diversity is reflected in the variety of ways plants are affected by fungal infec-

Table 10.2. *Examples of diseases caused by fungal pathogens*

Disease	Example of pathogen	Host	Organ or tissue infected	Main process disrupted
Root rot	*Phytophthora megasperma* f.sp. *glycinea*	soybean	root	Absorption of water and nutrients
Stem rot	*Phytophthora vignae*	cowpea	stems	Translocation of water and nutrients to the crown
Vascular wilts	*Fusarium solani* f.sp. *pisi*	pea	xylem	Translocation of water and nutrients to the crown
Powdery mildew	*Erysiphe graminis* f.sp. *hordei*	barley	leaves	Photosynthesis
Leaf rust	*Puccinia graminis* f.sp. *avenae*	oats	leaves	Photosynthesis
Covered smut or bunt	*Tilletia caries*	wheat	flower	Reproductive capacity

tion (Table 10.2). The terminology used to describe the host–pathogen interactions is summarized in Table 10.3.

Pathogenic fungi are usually classed within two broad categories based on the physiology of the host–fungus interaction: Biotrophic pathogens take nutrients only from living cells and hence establish a highly specialized relationship with the host plant. Some remain on the plant surface, or grow intercellularly, extending haustoria (modified feeding hyphae) into the host cells to absorb nutrients. These pathogens co-evolved with their plant hosts (Bailey, 1983), that is, they form gene-for-gene relationships, and generally will not grow in the absence of the host. They may not kill the host cells until several days after infection and can redirect host metabolism to suit their nutritional requirements. Examples of this group are the rusts, smuts and the powdery and downy mildews.

On the other hand, necrotrophic pathogens derive nutrients from dead cells. The host cell either does not survive long after hyphal invasion, or it is killed in advance of penetration by toxins or enzymes released by the fungus. Necrotrophic fungi generally do not form specialized relationships with the host cells. Pathogens in this category include soft rots of fruit, leaf spots, root- and stem-rots (for general introductory

Table 10.3. *Terminology used to describe host–pathogen cellular interactions between biotrophic pathogens and their hosts*

Host	Pathogen	Interaction
Resistant	Avirulent	Incompatible (low infection type)
Susceptible	Virulent	Compatible (high infection type)

An interaction between a resistant host and an avirulent pathogen is said to be incompatible (not successful), and an interaction between a susceptible host and a virulent pathogen is said to be compatible (successful). This terminology does not apply to interactions with necrotrophic pathogens, as virulent necrotrophs form relationships which are incompatible in cellular terms. The terms 'high and low infection type' apply to both biotrophic and necrotrophic pathogens.

plant pathology see Dickinson & Lucas, 1983). The division between biotrophs and necrotrophs is not clear-cut, as some pathogens can exhibit both characteristics in a single host interaction. The physiological requirements of the pathogen and the mechanisms it has evolved to penetrate and grow within the host tissue are interdependent; the success or failure of infection depends on the host response. If the host is sensitive to enzymes and toxins of necrotrophs or remains alive after invasion by the biotroph, then a pathogenic relationship may be established. However, if the host plant responds so that the required relationship is not established, the association is non-pathogenic.

There is no single response which correlates exclusively with resistance, and responses of one plant may vary depending upon the type of pathogen involved. The responses can be treated at three levels: (i) tissue; (ii) cellular (including cell wall); and (iii) cytoplasmic.

(i) *Tissue responses*: Examples of this type of response are the formation of cork layers to contain the pathogen in a limited area, and the formation of abscission layers, particularly in young leaf tissue, which results in the infected area being isolated and discarded from the plant. Such responses involve the dedifferentiation of plant cells to form a layer of meristematic cells surrounding the area of pathogen invasion. Phenolic derivatives, such as lignin and suberin, usually accumulate in the walls of cells newly formed from the meristem, creating a barrier to the pathogen (Beckman, 1980; Aist, 1983).

Many pathogens which invade the vascular tissue of the host (e.g. *Fusarium* sp., *Verticillium* sp.) induce the formation of tyloses. These are outgrowths of adjacent living parenchyma cells which protrude into xylem vessels through pits in the vessel walls, clogging the vessel. If tyloses form abundantly and ahead of the pathogen, they may stop the spread of the pathogen throughout the vascular system (Beckman & Talboys, 1981). Gums also accumulate within the cells and in intercellular spaces in response to infection, possibly to localize the area of infection (Beckman, 1980; Beckman & Talboys, 1981). Phenolic compounds are also components of some secreted gums (Beckman & Talboys, 1981).

These responses are part of the general wound response of the plant, and there is no unequivocal information on their role as resistance mechanisms, or on the recognition events leading to their formation.

(ii) *Cellular* (including cell wall): Alteration in the appearance of the cell wall is often an early response of plant cells to invasion by a pathogen (for reviews, see Heath, 1980; Beckman, 1980; Aist, 1983). One response is the incorporation of phenolic derivatives, such as lignin and suberin into the cell wall, leading to increased mechanical strength and perhaps decreased permeability of the wall, or decreased susceptibility to fungal enzymes (Vance, Kirk & Sherwood, 1980; Ride, 1980; Kolattukudy, 1981, 1984). This response is also elicited by wounding, but there are cases where lignification occurs to a greater degree after infection by a non-pathogen, or an avirulent pathogen race (Hammerschmidt, 1984; Hammerschmidt, Lamport & Muldoon, 1984). Lignification can also be induced by defined components of host or pathogen origin (Pearce & Ride, 1980, 1982) suggesting that mechanisms exist to regulate the lignification response to a pathogen.

The level of hydroxyproline-rich glycoprotein (HPRGP) also changes in some plants in response to fungal invasion (Clarke, Lisker, Lamport & Ellingboe, 1981; Toppan, Roby & Esquerre-Tugaye, 1982; Kratka & Kudela, 1984; Hammerschmidt *et al.* 1984). An increase in HPRGP is associated with resistance in one system (Hammerschmidt *et al.* 1984), but the general role of HPRGP in response to infection is not known; it may function in cell wall growth (Touze & Esquerre-Tugaye, 1982) or act as a matrix for lignification (Whitmore, 1978).

Localized cell wall modifications are another common response to infection. Wall modifications result from (i) alteration of the existing cell wall to form 'haloes' around the site of pathogen penetration – these are usually visualized by changes in the staining properties of the wall (Russo & Pappelis, 1981; Sargent & Gay, 1977), and (ii) deposition of material between the plasma membrane and the existing cell wall to form thickened appositions, or 'papillae' (Aist, 1976, 1983). These responses are usually preceded by cytoplasmic aggregation adjacent to the area of pathogen penetration (Aist, 1983).

The composition of 'halo regions' is diverse; silicon deposits (Heath, 1980; Kunoh & Ishizaki, 1975; Sargent & Gay, 1977) and lipid material (Sargent & Gay, 1977) have been reported. Papillae also vary in composition. The most commonly identified component is callose, which is a β-glucan with predominantly (1,3)-linkages, although (1,4)-linkages may also be present. It is detected by its fluorescence after staining with decolorized aniline blue (Eschrich & Currier, 1964). Some papillae contain lignin, cellulose and silicon (Aist, 1976, 1983). Papilla formation is correlated with non-host resistance (Heath, 1977; Hinch & Clarke, 1982; Sherwood & Vance, 1976) and is also associated with gene-for-gene interactions (Stossel, Lazarovits & Ward, 1981; Bird & Ride, 1981; Skou, Helms Jorgensen & Lilholt, 1984). The evidence that wall modifications form part of the plants' resistance mechanism is based on the correlation between resistance (or penetration failure) and the presence of haloes or papillae (Aist, 1983). Although these modifications may form in both susceptible and resistant hosts, papillae are not involved in at least one interaction (*Olpidium brassicae* and Kohlrabi root hairs (Aist & Israel, 1977)). The effectiveness of papillae as infection barriers appears to depend upon their frequency, size and speed of development (Skou *et al.* 1984; Hachler & Hohl, 1984).

Papillae may function as mechanical barriers to penetration; the acoustic reflectance (and hence the viscosity, density and elasticity) of preformed barley papillae which are resistant to fungal penetration is greater than that of papillae susceptible to penetration (Israel, Wilson, Aist & Kunoh, 1980). The failure of intracellular hyphae (e.g. haustoria) to develop fully in incompatible host cells is frequently associated with their complete encasement by papilla-like structures. The papillae may therefore play a role in preventing haustorial development although it is possible that the intracellular hyphae are already dying and that encasement is the host cell response to the presence of necrotic hyphae (Hickey & Coffey, 1980), It has also been proposed that papillae act as a permeability barrier between host and fungus. In one study, papillae formed in barley in response to *Erysiphe graminis* f.sp. *hordei*, were permeable to ions and molecules up to 416 daltons, but not above 500 daltons (Smart, Aist & Israel, 1984).

Little is known about the molecular basis of halo or papilla induction, although, as for lignification, there is some evidence for regulation by material of both plant and fungal origin. Several lines of evidence suggest that papilla formation is modulated by Ca^{2+} (Aist, 1983; Tighe & Heath, 1982).

(iii) *Cytoplasmic*: The hypersensitive response (HR) is a major cytoplasmic response of plant cells to invasion by microorganisms. Hypersensitivity means that the plant cells are more than normally sensitive, and rapidly become necrotic following contact with the potential pathogen. It was first described as a resistance mechanism of wheat to stem rust (Stakman, 1915). By sacrificing a few cells, the plant effectively deprives the invading biotrophic fungus of contact with living tissue, thus leading to a low infection type or failure of pathogenesis (incompatible interaction). However, this would not necessarily be an effective response against necrotrophic pathogens. The reaction is usually localized to a small number of cells in the immediate vicinity of the penetrating pathogen and is often associated with changes in the permeability of the cell membrane, leading to loss of water and electrolytes from the cell (Kiraly, 1980). Respiration rates, polyphenol oxidase and peroxidase activities also increase and there are associated increases in phenol and flavanoid compounds which lead to dark pigmentation of the infected tissue (Kiraly, 1980).

Hypersensitivity coincides with resistance in a number of systems; both in non-host responses, for example in the interaction of *Acacia pulchella* with *Phytophthora cinnamomi* (Tippett & Malajczuk, 1979), and as a response in gene-for-gene interactions such as that between potato (*Solanum tuberosum*) and *Phytophthora infestans* (Yamamoto, 1982). In the latter case hypersensitivity is influenced by the host R-genes. In the incompatible interaction, cytoplasmic streaming stops within 30 min of penetration and by 24 h the invaded cells have brown granular cytoplasm. In contrast, potato cells infected by compatible races of the fungus coexist with the pathogen for at least three days (Yamamoto, 1982). A major problem in assessing the contribution of HR to resistance has been the difficulty of defining cell death (Heath, 1976; Bailey, 1983); it is also difficult to establish at what stage during the HR the pathogen dies, if at all (Bailey, 1983). There are systems where HR is apparently not the primary determinant of resistance, and, like papilla formation, it is possible that HR is a consequence, rather than the cause, of host resistance (Kiraly, Barna & Ersek, 1972; Mayama, Daly, Rehfeld & Daly, 1975a; Mayama, Rehfeld & Daly, 1975b). The contribution of HR to resistance probably varies with each host–pathogen interaction, and may act in concert with other responses, such as the formation of structural barriers (Vance et al. 1980) or phytoalexin production.

Phytoalexins are low molecular weight antimicrobial compounds, synthesized by plants, which accumulate in localized areas of plant tissue following pathogen invasion (Paxton, 1981; Deverall, 1982). Phytoalexins, their mode of elicitation and their role as a defence mechanism have been reviewed extensively (Bailey & Mansfield, 1982; Kuc, 1976; derable structural diversity between different plant families, the phyto-alexins produced by members of a single family may be closely related, lexins produced by members of a single family may be closely related, e.g. isoflavanoid compounds produced by the Leguminoseae, and sesqui-terpenes by the Solanaceae. Phytoalexins are frequently associated with HR as they accumulate in the dead (or dying) host cells. As well as being toxic to microbial pathogens, they are phytotoxic; they are thought to be synthesized in adjacent healthy cells and transported by unknown mechanisms to the area around the invading pathogen (Keen, 1982). Although phytoalexins can accumulate in both resistant and susceptible necrotic tissues, resistance occurs when they accumulate rapidly and are present locally in concentrations high enough to retard pathogen growth (Bailey, 1983). As with hypersensitivity, there is no conclusive evidence that phytoalexins are lethal to the invading pathogen. They may act by halting further fungal growth (Yoshikawa, Yamanchi & Masago, 1978*a*).

There are several proposed mechanisms by which the pathogen may overcome the detrimental effects of HR and phytoalexin accumulation (Darvill & Albersheim, 1984). Some fungal pathogens are virulent because they have mechanisms to detoxify or tolerate certain phytoalexins (Matthews & Van Etten, 1983; Denny & Van Etten, 1983), thus render-ing a host species susceptible. Pathogens involved in gene-for-gene relationships are generally unable to detoxify phytoalexins (Keen, 1982). Compatible or high infection relationships are established when the pathogen avoids recognition mechanisms of the host, or actively suppresses phytoalexin accumulation (Keen, 1982). Phytoalexins are products of secondary metabolism and their accumulation is generally considered remote from the initial recognition events which determine the outcome of the host–pathogen interaction. Nevertheless, there is evidence that phytoalexin accumulation involves *de novo* synthesis of enzymes involved in their production (Yoshikawa, Yamauchi & Masago, 1978*b*; Bell *et al.* 1984) and that the rate of synthesis is controlled at the transcriptional level. Their role in plant defence is also indicated by the discovery of molecules of both plant and pathogen origin that elicit phytoalexin production either specifically or non-specifically. Examples of these elicitors are summarized in Table 10.4, and their possible involve-ment in plant–pathogen recognition is discussed further in Section 10.4.

Interpretation of host responses in genetic terms is difficult as many biochemical studies of resistance have been performed on systems with

Table 10.4. *Examples of substances ('elicitors') which elicit production of phytoalexins in higher plants*

Elicitors	Source	Experimental plant	Reference
Non-specific elicitors			
Abiotic elicitors			
Actinomycin D	Commercial	pea endocarp	Loschke *et al.* 1983
psoralens			
UV light			
CdCl$_2$			
Autoclaved ribonuclease A		french bean suspension-cultured cells and hypocotyl sections	Dixon *et al.* 1983*a*
Fungal components			
Hepta-β-glucoside containing 1,6- and 1,3- linkages	*Phytophthora megasperma* var. *sojae*	soybean cotyledons, hypocotyls and suspension-cultured cells	Darvill & Albersheim, 1984
Chitosan heptamer of β-1,4-glucosamine	Commercial (but a known component of fungal cell walls)	pea endocarp	Loschke *et al.* 1983; Hadwiger & Beckman, 1980
		soybean suspension-cultured cells	Kohle *et al.* 1984
Glucan	*Uromyces phaseoli*	french bean	Hoppe *et al.* 1980
Polysaccharide containing glc, man, ara	*Colletotrichum lindemuthianum*	french bean suspension-cultured cells	Dixon *et al.* 1981; Dixon *et al.* 1983*b*
Eicosapentaenoic acid and arachidonic acid; β-glucan and arachidonic acid	*Phytophthora infestans*	potato tuber tissue	Bostoc *et al.* 1982
	Phytophthora infestans	potato tuber tissue	Maniara *et al.* 1984
Plant components			
Trideca-α-1,4-D-galacturonide	Commercial polygalacturonic acid or castor bean cell walls digested by endopolygalacturonase from *Rhizopus stolonifer*	castor bean	Bruce & West, 1982; Lee & West, 1981; Jin & West, 1984
Dodeca-α-1,4-D-galacturonide ('endogenous elicitor')	Plant cell walls (soybean, tobacco, sycamore, wheat, citrus pectin)	soybean cotyledon	Nothnagel *et al.* 1983; Hahn *et al.* 1981
Proteinase inhibitor inducing factor (PIIF) (pectic polysaccharides)	Tomato and potato leaves	pea endocarp	Walker-Simmons *et al.* 1983
		castor bean seedings	Bruce & West, 1982
Race-specific elicitors			
High mol. wt glucan with (1,3)- and (1,4)-β-linkages + rha, man, gal and protein	*Colletotrichum* spp.	red kidney bean	Anderson, 1980
Glucomannan	*Phytophthora megasperma* f.sp. *glycinea*	soybean	Keen & Yoshikawa, 1983; Keen *et al.* 1983
Glycoprotein	*Phytophthora megasperma* f.sp. *glycinea*	soybean	Keen & Legrand, 1980
Peptidogalactoglucomannan	*Cladosporium fulvum*	tomato	de Wit & Roseboom, 1980; de Wit *et al.* 1984

poorly defined genetics making interpretation of the results difficult; furthermore unnatural inoculation techniques, which may involve wounding, are often used to obtain large amounts of uniformly infected host tissue and may bypass natural recognition barriers.

10.3   Principles of cell–cell recognition

Cell recognition may be defined as the initial event of cell–cell communication which elicits a defined biochemical, physiological or morphological response (Clarke & Knox, 1978). Many of our approaches to understanding cell–cell recognition in plant systems are based on experience from investigations of animal cell interactions. In animal systems recognition involves specific binding of signals to complementary membrane-bound receptor proteins, and there is a belief that this will also apply to plant cell–cell recognition, notwithstanding the differences between plant and animal cells. Before considering the role of receptors in plant–pathogen interactions, it is useful to review briefly the principles of cell recognition involving animal cells.

10.3.1   Recognition between animal cells

There are three main mechanisms by which animal cells are believed to communicate:

(1) One cell secretes a chemical signal which is received by a specific receptor on a target cell some distance away.

(2) Cells are in direct contact, and specific interactions take place between complementary, plasma-membrane bound molecules on the surfaces of the interacting cells.

(3) Direct cytoplasmic communication between cells via gap junctions.

The first two mechanisms can be regarded as extracellular as they depend on the target cells receiving extracellular signals, either secreted from the signalling cell (1) or bound to the surface of the signalling cell (2). The third mechanism is intracellular (Alberts *et al.* 1983).

Because of the difficulties of studying membrane-bound molecules, relatively little is known of the way in which recognition between animal cells making direct contact is mediated. There is also little information regarding the molecular basis of communication via gap junctions. On the other hand, a great deal is known about the mechanisms by which secreted chemical signals are received by target cells. Much of the information comes from work on the water-soluble peptide growth hormones and neurotransmitters. The main points which have emerged from studies of such recognition systems are:

(1) The secreted signal molecules bind specifically, reversibly and saturably with high affinity to specific receptor proteins (association constant K_a $10^8 M^{-1}$) on the surface of the target cells. This signal–receptor binding initiates intracellular signals (second messengers) which alter the behaviour of the target cells.

(2) Cyclic AMP (cAMP) is one of two known second messengers. The biological effect of specific binding of signal molecules to the target cells may be achieved by regulation of the intracellular cAMP concentration via activation of adenylate cyclase. Cyclic GMP can act in a similar way. It produces its effect by activating specific cAMP, Ca^{2+}-dependent protein kinases which in turn phosphorylate and thereby activate specific enzymic proteins. These activated enzymes often produce the biological effect directly, but in some cases they initiate another reaction which produces the biological effect.

(3) Ca^{2+} is the other known second messenger. In this case the cell-surface receptors are functionally coupled to Ca^{2+} channels. Binding of signal molecules to the receptors alters gated ion channels in the plasma membrane; this results either in an ion flux across the plasma membrane or an influx of Ca^{2+} into the cytosol. In both cases the change in Ca^{2+} concentration is transient. Calmodulin is a ubiquitous intracellular receptor for Ca^{2+} and a number of cellular proteins are regulated by Ca^{2+}/calmodulin in a Ca^{2+}-dependent way. These proteins include protein kinases, adenylate cyclase and cyclic nucleotide phosphodiesterase. There is overlap between the activities regulated by cAMP and Ca^{2+}/calmodulin to the extent that the same protein kinase may be activated both by the cAMP-dependent kinase and by the binding of Ca^{2+} to calmodulin (Nestler & Greengard, 1983; Alberts et al. 1983).

Another mechanism by which signal–receptor binding may initiate an intracellular signal is receptor-mediated endocytosis: the receptor molecules with bound signal molecules move laterally through the membrane into a zone on the cell surface (coated pit) which then undergoes endocytosis to form a vesicle. This vesicle then fuses with a lysosome to form a secondary lysosome in which various changes to either the receptor molecule or the signal molecule (or both) may occur. This receptor-mediated endocytosis may or may not occur in reactions involving enzyme activation (Brown, Anderson & Goldsteïn, 1983). For example, the action of epidermal growth factor involves the activation of both protein kinases and endocytosis of the receptor–epidermal growth factor complex (Hunter, 1984).

10.3.2 How far can principles of cell–cell recognition in animals be applied to plants?

In considering the possible mechanism for movement of signals between plant cells, we come to a major difference between plants and animals, that is, the presence of the cell wall which encases and protects the protoplast. In some way signals get from the extracellular medium across the cell wall; how they are received and transduced to give intra-cellular cytoplasmic responses is not understood. Cyclic AMP, protein kinases, Ca^{2+} channels, calmodulin and coated pits have all been described in plant cells but their role, if any, in plant cell recognition is not yet defined. The principles for extracellular recognition signals between animal and plant target cells are compared in Fig. 10.1. No

Fig. 10.1. Schematic representation of the events involved in recognition of an extracellular signal by an animal cell (a) and a plant cell (b).

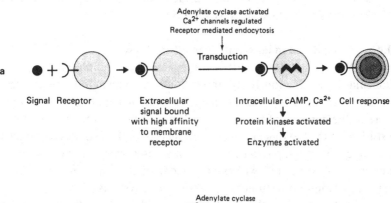

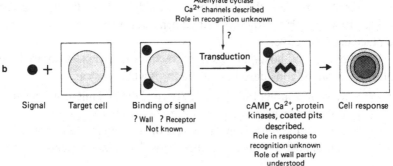

plant plasma membrane receptor for a particular signal has yet been isolated and fully characterized, although evidence for the presence of plasma membrane-associated receptors for β-glucans (Yoshikawa, Keen & Wang, 1983), auxins (Jacobs & Gilberts, 1983) and abscisic acid (Hornberg & Weiler, 1984) has been presented. Indeed, our understanding of the molecular structure of plant plasma membranes is poor compared with that of animal plasma membranes (Bowles, 1982). Chemical signals (for example, auxins, cytokinins, gibberellins) are secreted in plants, but we have a poor understanding of the way in which these and other molecules move extracellularly.

There are several other plant and microbial interactions leading to a defined response, recognition, which are simpler and better defined than recognition in host–fungal pathogen interactions. For example yeast mating is understood in some detail at the level of molecular genetics and chemistry of the interacting surfaces; in algae the *Chlamydomonas* mating interactions are also relatively simple and have been studied intensively (for list of interactions and details of references see Harris, Anderson, Bacic & Clarke, 1984). In higher plants, self-incompatibility has a relatively simple genetic basis and there is now some information on the nature of the interacting surfaces and the recognition events (Clarke, Anderson, Bacic & Harris, 1985).

10.4   Recognition in host–pathogen interactions

We know little about host–fungal pathogen interactions at the molecular level (Section 10.2), and virtually nothing about how the responses which are observed in compatible and incompatible interactions relate to the genetics of the systems (Section 10.6). There are a number of studies on the biologically active products of both hosts and pathogens (e.g. fungal elicitors, phytoalexins) but few cases where heritable disease phenotype is correlated with involvement of these products. Indeed, there are few in-depth studies of gene-for-gene systems (Day *et al.* 1983). An attempt to present the available knowledge schematically is given in Fig. 10.4 (p. 306), although our information is patchy and difficult to place in perspective.

The stages of the host–pathogen interactions leading to a compatible interaction can be considered as mechanisms which ensure:

(1) effective contact (adhesion) between host and microorganism,
(2) successful growth and penetration of the microorganism, and
(3) successful suppression or bypass of the diverse range of detrimental host responses, to allow growth of the pathogen.

Failure at any of these stages will result in failure to establish infection. Each stage could be considered to involve a specific recognition event, or events, the outcome of which determines whether the next interaction can occur.

We will review the information available for specific recognition at the three stages set out in Table 10.5. To keep it in perspective, we should stress:

(1) The highly variable and complex nature of the interacting systems.

(2) The relatively narrow data base; few systems have been studied in depth from all aspects. Responses which seem to be involved in several systems may not necessarily apply to all systems.

(3) The poor interface of information which relates the genetics of the systems to the specific responses observed.

10.4.1 Specific adhesion

Adhesion of the pathogen to the host is essential for infection. This is generally considered to be a non-specific interaction, and few systems have been studied at the molecular level. There is some preliminary information for two systems, which indicates that saccharide components of root surfaces are determinants, but not necessarily the sole determinants, of this type of interaction. These are the interactions between *Phytophthora cinnamomi* and corn (Hinch & Clarke, 1980) and *Pythium aphanidermatum*-cress (Callow, 1984) which apparently involve fucosyl residues of the root surface.

The experiments in these interactions are based on inhibition of binding of zoospores to roots by saccharide-specific lectins, and enzymatic removal of root-surface fucosyl residues. In neither case has either the root surface component containing the fucosyl residues been characterized, or the presumptive zoospore receptor for fucose isolated. These are necessary, but technically difficult, tasks required to define the system. There is little detailed information on these or other systems involving adhesion of fungi to plant surfaces, although there are a number of studies on adhesion of bacteria to plant surfaces (Lippincott & Lippincott, 1984; Pueppke, 1984).

10.4.2 Germination and penetration

Fungal pathogens enter the host by direct penetration of the epidermis, or through stomata, wounds, lenticels etc. The interaction between germ tubes and epidermal cell layers of the host is one point

Table 10.5. *Recognition events in the infection of a root by a host-specific pathogen of the Oomycetes*

	Event	Interacting surfaces	Result of successful interaction	Result of unsuccessful interaction
I	Adhesion	Zoospore surface Root surface	Adhesion	No adhesion
II	Germination and penetration	Germ tube surface Epidermal cells (intercellular, and/or intracellular growth?)	Penetration	No penetration
III	Continued growth of fungus through the host tissues	Hyphal surface cortex, vascular tissue (intercellular and/or intracellular growth?)	Growth through tissues, establishment of nutrient supply; successful sporulation	Reduced growth or no growth Barriers to growth such as wall modifications and hypersensitive response induced

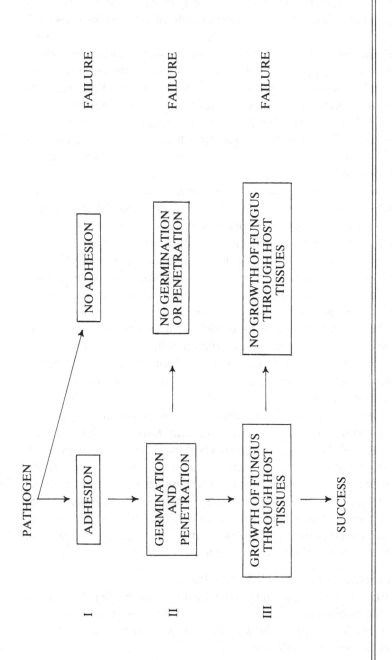

in the contact where specific recognition events may occur and the option for continued or arrested fungal growth may be exercised. For example, during rust infection, the germ-tube of the germinating spore grows and forms an appressorium over a stoma. An infection peg grows down through the stomatal aperture to form a sub-stomatal vesicle, from which infection hypha(e) grow intercellularly and adhere to the host mesophyll cell wall. This adherence appears to be a major factor determining subsequent haustorial mother cell formation, and haustorium production (Staples & Macko, 1980; Heath, 1974). The stages of infection are not random and all require directional growth by the fungus. There are a number of stages in this process where success or failure of the interaction could be determined.

While these events are described in detail at the electron-microscope level for a number of interactions, there is little molecular information available on the recognition events. One study which describes a system for exploring the basis of specificity is based on the observation that bean rust germ tubes (labelled with fluorescein isothiocyanate (FITC)) bind to host tissue cell walls, but not to cell walls of non-host tissue (Mendgen, 1978). The specificity implicit in both interacting partners would be amenable to study with monoclonal antibodies. One study which illustrates the complexity of the events occurring prior to host penetration is an ultrastructural description of the process of encystment of *Phytophthora cinnamomi* zoospores which precedes germination (Hardham, 1985*a,b*). This process can be triggered non-specifically, and hence may not involve specific recognition.

Cutin and suberin are the structural components of the major outer barriers which fungi must penetrate to gain entry into plants. Direct penetration of the cuticular barrier involves an extracellular enzyme, cutinase, produced by the germinating fungal spores. Inhibition of this enzyme prevents fungal penetration. Cutinase has recently been cloned and sequenced from *Fusarium solani* f.sp. *pisi* by Soliday, Flurkey, Okita & Kolattukudy (1984). The gene is only transcribed in the presence of either intact or partially hydrolysed cutin. This is the first fungal gene directly associated with pathogenicity to be cloned.

Growth of the fungus through the host tissues
The pattern of fungal growth through the host tissue is often determined by; (i) the requirements of the pathogen; (ii) the resistance or susceptibility of the host; (iii) the host tissue or organ infected; and (iv) host cell type in contact with the pathogen.

With *Phytophthora* spp., initial penetration and growth through the epidermal layers are usually intercellular in all tissue of resistant or susceptible plants (Coffey & Wilson, 1983; Stossel, Lazarovits & Ward, 1981; Miller & Maxwell, 1984a,b; Wetherbee *et al.* 1985), but after 4 h growth is more rapid in the susceptible host. As the invading hyphae reach the underlying tissue (parenchyma cells of the leaf mesophyll, or the stem and root cortex), the preferred infection pattern varies between intercellular, intercellular with haustoria, or inter- and intracellular. In non-host plants (Wetherbee *et al.* 1985) and in resistant hosts (Miller & Maxwell, 1984a,b) hyphae often grow intercellularly, through the middle lamella, or within the cell wall (Stossel *et al.* 1980, 1981).

In the compatible interaction, the barrier of the cell wall is breached and extensive inter- and intracellular hyphal growth often occurs. Logically we would expect the cell wall to play an important role in this interaction and the experimental evidence available illustrates that this is indeed the case.

The role of the plant cell wall

Our understanding of the structure and organization of the individual wall components of plant cells is far from complete. The proportion and detailed structures of the major groups of cell wall matrix polysaccharides, the pectic polymers and the other non-cellulosic polysaccharides can differ markedly in wall preparations from the same cell type of different species and in wall preparations from different cell types. These matrix polysaccharides and the wall-associated proteins including the hydroxyproline-rich glycoprotein, have all been implicated in particular recognition phenomena (for reviews see West, 1981; Keen, 1982; Bacic & Clarke, 1983; Darvill & Albersheim, 1984; Sequiera, 1984).

There are several ways in which the cell wall as a whole may act in the recognition process:

(1) The wall as a molecular sieve may restrict access of extracellular material to receptors associated with the plasma membrane

The matrix of primary cell walls is effectively a negatively charged porous gel. The nett charge may vary with the proportion of acidic polysaccharides and the degree of methylation of these components, as well as the content of high isoelectric point proteins and glycoproteins. Both the charge and porosity of the gel matrix will influence the movement of extracellular material through the wall to the plasma membrane; both these characteristics may differ for different cell types and for cells at different stages of growth. The effective porosity of the cell wall is controlled, at least in part, by the formation of 'junction zones' between unsubstituted regions of different wall polymers (Rees, 1977). Estimates of the molecular weight of globular proteins which can move from the external medium through primary cell walls to the plasma membrane vary from 17 kdaltons (Carpita, Dubularse, Montezinos & Delmer, 1979) to 60 kdaltons (Tepfer & Taylor 1981). (A molecular weight of 17 kdaltons for a globular protein corresponds to an effective cell wall pore size of 5 nm.) Cell walls may also be penetrated by cross-channels which are seen in some freeze-fracture studies. For example, morphological channels (50 nm diameter) have been observed in the cell walls of stigmatic papillae of *Gladiolus gandavensis* (Clarke, Abbott, Mandel & Pettitt, 1980). Interpretation of all these attempts to describe the capacity of the cell wall to allow movement of molecules from the external medium to the plasma membrane is difficult because of the possibility of artefacts induced by the experimental techniques. Movement of material in the opposite direction, from the plasma membrane through the wall to the external medium, may involve a different pathway as high molecular weight glycoconjugates (>100 kdaltons) such as arabinogalactan–proteins are secreted by, for example, suspension-cultured cells (Fincher, Stone & Clarke, 1983) and transmitting tissue cells of mature styles (Clarke *et al.* 1985).

(2) The wall as the primary receptor from which secondary signals are generated

Wall fragments released by enzymic signals: If, for instance, the primary signal from one cell is an enzyme such as a hydrolase or lyase, it may release fragments from the cell wall of the target cell. These wall fragments could be involved in the recognition process by interacting directly with a plasma membrane receptor. Alternatively, the primary signal (e.g. enzyme), by modifying the number of 'junction zones' between wall components may change the gel properties and effective porosity of the wall and thereby alter the accessibility of the primary and/or secondary signals to membrane-associated receptors.

Studies on interactions between pathogenic fungi and their host plants indicate that host cell wall fragments released by fungal enzymes can elicit cytoplasmic changes (for reviews see West, 1981; Bell, 1981; Darvill & Albersheim, 1984). One well studied mechanism of defence involves the accumulation of phytoalexins, at the site of infection (Deverall, 1977). West and coworkers (Lee & West, 1981; Bruce & West, 1982) isolated a heat-labile endopolygalacturonase from culture filtrates of *Rhizopus stolonifer* which can elicit accumulation of the phytoalexin casbene in castor bean (*Ricinus communis*). This endopolygalacturonase releases heat-stable, water-soluble cell wall fragments from the host plant. These wall fragments contain 70% of the carbohydrate as galacturonic acid or its methyl ester, indicating their derivation from pectic polysaccharides. Partial hydrolysis of polygalacturonic acid by the fungal enzyme also releases elicitor-active fragments; a minimum degree of polymerization of 10–11 units is required for activity (Jin & West, 1984). Oligosaccharides derived from pectic fragments are also active in inducing production of proteinase-inhibitors (Ryan, 1984). Presumably these low molecular weight cell-wall fragments, which could permeate through 'pores' in the wall, interact with primary receptors at the plasma membrane to elicit the cytoplasmic response. The nature of the presumptive receptors and the precise structural requirements of the pectic fragments remain undefined.

Wall-associated enzymes, lectins and agglutinins alter the extracellular signals: Another potential mechanism for modification of a primary extracellular signal is through interaction with wall-associated enzymes. This could generate a secondary signal which is accessible to the putative plasma membrane receptor. Alternatively, wall-bound lectins and agglutinins may bind and hence alter or immobilize extracellular signals. For instance, glycoproteins related to the hydroxyproline-rich cell wall glycoprotein, which are not lectins, but have been isolated from extracts of potato tubers (Leach, Cantrell & Sequiera, 1982), tobacco callus (Mellon & Helgeson 1982) and carrot (Van Holst & Varner, 1983). These glycoproteins are effective nonspecific bacterial agglutinins by virtue of their high pI. For a review of the role of cell wall components in microbial attachment see Whatley & Sequiera (1981). Another example of agglutination by cell wall components is the observation that a polygalacturonic acid-containing fraction from sweet potato agglutinates germinated spores of *Ceratocystis fimbriata* with some specificity in the presence of Ca^{2+} (Kojima, Kawakita & Uritani, 1982).

The most detailed illustrations of the importance of wall components in the receipt and modification of signals come from studies of host–pathogen interactions by Albersheim, Darvill and co-workers at the University of Colorado on the structure of the fungal β-glucan elicitor of phytoalexin accumulation in soybean (for review see Darvill & Albersheim, 1984). These neutral glucans are derived from hyphal cell walls and culture filtrates of the Oomycete *Phytophthora megasperma* f.sp. *glycinea*. The smallest elictor-active oligosaccharide purified from a partial acid hydrolysate of hyphal cell walls is a hepta-β-glucoside alditol (Fig. 10.2). This oligoglucoside is effective at eliciting phytoalexin production at concentrations of 10^{-9} to 10^{-10} M. Seven other hepta-β-glucoside alditols were inactive as elicitors of phytoalexins at concentrations up to 20 ng/cotyledon. The minimum structure required for biological activity is a (1,6)-β-glucosyl backbone substituted by single glucosyl residues at the C(0)3 position at two of the glucosyl backbone residues. The two substituted glucosyl backbone residues are separated by a single unbranched glucosyl residue. Darvill & Albersheim (1984) suggest that β-glucan hydrolases associated with the soyabean cell walls degrade the high molecular weight β-glucan elicitors to produce the most active elicitor, a hepta-β-glucoside.

The absolute requirement for a particular glucan structure emphasizes the informational potential of complex carbohydrates. It also implies that there must be complementary receptors for the active saccharide which, if protein in nature, would in effect be lectins. The importance of the binding requirements of lectins for particular saccharide sequences (Kauss, 1985) which allow them to discriminate between closely related saccharide sequences becomes apparent in this context. The host cell wall saccharides are not only involved in responses of the plant to patho-

Fig. 10.2. Structures of eight hepta-β-glucoside alditols purified from a partial acid hydrolysate of *Phytophthora megasperma* cell walls.* Indicates capacity to elicit phytoalexin accumulation.

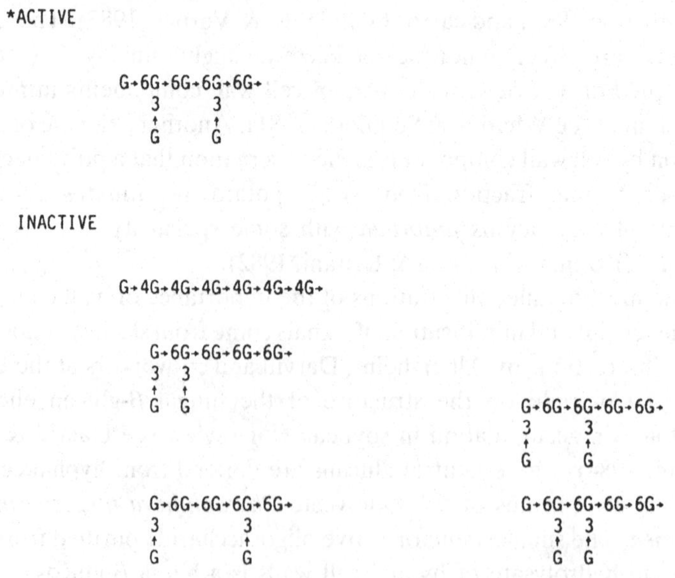

gen attack, but they also have a range of regulatory functions in growth and development (Darvill & Albersheim, 1984).

Another fungal wall polysaccharide which is an effective elicitor of phytoalexin production in peas is chitosan (polymeric $(1,4)$-β-glucosamine) (Kohle, Young & Kauss, 1984). In this case a heptamer is the smallest oligosaccharide active in eliciting production of pisatin in peas (Kendra & Hadwiger, 1984).

(3) The involvement of the cell wall in response to a recognition event
The response to reception of an extracellular signal in plant cells is not restricted to the cytoplasm, as the cell wall itself may be altered dramatically, for example, by laying down of papillae, lignification, suberization and increased production of the hydroxyproline-rich cell wall glycoprotein (for reviews see Aist, 1983; Heath, 1980). Studies of production of the polysaccharide components of papillae are characteristically associated with the plasma membrane (Henry, Schibeci & Stone, 1983); the plasma membrane followed by laying down of a core of electron-dense material and an overlaying with electron-lucent material. The plasma membrane proliferation is preceded by changes in the host cell wall at the point of contact with the fungal hyphae (Hinch, Wetherbee, Mallett & Clarke, 1985; Wetherbee, Hinch, Bonig & Clarke, 1985). $(1,3)$-β-glucan synthases which would be involved in biosynthesis of one of the polysaccharide components of papillae are characteristically associated with the plasma membrane (Henry, Schibeci & Stone, 1983); it may be that the proliferation of the plasma membrane observed is a mechanism for increasing the localized activity of these enzymes. In this regard, the observation that a Ca^{2+}-dependent $(1,3)$-β-glucan synthase in soybean suspension cultured cells is activated by a Ca^{2+} influx, which can be induced non-specifically by, for example, chitosan, or polylysine may be important. This suggests that a signal may be transduced by activation of Ca^{2+} channels to result in a change in the cell wall composition.

Possible pathways by which a signal which has glycan hydrolase activity may generate effects in the cell wall and the cytoplasm of a target cell are shown schematically in Fig. 10.3(IV).

Role of plasma membrane
It is believed, by analogy with animal cell recognition systems, that the plant plasma membrane plays a critical role in the recognition events between host and pathogen. There is however only indirect evidence to support this because of the difficulty of obtaining purified plasma membrane preparations from plants. One approach is to use isolated

protoplasts, which are agglutinated by lectins (Larkin, 1981; Fowke, this volume) and by a fungal β-glucan preparation (Peters, Cribbs & Stelzig, 1978). Fungal cell wall preparations also bind to membrane fractions (Yoshikawa *et al.* 1983). Further evidence for the interaction of plant plasma membranes and fungal hyphae is the persistent adherence of plasma membranes to intracellular hyphae following plasmolysis of infected tissue (Nozue, Tomiyama & Doke, 1979) and the hypersensitive-like response of protoplasts to hyphal wall components (Doke & Tomiyama, 1980).

Other responses

There are a number of other responses which show some specificity in some systems; production of host-specific toxins, (Gilchrist, 1983; Yoder, 1980). In this case the specificity depends on the host having

Fig. 10.3. Schematic summary of events of interaction between a resistant host plant and an avirulent pathogen.

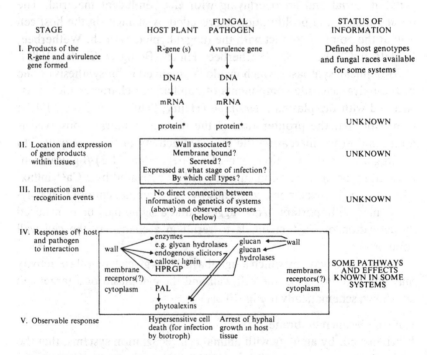

* Protein gene products may be:
  (1) Modified post-translationally (e.g. glycosylation)
  (2) Enzymically active (e.g. hydrolases)
  (3) Enzymes which activate another protein (e.g. glycosyl transferases, protein kinases)
† Different systems have different responses. In this panel some responses which occur in different systems are placed together to give an idea of the possible interactions.

complementary receptors for the toxin; the host also must lack the capacity to detoxify the toxin for a successful interaction to occur. Fungal enzymes which facilitate penetration through the host tissues may also be involved in specific interactions.

10.5 Requirements for defining host–pathogen relationships in molecular terms

Progress in understanding host–pathogen relationships in molecular terms will be made as particular interacting systems are studied in detail. The experimental systems which are most amenable to study are those in which the phenotypes of compatible and incompatible interactions are distinct, and thus easily recognizable. Information on the classical genetics, physiology and cytology of the interaction is also essential and it is helpful if derived callus lines of the resistant cultivars also express resistance.

10.5.1 Classical genetics

Capacity to manipulate the genetics of both the host and the pathogen is desirable. Ideally, both interacting partners (plant and pathogen) should have relatively short life cycles which include a diploid and a sexual phase so that the pattern of inheritance of resistance in the plant and avirulence in the pathogen can be studied. In many 'gene-for-gene' systems the inheritance of avirulence in the pathogen has not been followed; this information is crucial to any study of avirulence genes and their products.

If resistance to a particular pathogen is controlled by a single dominant gene (R-gene), then the gene may be transferred in a classical breeding program to a susceptible cultivar of the same plant, by selecting for resistance in successive backcross generations (Fig. 10.4). In this way, two cultivars with near-isogenic backgrounds, differing primarily in that one cultivar will contain the resistance gene, are available for study. The resistance gene and its products can be examined by comparing mRNA and/or proteins from both cultivars (see Section 10.6). If this type of approach is used, it is essential to have at least five back-crosses so that the background of the two cultivars is 98.4% isogenic. A major difficulty with this approach is that it is not known whether this type of resistance is constitutive or inducible. It is quite possible that the genes are expressed only during interaction with the fungus. This necessarily complicates the search for R-gene associated mRNA or protein products in the resistant cultivar.

The 'resistance gene' defined by classical genetics will not necessarily be a single gene in the sense of being DNA coding for a single protein. It will be DNA which is transferred during successive sexual crosses and which will bear the traits for disease resistance but may also bear other characteristics which are tightly linked and segregate with the resistance gene.

It is also helpful to have an established map of genetic markers at known chromosomal locations so that the site of the resistance genes in the plant or avirulence genes in the pathogen can eventually be mapped to a particular site. In some plants, groups of genes on the same chromosome confer resistance to several diseases (Day *et al.* 1983). For example barley chromosome 5 (short arm) bears several loci controlling resistance to *Erysiphe graminis* f. sp. *hordei*, a gene for resistance to *Puccinia hordei* and a gene for resistance to *P. striiformis*. This clustering of resistance genes at a known location greatly enhances the chance of identifying a single gene and its product in molecular terms.

Fig. 10.4. Scheme for back-crossing sexually compatible plants to transfer a particular trait (e.g. resistance) from one plant into the genetic background of a desirable cultivar.

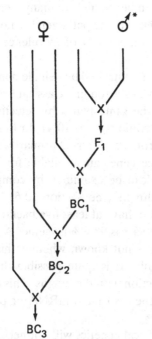

Select for
*trait in each
back cross (BC)

10.5.2 *Physiology and cytology*

It is important to understand the developmental biology of the system – questions such as: At what stage in the life cycle is the pathogen infective? What are the sites and times of expression of susceptibiity or resistance in the host plant? What tissues and cell types are involved in the interaction? Is the response temperature dependent? Information regarding the interaction requires experimentation at the whole plant level in conjunction with detailed light- and electron-microscopic studies of inoculated tissues of resistant and susceptible cultivars.

10.5.3 *Tissue culture systems*

If cultivar resistance to a particular pathogen is preserved in derived callus lines then large amounts of material suitable for analysis can be produced, thus obviating the need for continual seed production programs. Tissue culture is also necessary as a part of any gene transfer program. If genes associated with expression of resistance are cloned, their identity as R-genes will depend ultimately on the transformation of cultivar phenotype from susceptible to resistant.

10.6 Approaches to defining resistance and avirulence gene products

To get molecular information regarding candidates for gene products of the resistance and avirulence genes, a supply of young, relatively unlignified plant tissue which can be reproducibly infected and which will respond in a predictable time course is required. This tissue is used to prepare extracts to examine proteins or mRNA. The procedure is ideally designed so that the inoculated tissue is harvested at the time at which resistance is expressed, and the tissue is selected so that adjacent non-responding or non-infected regions are excluded to minimize the possibility of 'diluting-out' the R-gene products in the extracts.

Gel electrophoresis, including isoelectric focusing and both single and two-dimensional sodium dodecyl sulphate–polyacrylamide gel electrophoresis (SDS–PAGE) are rapid techniques for obtaining information on the protein components of the extracts. This approach has been used by Gabriel & Ellingboe (1982) who examined 12 congenic wheat lines, each of which contains separate homozygous Pm genes for resistance to powdery mildew (*Erysiphe graminis* f.sp. *tritici*). Comparison of SDS–PAGE gels of leaf extracts failed to show any consistent differences in polypeptides which could be attributed to a resistance-gene product.

Immunological techniques including monoclonal antibody technology are also useful for identifying specific gene products. However a practical difficulty in applying immunological methods to plants is that plant polysaccharides and glycoconjugates are highly antigenic in mice so that a proportion of hybridomas secrete antibody to saccharides or saccharide sequences. These antibodies are not specific to a particular glycoconjugate as many different glycoconjugates may bear the same saccharide sequence (Anderson, Sandrin & Clarke, 1984). This technique has not yet been applied extensively to examine host–pathogen systems, but offers such selectivity that it will be extremely useful if the problems with cross-reactivity due to anti-glycosyl antibodies are avoided.

If specific proteins which segregate with disease resistance can be identified, it would then be possible to isolate them, obtain partial amino acid sequences, construct synthetic nucleotide probes and screen a cDNA library for the genes which code for these proteins. This is a classic strategy and can be carried out on nanomole quantities of protein which can be isolated directly from SDS–PAGE gels (Hunkapillar *et al.* 1984).

Another strategy is to isolate mRNA from inoculated plants at a time when resistance is expressed and to prepare a cDNA library which can then be probed with mRNA prepared before and after inoculation of resistant and susceptible cultivars. Comparison of these RNA populations by such differential hybridization should detect RNA molecules associated with resistance. This approach has been used successfully to examine developmentally regulated genes in slime moulds (Williams & Lloyd, 1979) and *Xenopus* (Dworkin & David, 1984) as well as inducible genes involved in plant—microbial interactions. These include fungal cutinase which digests plant cutin (Soliday *et al.* 1984) and nodulins, a class of proteins specifically expressed in nitrogen-fixing soybean root nodules (Fuller, Kunster, Nguyen & Verma, 1983).

Chemical and ultraviolet mutagenesis can be used to create pathogens or plants with changed virulence or resistance phenotypes. The mutant genes can then be mapped by complementation studies. Transposable elements also may be used to mutate plant or pathogen genes, and have been used to great advantage in identifying the genes in *Rhizobium* spp. and *Agrobacterium* spp. involved in specific interaction with their host plants (Long, Buikema & Ausubel, 1982; Klee *et al.* 1983). The mutant genes can be physically isolated and the wild type gene recovered by hybridization. Transposable elements have been demonstrated in plants such as maize and snapdragon as well as in *Drosophila*, bacteria and yeast (for review see Flavell, 1984) but not as yet in any pathogenic fungus. Ellingboe (1984) has used the transposable element (Mu-1) to mutate a gene in maize (Rp1) which confers resistance to *Puccinia sorghii*. Fifty thousand treated seedlings were screened and 36 susceptible seedlings recovered. The cloned mutant gene has not yet been isolated, but it is likely that this approach will lead to a cloned resistance gene.

An avirulence gene from the bacterium *Pseudomonas syringae* p.v. *glycinea* race 6 has recently been identified and cloned by complementation studies (Staskawicz, Dahlbeck & Keen, 1984). The gene confers avirulence on other bacterial races which are normally virulent on the same soybean cultivar. Transposon mutagenesis has also been used to identify the avirulence gene of the bacterium *Pseudomonas syringae* p.v. *savastanoi* and its product which causes gall disease of oleander (Comai & Kosuge, 1982). Avirulent strains lack ability to synthesize the hormone indoleacetic acid (IAA) while virulent strains produce high levels of IAA. This is correlated with the avirulent strains having defective genes coding for enzymes involved in IAA production.

A fungal disease for which specific probes are available is crown rust of oats. The Pc-2 gene in oats confers resistance to avirulent races of the crown rust fungus *Puccinia coronata* f.sp. *avenae*; in terms of classic genetics the same gene also confers susceptibility to a devastating disease caused by *Helminthosporum victoriae*, which produces a potent host-specific toxin, victorin (for review see Yoder, 1980). Furthermore victorin elicits production of avenalumins (phytoalexins) only in oats carrying the Pc-2 gene (Mayama & Tani, 1981). Victorin is a low molecular weight peptide (Keen, Midland & Sims, 1983) and promises to be a useful tool in elucidating the mechanism of action of this dual-function gene, Pc-2.

Transformation systems for the plant and pathogen are essential if putative resistance or avirulence genes are to be unequivocally identified (see section 5.3). Transformation of dicotyledonous plants is now possible using vectors derived from the Ti plasmid of *Agrobacterium tumefaciens* (for review see Depicker, van Montagu & Schell, 1983) and there is recent evidence that monocots can be transformed by Ti plasmid vectors (Hooykaas-von Slogteren, Hooykas & Schilperoort, 1984). Eukaryotic microorganisms that have been successfully transformed to date are yeast (Hinnen, Hicks & Fink, 1978), and *Dictyostelium* (Ratner, Ward & Jacobson, 1981) but progress has been reported recently with *Neurospora* (Kinniard *et al.* 1982), *Podospora* (Stahl, Tudzynski, Kuch & Esser, 1982), *Aspergillus* (Ballance, Buxton & Turner, 1983) and *Ustilago* (Banks, 1983) and *Chlamydomonas* (Rochaix & Van Dillewijn, 1982). The only plant pathogen is *Ustilago*. Since the classical genetics of most of the pathogenic fungi are poorly understood, it will be some time before molecular genetics can be used to identify genes involved in avirulence. Techniques for separating single chromosomes are being developed for plants (Griesbach, Malmberg & Carlson, 1982) and trypanosomes (Van der Ploeg, Schwartz, Cantor & Brost, 1984). This may simplify the task of identifying the site of the R-gene in the genome.

There is as yet no connecting pathway of information between the genetics of the system and the observed response. Although there is detailed information on some aspects of some responses, a complete understanding of the expression of resistance will depend on application of multidisciplinary experimental approaches to investigate individual genetically defined systems.

10.7   Conclusions

(1) Host–pathogen interactions are diverse in nature.
(2) Many host–pathogen interactions can be described by the 'gene-for-gene' concept which states that for every gene for resistance in the host there is a corresponding gene for avirulence in the pathogen.
(3) Specific interactions between gene products of the resistance and avirulence genes acting as receptors, or signals, or both are implicit in the hypothesis.
(4) Avirulence genes have been cloned from two bacterial pathogens, but not yet from fungal pathogens. The nature of the resistance genes is not yet understood.
(5) The putative host receptors for the avirulence gene products have not been identified.
(6) The host cell wall plays an active role in both recognition of, and response to, contact with fungal pathogens.
(7) Understanding of the molecular basis of host–pathogen interactions is being approached from two directions. Firstly, understanding the molecular genetics of the interaction and secondly observing host responses to infection. As yet there is no pathway connecting information obtained by these two approaches.
(8) The availability of genetically defined host–pathogen systems, monoclonal antibody and recombinant DNA technology is leading to an understanding of gene expression and specific interactions in these systems.

We thank Dr David Guest and Dr Michael Smart for critically evaluating the manuscript for us at short notice and for their valuable comments. Miss Genevieve Loy gave us excellent secretarial and editorial assistance for which we are most grateful.

10.8 References

Aist, J. R. (1976). Papillae and related wound plugs of plant cells. *Ann. Rev. Phytopathol.* **14:** 145–63.

Aist, J. R. (1983). Structural responses as resistance mechanisms. In *The Dynamics of Host Defence*, ed. J. A. Bailey & B. J. Deverall, pp. 33–70. London, New York: Academic Press.

Aist, J. & Israel, H. (1977). Effects of heat-shock inhibition of papilla formation on compatible host penetration by two obligate parasites. *Physiol. Plant Pathol.* **10:** 13–20.

Alberts, B., Bray, D., Lewis, J., Raff, M., Robers, K. & Watson, J. D. (1983). *Molecular Biology of the Cell.* New York: Garland.

Anderson, A. J. (1980). Differences in the biochemical compositions and elicitor activity of extracellular components produced by three races of a fungal plant pathogen *Collectotrichum lindemuthianum. Can. J. Microbiol.* **26:** 1473–9.

Anderson, M. A., Sandrin, M. S. & Clarke, A. E. (1984). A high proportion of hybridomas raised to a plant extract secrete antibody to arabinose or galactose. *Plant Physiol.* **75:** 1013–16.

Bacic, A. & Clarke, A. E. (1983). The cell surface in plant recognition phenomena. In *Vegetative Compatibility Responses in Plants*, ed. R. Moore, pp. 139–60. Baylor University Press.

Bailey, J. A. (1983). Biological perspectives of host-pathogen interactions. In *The Dynamics of Host Defence*, ed. J. A. Bailey & B. J. Deverall, pp. 1–32. London, New York: Academic Press.

Bailey, J. A. & Mansfield, J. W. (eds.) (1982). *Phytoalexins.* Glasgow and London: Blackie.

Ballance, D. J., Buxton, F. P. & Turner, G. (1983). Transformation of *Aspergillus nidulans* by the orotidine 5' phosphate decarboxylase gene of *Neurospora crassa. Biochem. Biophys. Res. Comm.* **112:** 284–9.

Banks, G. (1983). Transformation of *Ustilago maydis* by a plasmid containing yeast 2-micron DNA. *Current Genetics* **7:** 79–83.

Beckman, C. H. (1980). Defenses triggered by the invader: physical defenses. In *Plant Disease: An Advanced Treatise*, vol. 5, ed. J. G. Horsfall & E. B. Cowling, pp. 225–45. London, New York: Academic Press.

Beckman, C. H. & Talboys, P. W. (1981). Anatomy of resistance. In *Fungal Wilt Diseases of Plants*, ed. M. E. Mace, A. A. Bell & C. H. Beckman, pp. 487–518. New York, London: Academic Press.

Bell, A. A. (1981). Biochemical mechanisms of disease resistance. *Ann. Rev. Plant Physiol.* **32:** 21–81.

Bell, J. N., Dixon, R. A., Bailey, J. A., Rowell, P. M. & Lamb, C. J. (1984). Differential induction of chalcone synthase mRNA activity at the onset of phytoalexin accumulation in compatible and incompatible plant–pathogen interactions. *Proc. Natl. Acad. Sci. U.S.A.* **81:** 3384–8.

Bird, P. M. & Ride, J. P. (1981). The resistance of wheat to *Septoria nodorum*: fungal development in relation to host lignification. *Physiol. Plant Pathol.* **19:** 289–99.

Bowles, D. J. (1982). Membrane glycoproteins. In *Encyclopaedia of Plant Physiology*, ed. A. Pirson & M. H. Zimmerman, vol. 13A, pp. 584–97. Berlin: Springer-Verlag.

Bostoc, R. M., Laine, R. A. & Kuc, J. (1982). Factors affecting the elicitation of sesquiterpenoid phytoalexin accumulation by eicosapentaenoic and arachidonic acids in potato. *Plant Physiol.* **70:** 1417–24.

Brown, M. S., Anderson, R. G. W. & Goldstein, J. L. (1983). Recycling receptors: the round-tip itinerary of migrant membrane proteins. *Cell* **32:** 663–7.

Bruce, R. J. & West, C. A. (1982). Elicitation of casbene synthetase activity in castor bean. The role of pectic fragments of the plant cell wall in elicitation by a fungal endopolygalacturonase. *Plant Physiol.* **69:** 1181–8.

Bushnell, W. R. & Rowell, J. B. (1981). Suppressors of defense reactions: a model for roles in specificity. *Phytopathology* **71**: 1012–24.

Callow, J. (1984). Cellular and molecular recognition between higher plants and fungal pathogens. In *Encyclopaedia of Plant Physiology*, H. F. Linskens & J. Heslop-Harrison, vol. 17, pp. 212–32. Heidelberg: Springer-Verlag.

Caplan, A., Herrera-Estrella, L., Inze, D., Van Haute, E., Van Montague, M., Schell, J. & Zambryski, P. (1983). Introduction of genetic material into plant cells. *Science* **222**: 815–21.

Carpita, N., Dubularse, D., Montezinos, D. & Delmer, D. P. (1979). Determination of the pore size of cell walls of living plant cells. *Science* **205**: 1144–7.

Clarke, A. E., Abbott, A., Mandel, T. & Pettitt, J. (1980). Organization of the wall layers of the stigmatic papilli of *Gladiolus gandavensis*: a freeze fracture study. *J. Ultrastruct. Res.* **73**: 269–81.

Clarke, A. E., Anderson, M. A., Bacic, T. & Harris, P. J. (1985). Molecular basis of cell recognition during fertilization in higher plants. *J. Cell Sci.* Suppl. 1 (in press).

Clarke, A. E. & Knox, R. B. (1978). Cell recognition in plants. *Quart. Rev. Biol.* **53**: 3–28.

Clarke, J. A., Lisker, N., Lamport, D. T. A. & Ellingboe, A. H. (1981). Hydroxyproline enhancement as a primary event in the successful development of *Erysiphe graminis* in wheat. *Plant Physiol.* **67**: 188–9.

Coffey, M. D. & Wilson, U. E. (1983). An ultrastructural study of the late-blight fungus *Phytophthora infestans* and its interaction with the foliage of two potato cultivars possessing different levels of general (field) resistance. *Can. J. Bot.* **61**: 2669–85.

Comai, L. & Kosuge, T. (1982). Cloning and characterisation of iaaM, a virulence determinant of *Pseudomonas syringae* pv. *savastanoi. J. Bacteriol.* **149**: 40–6.

Darvill, A. G. & Albersheim, P. (1984). Phytoalexins and their elicitors – a defense against microbial infection in plants. *Ann. Rev. Plant Physiol.* **35**: 243–75.

Day, P. R., Barrett, J. A. & Wolfe, M. S. (1983). The evolution of host-parasite interactions. In *Genetic Engineering of Plants*, ed. T. Kosuge, C. P. Meredith & A. Hollaender, pp. 419–30. New York and London: Plenum Press.

Denny, T. P. & Van Etten, H. D. (1983). Tolerance of *Nectria haematococca* MP VI to the phytoalexin pisatin in the absence of detoxification. *J. Gen. Microbiol.* **129**: 2893–901.

Depicker, A., van Montagu, M. & Schell, J. (1983). Plant cell transformations by *Agrobacterium* plasmids. In *Genetic Engineering of Plants*, ed. T. Kosuge, C. M. Meredith & A. Hollaender, pp. 143–76. New York and London: Plenum Press.

Deverall, B. J. (1977). *Defence Mechanisms of Plants*. Cambridge Monographs in Experimental Biology, No. 19. Cambridge University Press.

Deverall, B. J. (1982). Introduction. In *Phytoalexins*, ed. J. A. Bailey & J. W. Mansfield, pp. 1–20. Glasgow and London: Blackie.

Dickinson, C. H. & Lucas, J. A. (1983). *Plant Pathology and Plant Pathogens. Basic Microbiology*, vol. 6. London: Blackwell Scientific Publications.

Dixon, R. A., Dey, P. M., Lawton, M. A. & Lamb, C. J. (1983*a*). Phytoalexin production in French Bean. Intercellular transmission of elicitation in cell suspension cultures and hypocotyl sections of *Phaseolus vulgaris. Plant Physiol.* **71**: 251–6.

Dixon, R. A., Dey, P. M., Murphy, D. L. & Whitehead, I. M. (1981). Dose responses for *Colletrichum lindemuthianum* elicitor-mediated enzyme induction in french bean cell suspension cultures. *Planta* **151**: 272–80.

Dixon, R. A., Gerrish, C., Lamb, C. J. & Robbins, M. P. (1983*b*). Elicitor-mediated induction of chalcone isomerase in *Phaseolus vulgaris* cell suspension cultures. *Planta* **159**: 561–9.

Doke, N. & Tomiyama, K. (1980). Effect of hyphal wall components from *Phytophthora infestans* on protoplasts of potato tuber tissues. *Physiol. Plant Pathol.* **16**: 169–76.

Dworkin, H. B. & David, I. B. (1984). Use of a cloned library for the study of abundant poly(A)$^+$ RNA during *Xenopus laevis* development. *Dev. Biol.* **76**: 449–4.

Ellingboe, A. H. (1981). Changing concepts in host–pathogen genetics. *Ann. Rev. Phytopathol.* **19**: 125–43.

Ellingboe, A. H. (1984). Genetics of resistance in plants. *Proc. Molecular Basis of Plant Disease Conference*, Davis, Ca., U.S.A.

Eschrich, W. & Currier, H. B. (1964). Identification of callose by its diachrome and fluorochrome reactions. *Stain Technology* **39**: 303–7.

Fincher, G. B., Stone, B. A. & Clarke, A. E. (1983). Arabinogalactan-proteins – structure, biosynthesis and function. *Ann. Rev. Plant Physiol.* **34**: 47–70.

Flavell, R. B. (1984). Transposable elements. In *Oxford Surveys of Plant Molecular and Cellular Biology*, ed. B. J. Miflin, vol. 1, pp. 207–10. Oxford: Oxford University Press.

Flor, H. H. (1955). Host–parasite interaction in flax rust – its genetics and other implications. *Phytopathology* **45**: 680–5.

Fuller, F. F., Kunstner, P., Nguyen, T. & Verma, D. P. S. (1983). Soybean nodulin genes: analysis of cDNA clones reveals four plant genes which are expressed specifically in nitrogen fixing root nodules. *Proc. Nat. Acad. Sci. U.S.A.* **80**: 2594–8.

Gabriel, D. W. & Ellingboe, A. H. (1982). High resolution two-dimensional electrophoresis of proteins from congenic wheat lines. *Physiol. Plant Pathol.* **20**: 349–57.

Gilchrist, D. J. (1983). Molecular modes of action. In *Toxins and Plant Pathogenesis*, ed. J. M. Daly & B. J. Deverall, pp. 81–130. Australia: Academic Press.

Gilchrist, D. G. & Yoder, O. C. (1984). Genetics of host–parasite systems: a prospectus for molecular biology. In *Plant-Microbe Interactions – Molecular and Genetic Perspectives*, vol. 1, ed. T. Kosuge & E. W. Nester, pp. 69–92. New York: Macmillan.

Gould, A. R. & Symons, R. H. (1983). A molecular biological approach to relationships among viruses. *Ann. Rev. Phytopathol.* **21**: 179–242.

Grisebach, R. J. Malmberg, R. L. & Carlson, P. S. (1982). An improved technique for the isolation of higher plant chromosomes. *Plant Sci. Letters* **24**: 55–60.

Hachler, H. & Hohl, H. R. (1984). Temporal and spatial distribution patterns of collar and papillae wall appositions in resistant and susceptible tuber tissue of *Solanum tuberosum* infected by *Phytophthora infestans*. *Physiol. Plant Pathol.* **24**: 107–18.

Hadwiger, L. A. & Beckman, J. M. (1980). Chitosan as a component of pea–*Fusarium solani* interactions. *Plant Physiol.* **66**: 205–11.

Hahn, M. G., Darvill, A. G. & Albersheim, P. (1981). Host–pathogen interactions. XIX. The endogenous elicitor, a fragment of a plant cell wall polysaccharide that elicits phytoalexin accumulation in soybeans. *Plant Physiol.* **68**: 1161–9.

Hammerschmidt, H. (1984). Rapid deposition of lignin in potato tuber tissue as a response to fungi non-pathogenic on potato. *Physiol. Plant Pathol.* **24**: 33–42.

Hammerschmidt, R., Lamport, D. T. A. & Muldoon, E. P. (1984). Cell wall hydroxyproline enhancement and lignin deposition as an early event in the resistance of cucumber to *Cladosporium cucumerium*. *Physiol. Plant Pathol.* **24**: 43–7.

Hardham, A. R. (1985*a*). Studies on the cell surface of zoospores and cysts of the fungus, *Phytophthora cinnamomi*. The influence of fixation on the pattern of lectin binding. *J. Histochem. Cytochem.* **33**: 110–18.

Hardham, A. R. (1985*b*). Studies on the cell surface of zoospores and cysts of the fungus, *Phytophthora cinnamomi*. Localization of lectin receptors. *J. Histochem. Cytochem.* (in preparation).

Harris, P. J., Anderson, M. A., Bacic, A. & Clarke, A. E. (1984). Cell–cell recognition in plants with special reference to the pollen–stigma interaction. In *Oxford Surveys of Plant Molecular and Cellular Biology*, vol. 1, ed. B. J. Miflin, pp. 161–203. Oxford: Oxford University Press.

Heath, M. C. (1974). Light and electron microscopic studies of the interactions of host and non-host plants with cowpea rust – *Uromyces phaseoli* var. *vignae*. *Physiol. Plant Pathol.* **4:** 403–14.

Heath, M. C. (1976). Hypersensitivity, the cause or the consequence of rust resistance? *Phytopathology* **66:** 935–6.

Heath, M. C. (1977). A comparative study of non-host interactions with rust fungi. *Physiol. Plant Pathol.* **10:** 73–88.

Heath, M. C. (1980). Reactions of nonsuscepts to fungal pathogens. *Ann. Rev. Phytopathol.* **18:** 211–36.

Heath, M. C. (1981). A generalized concept of host–parasite specificity. *Phytopathology* **71:** 1121–3.

Henry, R. J., Schibeci, A. & Stone, B. A. (1983). Localization of β-glucan synthases on the membranes of cultured *Lolium multiflorum* (ryegrass) endosperm cells. *Biochem. J.* **209:** 627–33.

Hickey, E. L. & Coffey, M. D. (1980). The effects of Ridomil on *Peronospora pisi* parasitizing *Pisum sativum*: an ultrastructural investigation. *Physiol. Plant Pathol.* **17:** 199–204.

Hinch, J. M. & Clarke, A. E. (1980). Adhesion of fungal zoospores to root surfaces is mediated by carbohydrates of the root slime. *Physiol. Plant Pathol.* **16:** 305–7.

Hinch, J. M. & Clarke, A. E. (1982). Callose formation in *Zea mays* as a response to infection with *P. cinnamomi*. *Physiol. Plant Pathol.* **21:** 113–24.

Hinch, J. M., Wetherbee, R., Mallett, J. & Clarke, A. E. (1985). Response of *Zea mays* roots to infection with *Phytophthora cinnamomi*: I. The epidermal layer. *Protoplasma* **126:** 178–87.

Hinnen, A., Hicks, J. B. & Fink, G. R. (1978). Transformation of yeast. *Proc. Nat. Acad. Sci. U.S.A.* **75:** 1919–33.

Hooykaas-van Slogteren, G. M. S., Hooykaas, P. J. J. & Schilperoort, R. A. (1984). Expression of Ti plasmid genes in monocotyledonous plants infected with *Agrobacterium tumefaciens*. *Nature* **311:** 763–4.

Hoppe, H. H., Humme, B. & Heitefuss, R. (1980). Elicitor induced accumulation of phytoalexins in healthy and rust infected leaves of *Phaseolus vulgaris*. *Phytopathol. Z.* **97:** 85.

Hornberg, C. & Weiler, E. W. (1984). High affinity sites for abscisic acid on the plasmalemma of *Vicia faba* guard cells. *Nature* **310:** 321–4.

Hunkapillar, M., Kent, S., Caruthers, M., Dreyer, W., Firea, J., Giffin, C., Horrath, S., Hunkapillar, J., Tempst, P. & Hood, L. (1984). A microchemical facility for the analysis and synthesis of genes and proteins. *Nature* **310:** 105–11.

Hunter, T. (1984). Growth factors: the epidermal growth factor receptor gene and its product. *Nature* **311:** 414–16.

Israel, H. W., Wilson, R. G., Aist, J. R. & Kunoh, H. (1980). Cell wall appositions and plant disease resistance. Acoustic microscopy of papillae that block fungal ingress. *Proc. Nat. Acad. Sci. U.S.A.* **77:** 2046–9.

Jacobs, M. & Gilberts, S. F. (1983). Basal localisation of the presumptive auxin carrier in pea stem cells. *Science* **220:** 1297–1300.

Jacobs, S., Hazum, Y., Schechter, P. & Cuatrecasas, P. (1979). Insulin receptor: covalent labelling and identification of subunits. *Proc. Nat. Acad. Sci. U.S.A.* **76:** 4918–21.

Jin, D. F. & West, C. A. (1984). Characteristics of galacturonic acid oligomers as elicitors of casbene synthetase activity in castor bean seedlings. *Plant Physiol.* **74:** 989–92.

Kauss, H. (1985). Role of Ca^{2+} in callose formation. *Proc. John Innes Symp.* (in press).

Keen, N. T. (1982). Specific recognition in gene-for-gene host-parasite systems. In *Advances in Plant Pathology*, vol. 1, ed. D. S. Ingram & P. H. Williams, pp. 35–82. London: Academic Press.

Keen, N. T. & Legrand, M. (1980). Surface glycoproteins: evidence that they may function as the race specific phytoalexin elicitors of *Phytophthora megasperma* f.sp. *glycinea. Physiol. Plant Pathol.* **17:** 175–92.

Keen, N. T., Midland, S. & Sims, J. J. (1983*a*). Purification of victorin. *Phytopathology* **73:** 830 Abs.

Keen, N. T. & Yoshikawa, M. (1983). 1,3-endoglucanase from soybean releases elicitors-active carbohydrates from fungus cell walls. *Plant Physiol.* **71:** 460–5.

Keen, N. T., Yoshikawa, M. & Wong, M. C. (1983*b*). Phytoalexin elicitor activity of carbohydrates from *Phytophthora megasperma* f.sp. *glycinea* and other sources. *Plant Physiol.* **71:** 466–71.

Kendra, D. F. & Hadwiger, L. E. (1984). Characterization of the smallest chitosan oligomer that is maximally antifungal to *Fusarium solani* and elicits pisatin formation in *Pisum sativum. Exp. Mycol.* **8:** 276–81.

Kinniard, J. H., Keighren, M. A., Kinsey, J. A., Eaton, M. & Fincham, J. R. S. (1982). Cloning of the an (glutamate dehydrogenase) gene of *Neurospora crassa* through the use of a synthetic DNA probe. *Gene* **20:** 387–96.

Kiraly, Z. (1980). Defenses triggered by the invader: hypersensitivity. In *Plant Disease: An Advanced Treatise,* vol. 5, ed. J. G. Horsfall & E. G. Cowling, pp. 201–24. New York: Academic Press.

Kiraly, Z., Barna, B. & Ersek, T. (1972). Hypersensitivity as a consequence, not the cause, of plant resistance to infection. *Nature* **239:** 456–567.

Klee, H. J., White, F. F., Iver, V. N., Gordon, M. P. & Nester, E. W. (1983). Mutational analysis of the virulence region of an *Agrobacterium tumefaciens* Ti plasmid. *J. Bacteriol.* **153:** 879–83.

Kohle, H., Young, D. H. & Kauss, H. (1984). Physiological changes in suspension-cultured soybean cells elicited by treatment with chitosan. *Plant Science Letters* **33:** 221–30.

Kojima, M., Kawakita, K. & Uritani, I. (1982). Studies on a factor in sweet potato root which agglutinates spores of *Ceratocystis fimbriata*, Black rot fungus. *Plant Physiol.* **69:** 474–8.

Kolattukudy, P. E. (1981). Structure, biosynthesis and biodegradation of cutin and suberin. *Ann. Rev. Plant Physiol.* **32:** 539–67.

Kolattukudy, P. E. (1984). Fungal penetration of defensive barriers of plants. In *Structure, Function and Biosynthesis of Plant Cell Walls*, ed. W. M. Dugger & S. Bartnicki-Garcia, pp. 302–43. Baltimore: Waverley Press.

Kratka, J. & Kudela, V. (1984). Effect of alfalfa wilts on the hydroxyproline content in cell wall. *Phytopath. Z.* **110:** 127–33.

Kuc, J. (1976). Phytoalexins. In *Physiological Plant Pathology, Encyclopedia of Plant Physiology*, ed. R. Heitefuss & P. H. Williams, pp. 632–52. Berlin and New York: Springer-Verlag.

Kunoh, H. & Ishizaki, H. (1975). Silicon levels near penetration sites of fungi on wheat, barley, cucumber and morning glory leaves. *Physiol. Plant Pathol.* **5:** 283–7.

Larkin, P. J. (1981). Plant protoplast agglutination and immobilization. In *Recent Advances in Phytochemistry*, vol. 15: *The Phytochemistry of Cell Recognition and Cell Surface Interactions*, ed. F. A. Loewus & C. A. Ryan, pp. 135–60. New York and London: Plenum Press.

Leach, J. E., Cantrell, M. A. & Sequiera, L. (1982). Hydroxyproline-rich bacterial agglutinin from potato. *Plant Physiol.* **70:** 1353–8.

Lee, S. C. & West, C. A. (1981). Polygalacturonase from *Rhizopus stolonifer*, an elicitor of casbene synthetase activity in castor bean (*Ricinus communis* L.) seedlings. *Plant Physiol.* **67:** 633–9.

Lippincott, J. A. & Lippincott, B. B. (1984). Concepts and experimental approaches in host-microbe recognition. In *Plant–Microbe Interactions: Molecular and Genetic Perspectives*, vol. 1, pp. 195–214. New York: Macmillan.

Long, S. R., Buikema, W. J. & Ausubel, F. M. (1982). Cloning of *Rhizobium meliloti* nodulation genes by direct complementation of Nod-mutants. *Nature* **298**: 485–8.

Loschke, D. C., Hadwiger, L. A. & Wagoner, W. (1983). Comparison of mRNA populations coding for phenylalanine ammonia lyase and other peptides from pea tissue treated with biotic and abiotic phytoalexin inducers. *Physiol. Plant Pathol.* **23**: 163–73.

Maniara, G., Laine, R. & Kuc, J. (1984). Oligosaccharides from *Phytophthora infestans* enhance the elicitation of secquiterpenoid stress metabolites by arachidonic acid in potato. *Physiol. Plant Pathol.* **24**: 177–86.

Matthews, D. E. & Van Etten, H. D. (1983). Detoxification of the phytoalexin pisatin by a fungal cytochrome P-450. *Arch. Biochem. Biophys.* **224**: 494–505.

Matzke, M. A., Susani, M., Binns, A. N., Lewis, E. D., Rubenstein, I. & Matzke, A. J. M. (1984). Transcription of a zein gene introduced into sunflower using a Ti plasmid vector. *EMBO J.* **3**: 1525–31.

Mayama, S., Daly, J. M., Rehfeld, D. W. & Daly, R. C. (1975a). Hypersensitive response of near-isogenic wheat carrying the temperature-sensitive Sr6 allele for resistance to stem rust. *Physiol. Plant Pathol.* **7**: 35–47.

Mayama, S., Rehfeld, D. W. & Daly, R. C. (1975b). The effect of detachment on the development of rust disease and the hypersensitive response on wheat leaves infected with *Puccinia graminis tritici. Phytopathology* **65**: 1139–42.

Mayama, S. & Tani, T. (1981). Relationship between the production of phytoalexin avenalumin and host sensitive infection of oat Pc lines by *Helminthosporium victoriae. Ann. Phytopath. Soc. Jpn* **47**: 124.

Mellon, J. E. & Helgeson, J. P. (1982). Interactions of a hydroxyproline-rich glycoprotein from tobacco callus with potential pathogens. *Plant Physiol.* **70**: 401–5.

Mendgen, K. (1978). Attachment of bean rust cell wall material to host and non-host plant tissue. *Arch. Microbiol.* **119**: 113–17.

Miller, S. A. & Maxwell, D. P. (1984a). Light microscope observations of susceptible, host resistant, and nonhost resistant interactions of alfalfa with *Phytophthora megasperma. Can. J. Bot.* **62**: 100–16.

Miller, S. A. & Maxwell, D. P. (1984b). Ultrastructure of susceptible, host resistant, and nonhost resistant interactions of alfalfa with *Phytophthora megasperma. Can. J. Bot.* **62**: 117–28.

Nestler, E. J. & Greengard, P. (1983). Protein phosphorylation in the brain. *Nature* **305**: 583–8.

Nothnagel, E. A., McNeil, M., Dell, A. & Albersheim, P. (1983). Host–pathogen interactions. XXII. A galacturonic acid oligosaccharide from plant cell walls elicits phytoalexins. *Plant Physiol.* **71**: 916–26.

Nozue, M., Tomiyama, K. & Doke, K. (1979). Evidence for adherence of host plasmalemma to infecting hyphae of both compatible and incompatible races of *Phytophthora infestans. Physiol. Plant Pathol.* **15**: 111–15.

Paxton, J. D. (1981). Phytoalexins – a working redefinition. *Phytopathol. Z.* **101**: 106–9.

Pearce, R. B. & Ride, J. P. (1980). Specificity of induction of the lignification response in wounded wheat leaves. *Physiol. Plant Pathol.* **16**: 197–204.

Pearce, R. B. & Ride, J. P. (1982). Chitin and related compounds as elicitors of the lignification response in wounded wheat leaves. *Physiol. Plant Pathol.* **20**: 119–23.

Peters, B. M., Cribbs, D. H. & Stelzig, D. A. (1978). Agglutination of plant protoplasts by fungal cell-wall glucans. *Science* **201**: 364–5.

Pueppke, S. G. (1984). Adsorption of bacteria to plant surfaces. In *Plant–Microbe Interactions: Molecular and Genetic Perspectives*, vol. 1, ed. T. Kosuge & E. W. Nester, pp. 215–64. New York: Macmillan.

Ratner, D. I., Ward, T. E. & Jacobson, A. (1981). Evidence for the transformation of *Dictyostelium discoideum* with homologous DNA. In *Developmental Biology Using Purified Genes*, ed. D. D. Brown & C. F. Fox, pp. 595–605. New York: Academic Press.

Rees, D. A. (1977). *Polysaccharide Shapes. Outline Studies in Biology.* London: Chapman and Hall.

Ride, J. P. (1980). The effect of induced lignification on the resistance of wheat cell walls to fungal degradation. *Physiol. Plant Pathol.* **16**: 187–96.

Rochaix, J. D. & van Dillewijn, N. (1982). Transformation of the green alga *Chlamydomonas reinhardii* with yeast DNA. *Nature* **296**: 70–2.

Russo, V. M. & Pappelis, A. J. (1981). Observations of *Colletotrichum dematium* f. *circinans* on *Allium cepa*: halo formation and penetration of epidermal walls. *Physiol. Plant Pathol.* **19**: 127–36.

Ryan, C. A. (1984). Systemic responses to wounding. In *Plant–Microbe Interactions: Molecular and Genetics Perspectives*, vol. 1, ed. T. Kosuge & E. W. Nester, pp. 307–20. New York: Macmillan.

Sargent, C. & Gay, J. L. (1977). Barley and epidermal apoplast structure and modification by powdery mildew contact. *Physiol. Plant Pathol.* **11**: 195–205.

Sequiera, L. (1984). Plant bacterial interactions. In *Encyclopaedia of Plant Physiology*, vol. 17, ed. H. F. Linskens & J. Heslop-Harrison, pp. 187–207. Heidelberg: Springer-Verlag.

Sherwood, R. T. & Vance, C. P. (1976). Histochemistry of papillae formed in reed canary grass leaves in response to non-infecting pathogenic fungi. *Phytopathology* **66**: 503–10.

Skou, J. P., Helms Jorgensen, J. & Lilholt, U. (1984). Comparative studies on callose formation in powdery mildew compatible and incompatible barley. *Phytopath. Z.* **109**: 147–68.

Smart, M. G., Aist, J. R. & Israel, H. W. (1984). Are papillae in barley coleoptiles impermeable to small molecules? *Abstract A494, Phytopathology* **74**: 851.

Soliday, C. L., Flurkey, W. H., Okita, T. W. & Kolattukudy, P. E. (1984). Cloning and structure determination of cDNA for cutinase, an enzyme involved in fungal penetration of plants. *Proc. Nat. Acad. Sci. U.S.A.* **81**: 3939–43.

Stahl, U., Tudzynski, P., Kuch, U. & Esser, K. (1982). Replication and expression of a bacterial–mitochondrial hybrid plasmid in the fungus *Podospora anserina*. *Proc. Nat. Acad. Sci. U.S.A.* **79**: 3641–5.

Stakman, E. C. (1915). Relation between *Puccinia graminis* and plants highly resistant to its attack. *J. Agric. Res.* **4**: 193–200.

Staples, R. C. & Macko, V. (1980). Formation of infection structures as a recognition response in fungi. *Ex. Mycol.* **4**: 2–16.

Staskawicz, B. J., Dahlbeck, D. & Keen, N. T. (1984). Cloned avirulence gene of *Pseudomonas syringae* pv. *glycinea* determines race-specific incompatibility on *Glycine max* (L.) Merr. *Proc. Nat. Acad. Sci. U.S.A.* **81**: 6024–8.

Stossel, P., Lazarovits, G. & Ward, E. W. B. (1980). Penetration and growth of compatible and incompatible races of *Phytophthora megasperma* var. *sojae* in soybean by hypocotyl tissue differing in age. *Can. J. Bot.* **58**: 2594–2601.

Stossel, P., Lazarovits, G. & Ward, E. W. B. (1981). Electron microscope study of race-specific and age-related resistant and susceptible reactions of soybeans to *Phytophthora megasperma* var. *sojae*. *Phytopathology* **71**: 617–23.

Stoessl, A. (1983). Secondary plant metabolites in preinfectional and postinfestional resistance. In *The Dynamics of Host Resistance*, ed. J. A. Bailey & B. J. Deverall, pp. 71–123. New York, London: Academic Press.

Tanabe, T., Pricer, W. E. & Ashwell, G. (1979). Subcellular membrane topology and turnover of a rat hepatic binding protein specific for asialoglycoproteins. *J. Biol. Chem.* **254**: 1038–43.

Tepfer, M. & Taylor, I. E. P. (1981). The permeability of plant cell walls as measured by gel filtration chromatography. *Science* **213**: 761–3.

Tighe, D. M. & Heath, M. C. (1982). Callose induction in cowpea by uridine diphosphate glucose and calcium phosphate–boric acid treatments. *Plant Physiol.* **69**: 366–70.

Tippett, J. & Malajczuk, N. (1979). Interaction of *Phytophthora cinnamomi* and a resistant host *Acacia pulchella*. *Cytol. Hist. Phytopathology* **69**: 764–72.

Toppan, A., Roby, D. & Esquerre-Tugaye, M. T. (1982). Cell surfaces in plant-microorganism interactions. III. In vivo effect of ethylene on hydroxyproline-rich glycoprotein accumulation in the cell wall of diseased plants. *Plant Physiol.* **70**: 82–6.

Touze, A. & Esquerre-Tugaye, M-T. (1982). Defense mechanisms of plants against varietal non-specific pathogens. In *Active Defense Mechanisms in Plants*, ed. R. K. S. Wood, pp. 103–17. New York and London: Plenum Press.

Vance, C. P., Kirk, T. K. & Sherwood, R. T. (1980). Lignification as a mechanism of disease resistance. *Ann. Rev. Phytopathol.* **18**: 259–88.

Vanderplank, J. E. (1982). *Host–Pathogen Interactions in Plant Disease*. London: Academic Press.

Van der Ploeg, L. H. T., Schwartz, D. C., Cantor, C. R. & Brost, P. (1984). Antigenic variation in *Trypanosoma brucei* analysed by electrophoretic separation of chromosome sized DNA molecules. *Cell* **37**: 77–84.

Van Holst, G. J. & Varner, J. E. (1984). Reinforced polyproline. II. Conformation in hydroxyproline-rich cell wall glycoprotein from carrot root. *Plant Physiol.* **74**: 247–51.

Walker-Simmons, M., Hadwiger, L. & Ryan, C. A. (1983). Chitosans and pectic polysaccharides both induce the accumulation of the antifungal phytoalexin pisatin in pea pods and antinutrient proteinase inhibitors in tomato leaves. *Biochem. Biophys. Res. Commun.* **110**: 94–9.

West, C. A. (1981). Fungal elicitors of the phytoalexin response in higher plants. *Naturwiss.* **68**: 447–57.

Wetherbee, R., Hinch, J. M., Bonig, I. & Clarke, A. E. (1985). Response of *Zea mays* roots to infection with *Phytophthora cinnamomi*: II. The cortex and stele. *Protoplasma* **126**: 188–97.

Whatley, M. H. & Sequiera, L. (1981). Bacterial attachment to plant cell walls. *Recent Adv. Phytochem.* **15**: 213–40.

Whitmore, F. W. (1978). Lignin–protein complex catalysed by peroxidase. *Plant Science Letters* **13**: 241–5.

Williams, J. G. & Lloyd, M. M. (1979). Changes in the abundance of polyadenylated RNA during slime mould development measured using cloned molecular hybridisation probes. *J. Mol. Biol.* **129**: 19–35.

de Wit, P. J. G. M. & Roseboom, P. H. M. (1980). Isolation, partial characterisation and specificity of glycoprotein elicitors from culture filtrates, mycelium and cell walls of *Cladosporium fulvum* (syn. *Fulvia fulva*). *Physiol. Plant Pathol.* **16**: 391–408.

de Wit, P. J. G. M., Hofman, J. E. & Aarts, J. M. M. J. G. (1984). Origin of specific elicitors of chlorosis and necrosis occurring in intercellular fluids of compatible interactions of *Cladosporium fulvum* (syn. *Fulvia fulva*) and tomato. *Physiol. Plant Pathol.* **24**: 17–23.

Yamamoto, M. (1982). Recent advance to studies of potato late blight, with special reference to the nature of resistance of potatoes to the invasion of *Phytophthora infestans*. *Shokubutsu Byogai Kenku* **9**: 1–37.

Yoder, O. C. (1980). Toxins in pathogenesis. *Ann. Rev. Phytopathol.* **18**: 103–29.

Yoshikawa, M., Yamauchi, K. & Masago, H. (1978*a*). Glyceollin: its role in restricting fungal growth in resistant soybean hypocotyls infected with *Phytophthora megasperma* var. *sojae*. *Physiol. Plant Pathol.* **12**: 73–82.

Yoshikawa, M., Yamauchi, K. & Masago, H. (1978*b*). De novo messenger RNA and protein synthesis are required for phytoalexin-mediated disease resistance in soybean hypocotyls. *Plant Physiol.* **61**: 314–17.

Yoshikawa, M., Keen, N. T. & Wang, M. C. (1983). A receptor on soybean membranes for a fungal elicitor of phytoalexin accumulation. *Plant Physiol.* **73**: 497–506.

11

Receptors for attachment of Rhizobium to legume root hairs

Frank B. Dazzo

11.1 Introduction

Cellular recognition between microorganisms and higher plants plays a role in plant morphogenesis, nutrition, and protection against infectious disease. Positive cellular recognitions are believed to arise from a specific union, reversible or irreversible, between chemical receptors on the surface of interacting cells (Burnet, 1971). This implies that cells recognize one another when they come into contact, and therefore the complementary components of the cell surfaces are a focus for many biochemical studies on cellular recognition. Such was the case for the studies by Hamblin & Kent (1973) and Bohlool & Schmidt (1974), which suggested that the specific interaction between legume lectins and polysaccharides on the *Rhizobium* symbiont serves as a basis of specificity in this plant–microorganism symbiosis.

The bacterium, *Rhizobium*, selectively infects legume roots and forms root nodules that fix atmospheric nitrogen into ammonia which can be assimilated for plant growth. The selectivity of rhizobia and legume host in the infection process is used to define the various species of *Rhizobium*. For example, *R. trifolii* infects clover root hairs and *R. meliloti* infects alfalfa root hairs. Neither species infects root hairs of the heterologous plant host. Successful infection of legume roots by these nitrogen-fixing bacteria in soil is of immense importance in the nitrogen cycle on earth.

There are many cellular recognition phenomena which occur during the infection of legume roots by *Rhizobium*. Lectin-mediated attachment of rhizobia to legume host root hairs is considered an early step of cellular recognition as illustrated by non-infective mutant strains which are defective in hapten-specific attachment steps (Dazzo, Napoli & Hubbell, 1976; Kato, Maruyama & Nakamura, 1981; Paau, Leps & Brill, 1981; Zurkowski, 1980). However, recent studies have shown clearly that there are other important keys needed to unlock the many doors blocking the way to successful infection. An understanding of the recognition code to host specificity would help not only to elucidate the developmental events in this plant–microorganism symbiosis, but also to indicate ways in which *Rhizobium* and the host plant may be manipulated genetically to broaden the range of agricultural crops which can enter efficient nitrogen-fixing symbioses. In this chapter, I review the biology and biochemistry of attachment stages in the *Rhizobium*–legume symbiosis.

11.2   Bacterial attachment

11.2.1   *Quantitation and electron microscopy*

Quantitative microscopic assays (Dazzo *et al.* 1976; Dazzo *et al.* 1984) and transmission electron-microscopic studies (Dazzo & Hubbell, 1975; Napoli & Hubbell, 1975; Kumarasinghe & Nutman, 1977) have revealed multiple mechanisms of rhizobial attachment to clover root hairs. A nonspecific mechanism allows all rhizobia to attach in low numbers (2–4 cells per 200 μm root hair length per 12 using low inoculum per seedling). In addition, a specific mechanism allows selective attachment in significantly ($P = 0.005$) larger numbers (22–27 cells per 200 μm root hair length per 12) under identical conditions (Dazzo *et al.* 1976). Electron microscopy (Dazzo & Hubbell, 1975) disclosed that the initial bacterial attachment step consisted of contact between the fibrillar capsule of *R. trifolii* and electron-dense globular aggregates lying on the outer periphery of the fibrillar clover root hair cell wall (Fig. 11.1). This 'docking' stage is the first point of physical contact between the microbe and the host (Phase I attachment), and occurs within minutes after inoculation of encapsulated cells of *R. trifolii* on the host clover.

11.2.2   *Phase I attachment*

To identify which cell surface molecules are involved in Phase I attachment, we are examining the surface components of the bacterium and the host that interact with the same order of specificity as is observed with the adhesion of the bacterial cells. Immunochemical and genetic studies have demonstrated that the surfaces of *R. trifolii* and clover epidermal cells contain a unique antigen that is immunochemically cross-reactive (Dazzo & Hubbell, 1975; Dazzo & Brill, 1979), suggesting its structural relatedness on both symbionts. This antigen contains receptors that bind reversibly to a multivalant clover lectin called trifoliin A (originally trifoliin) which has been isolated from seeds and seedling roots (Dazzo, Yanke & Brill, 1978; Dazzo & Brill, 1979). A specific hapten inhibitor of trifoliin A binding to these receptors is 2-deoxy-D-glucose

(Dazzo & Hubbell, 1975; Dazzo & Brill, 1977). The first clue that trifoliin A on the root may be involved in rhizobial attachment came from the observation that 2-deoxy-D-glucose specifically inhibited the attachment of *R. trifolii* to clover root hairs (Dazzo *et al.* 1976), reducing the high level of bacterial adhesion to that characteristic of background. Subsequent studies showed that 2-deoxy-D-glucose specifically facilitated the elution of trifoliin A from the intact clover root (Dazzo & Brill, 1977; Dazzo *et al.* 1978), and inhibited the binding of *R. trifolii* capsular polysaccharide to clover root hairs (Dazzo & Brill, 1977). As a negative control, 2-deoxy-D-glucose did not inhibit adsorption of *R. meliloti* or its capsular polysaccharide to alfalfa root hairs (Dazzo *et al.* 1976; Dazzo & Brill, 1977). Consistent with the above results in the *Rhizobium trifolii*–clover system, lectins on pea, alfalfa, and soybean roots accessible for binding symbiont rhizobia have been demonstrated (Kijne, van der Schaal & de Vries, 1980; Kijne, van der Schaal, Diaz & van Iren, 1982; Kato, Maruyama & Nakamura, 1980, 1981; Gatehouse & Boulter, 1980; van der Schaal & Kijne, 1981; Paau *et al.* 1981; Stacey, Paau & Brill,

Fig. 11.1. Docking stage of Phase 1a attachment of *Rhizobium trifolii* NA-30 to the clover root-hair cell wall (transmission electron micrograph). Note the fibrillar capsule which contacts electron-dense aggregates on the outer periphery of the root hair cell wall. (From Dazzo & Hubbell, 1975, and courtesy of the American Society for Microbiology.)

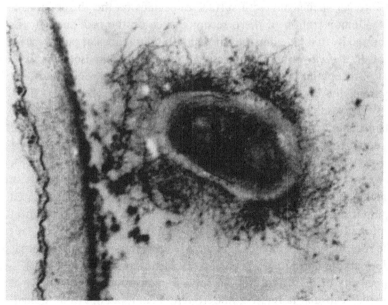

1980; Gade *et al.* 1981). In addition, specific hapten-facilitated elution of lectin from pea, alfalfa, and soybean roots have also been found (Kijne *et al.* 1980; van der Schaal & Kijne, 1981; Gade *et al.* 1981; W. Kamberger, personal communication). In such hapten-facilitated elution techniques, it is believed that the sugar acts by combining specifically with the site on the lectin which is normally occupied by the natural saccharide receptor. This implies a close but not necessarily identical structure of the hapten and the native determinant. However, some lectins undergo conformational changes when associated with saccharide binding (Reeke *et al.* 1975), and so this possibility must also be considered in the interpretation of hapten inhibition studies.

To explain this early recognition event of attachment on the clover root-hair surface prior to infection, we proposed (Dazzo & Hubbell, 1975) that the multivalent trifoliin A (it agglutinates cells) recognizes similar saccharide residues on *R. trifolii* and clover and cross-bridges them in a complementary fashion to form the correct molecular inter-facial structure that initiates the preferential and specific adsorption of the bacteria to the root-hair surface. By mediating the adhesion of these specific cells, the lectin may also function as a 'cell recognition molecule' since it could feasibly influence which cells associate in sufficient proximity to the root hairs to allow subsequent specific recognition steps to occur.

This model predicts that host-specific receptor sites on the legume root interact specifically with surface molecules on the rhizobial symbiont. Demonstration of these receptor sites on the root surface was performed by first labelling the trifoliin A-binding capsular polysaccharide of *R. trifolii* with the fluorescent dye fluorescein isothiocyanate (FITC), incubating the conjugate with sterile seedling roots, and then examining the roots by epifluorescence microscopy (Dazzo & Brill, 1977). The results are shown in Fig. 11.2. These receptor sites on clover roots accumulated at root-hair tips and diminished toward the base of the root hair. This unique location exactly matched the distribution of trifoliin A on the surface of the clover seedling root (Dazzo *et al.* 1978), and the sites which immediately bound encapsulated *R. trifolii* at Phase I attachment (Fig. 11.3; Dazzo & Brill, 1979). This result highlighted the importance of epidermal cell differentiation in the development of receptor sites that recognize rhizobia. Undifferentiated epidermal cells in the root-hair region did not bind the bacterial polysaccharide, whereas epidermal root hair primordia gained this surface property (Fig. 11.2).

Specificity of these receptor sites was demonstrated by the ability of unlabelled capsular polysaccharide from *R. trifolli*, but not from *R. meli-*

loti, to block the binding of the labelled polysaccharide. Similar specific binding of bacterial polysaccharides to legume host root hairs has been demonstrated in the *R. meliloti*–alfalfa (Dazzo & Brill, 1977), *R. legumi-nosarum*-pea (Kato, Maruyoma & Nakamura, 1980), and *R. japonicum*–soybean system (Hughes, Leece & Alkan, 1979; Hughes & Elkan, 1981).

The results of immunochemical and genetic studies suggest that trifo-liin A and the cross-reactive anti-clover root antibody bind to the same or similar overlapping saccharide determinants on *R. trifolii* (Dazzo & Brill, 1979). First, the antibody and the lectin bind specifically to the same isolated polysaccharides from *R. trifolii*. Second, this interaction is specifically inhibited by the hapten, 2-deoxy-D-glucose. Third, the genetic markers of *R. trifolii* that bind trifoliin A and the antibody cotransform into *Azotobacter vinelandii* with 100% frequency (Bishop *et al.* 1977). And fourth, monovalent Fab fragments of IgG from anti-clover root antiserum strongly block the binding of trifoliin A to *R. trifolii* (Dazzo & Brill, 1979). Considered collectively, these studies

Fig. 11.2. Specific binding of FITC-labelled capsular polysaccharide from *Rhizobium trifolii* 0403 to clover root hairs (epifluorescence micrograph). (From Dazzo & Brill, 1977, and courtesy of the American Society for Microbiology.)

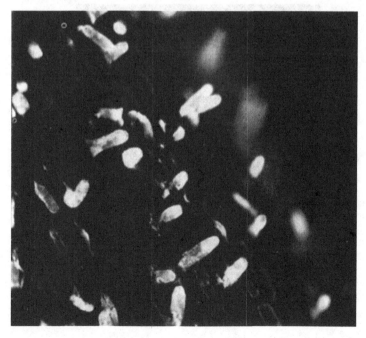

suggest that *R. trifolii* and clover roots have similar saccharide receptors for trifoliin A. However, the definitive test of their identity as antigenically related structures will require knowledge of the minimal saccharide sequence that binds the clover lectin.

Three experiments suggest that trifoliin A and antibody to the cross-reactive antigen bind to the same *R. trifolii* saccharide determinants that bind these bacteria to clover root hairs (Dazzo & Brill, 1979). First, Fab fragments of anti-clover root IgG blocked Phase I attachment of *R. trifolii* to clover root hairs. Second, only the *A. vinelandii* hybrid transformants that carried the trifoliin A receptor bound to clover root hairs in Phase I attachment assays. Third, competition assays using fluorescence microscopy indicated that the *R. trifolii* polysaccharides that bound trifoliin A had the highest affinity for clover root hairs.

Selectivity of rhizobial attachment to legume host root hairs has been found in many, but not all, cases. For instance, hapten-inhibitable and host-specific attachment has been demonstrated in *R. japonicum*-soybean (Stacey, Paau & Brill, 1980) and *R. leguminosarum*–pea systems

Fig. 11.3. Attachment of encapsulated *Rhizobium trifolii* 0403 to the tip of a clover root hair after short-term incubation (scanning electron micrograph). (From Dazzo & Brill, 1979, and courtesy of the American Society for Microbiology.)

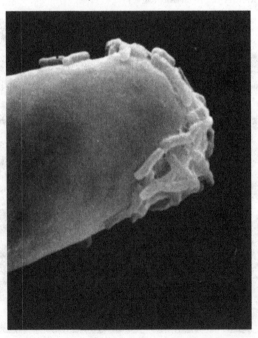

(Kato *et al.* 1980). *Rhizobium japonicum* also selectively attaches to soybean root cells in suspension culture (Reporter, Raveed & Norris, 1975). As with the *R. trifolii*—clover system (Dazzo *et al.* 1976; Zurkowski, 1980), non-nodulating mutant strains of *R. leguminosarum* which produce less extracellular/capsular polysaccharide (Saunders, Carlson & Albersheim, 1978; Napoli & Albersheim, 1980) or heterologous rhizobia (Kato *et al.* 1980) attach in smaller numbers to pea root hairs as compared with the wild-type nodulating strains (Kato *et al.* 1980, 1981; C. Napoli, personal communication). On the other hand, specificity in quantitative studies of root attachment is not found when enormously large bacterial densities are incubated with roots (e.g. 10^9–10^{10} cells per seedling: Chen & Phillips, 1976). Nevertheless, the latter study was significant in that it revealed the importance of root hair tips as sites for rhizobial attachment.

11.2.3 Phase II adhesion

Phase I attachment is followed by Phase II adhesion, characterized by the firm anchoring of the bacterial cell to the root-hair surface during later preinfective stages (Dazzo & Truchet, 1983; Dazzo *et al.* 1984). Phase II adhesion may be important in maintaining the first contact between the bacterium and the host root hair necessary for triggering the tight root-hair curling (shepherd's crook formation) and successful penetration of the root-hair cell wall during infection (Napoli, Dazzo & Hubbell, 1975). During Phase II adhesion, fibrillar materials, recognized by scanning electron microscopy, are characteristically found associated with the adherent bacteria. The nature of these microfibrils is unknown. One possibility is that they are bundles of cellulose microfibrils, known to be produced by many rhizobia (Deinema & Zevenhuizen, 1971; Napoli *et al.* 1975). Another possibility is that they are collections of pili, which have been demonstrated recently in *Rhizobium* (Kijne *et al.* 1982). Future studies should be directed to isolate and characterize these fibrils associated with the adherent bacteria in order to understand the Phase II adhesion process better.

Several lines of evidence indicate that attachment of rhizobia to root hairs is important but not the most crucial prerequisite for infection. First of all, very few root hairs to which infective rhizobia attach eventually become infected. This may be due to a transient susceptibility of the root hairs to infection by the rhizobial symbiont (Bhuvaneswari, Turgeon & Bauer, 1980; Bhuvaneswari, Bhagwat & Bauer, 1981). Secondly, genetic hybrids of *Azotobacter vinelandii* (which carry the trifoliin A-binding saccharide receptor on their surface as a result of

intergeneric transformation with DNA from *R. trifolii*: Bishop *et al.*
1977), have acquired the ability to attach specifically to clover root hairs
(Dazzo & Brill, 1979) but do not infect them. Finally, although mutant
strains which fail to bind the host lectin neither attach well to the host
root hairs nor infect them (Dazzo *et al.* 1976; Paau *et al.* 1981), another
class of non-infective mutant strains has been shown to bind the host
lectin and attach to the host root hairs (Kamberger, 1979; Paau *et al.*
1981). These results are consistent with lectin-mediated root-hair attachment, but indicate that other cell-recognition events must occur for the
bacterium to penetrate the root hair. Possibilities for other genes or
gene products which may not have been expressed in the above situations
include those controlling cell-wall hydrolytic enzymes (Hubbell, Morales
& Umali-Garcia, 1978; Martinez-Molina, Morales & Hubbell, 1979),
inducers of host polygalacturonase (Ljunggren & Fahraeus, 1961;
Palomares, Montega & Olivares, 1978), root-hair curling factors (Yao &
Vincent, 1976), and periplasmic extrinsic substance ES-6000 which pro-
motes root-hair infection (Higashi & Abe, 1980).

11.3   Regulation of the attachment process

11.3.1   Combined nitrogen supplied to growing roots

Combined nitrogen inhibits nodule formation. For instance,
nitrate supplied at critical concentrations inhibits all of the morphogenetic steps of the nodulation process known to involve the bacterial symbiont (Truchet & Dazzo, 1982).

We have studied the effect of nitrate supply on accumulation of trifoliin
A as a model for how this combined nitrogen ion regulates the infection
process. Microscopic assays indicated that the specific binding of *R. trifolii* 0403 to clover root hairs and the levels of trifoliin A on these epidermal
cells declined in parallel as the nitrate concentration was increased from
1 mM to 15 mM in the rooting medium (Dazzo & Brill, 1978). The inhibition was due specifically to nitrate ion, and 15 mM nitrate did not stunt
seedling growth. Nitrate does not bind directly to trifoliin A or its glycosylated receptors in a way which would reduce the levels of this lectin
on clover roots or block attachment of *R. trifolii* 0403 to root hairs.
Rather, some intervening process modulated by nitrate supply over-
periods greater than 1 h regulates these early recognition events of the
infection process (Dazzo & Hrabak, 1982).

Other studies have shown that nitrate supply affects root cell wall
composition (Dazzo *et al.* 1981; Diaz, Kijne & Quispel, 1981). For
instance, nitrate supply increases the levels of extensin, the hydroxyproline-rich glycoprotein in root cell walls (Dazzo *et al.* 1981). Since rhizobia

must penetrate the host cell wall, changes in the chemistry of the wall could have an important impact on the infection process. In addition, the accessibility of trifoliin A receptors on clover root cell walls is reduced when the plant is grown with nitrate. Isolated root cell walls were assayed for the ability to bind trifoliin A and reduce its agglutination titre with *R. trifolii* 0403. Walls from plants grown in nitrogen-free conditions adsorbed 3–4-fold more trifoliin A-agglutinating activity per mg dry wt of walls than walls of roots grown with 15 mM nitrate (Dazzo *et al.* 1981). Nitrate supply also seems to affect the accumulation of pea lectin on pea roots (Diaz *et al.* 1981).

Since trifoliin A did not accumulate on or bind well to root cell walls grown in 15 mM nitrate, we wondered whether trifoliin A would be released from the roots and accumulate in root exudate. Immunofluorescence assays of root exudate detected 'active' trifoliin A capable of binding to the bacterial symbiont *R. trifolii*; 30-fold higher levels of active trifoliin A (per constant total protein concentration) were detected from root exudate of clovers grown under nitrogen-free conditions than from exudate of root grown with 15 mM nitrate (Dazzo & Hrabak, 1981). The presence of trifoliin A in root exudate of two white clover varieties was confirmed by its purification using immunoaffinity chromatography (Dazzo & Hrabak, 1981; Dazzo *et al.* 1982). This work has recently been extended by demonstrating the *in vivo* synthesis of trifoliin A in seedling roots of white clover (Sherwood, Truchet, Hrabak & Dazzo, 1984). Axenic growth of seedlings with an excess supply of nitrate does not repress trifoliin A synthesis but does lower the levels of active trifoliin A on the root surface and the external root environment which could interact with the bacterial symbiont.

The presence in clover root exudate of trifoliin A which can bind to receptors on *R. trifolii* provides supporting evidence for a lectin-recognition model proposed by Solheim (1975). According to this model, a glycoprotein lectin excreted from the legume root binds to the rhizobia. This active complex then combines with a receptor site on the root. Thus, both partners in the symbiosis could benefit from the discriminatory reaction of a cross-bridging lectin which could be either bound to a glycosylated receptor on the root-hair cell wall (Dazzo & Hubbell, 1975) or released from the root to bind to the rhizobial cell (Solheim, 1975). This event would help to ensure that only the symbiotic bacterium could establish the proper intimate contact with the host cell required to trigger other recognition events that lead to successful infection. Combined nitrogen (e.g., nitrate) would play a role in regulating the recognition process as proposed by Solheim (1975).

11.3.2   *Transient appearance of lectin receptors on* Rhizobium

The selective ability of *R. trifolii* to attach to clover root hairs is controlled by the transient accumulation of the saccharide receptor for trifoliin A on the bacterium. For instance, the transient appearance of trifoliin A receptors on *R. trifolii* may influence the ability of these bacteria to attach to clover root hairs (Dazzo, Urbano & Brill, 1979; Sherwood, Vasse, Dazzo & Truchet, 1984). Cells grown on agar plates of a defined medium were most susceptible to agglutination by trifoliin A when they were harvested at 5 days of growth, and this is when they attached to clover roots in greatest quantity (Dazzo *et al.* 1979; Sherwood *et al.* 1984). In broth cultures, these receptors were detected as cultures left their lag phase of growth and again as they entered stationary phase. Clover roots adsorbed the bacteria in greatest quantity when the cells were in early stationary phase (Dazzo *et al.* 1979).

The dominant receptor for trifoliin A on *R. trifolii* is the capsular polysaccharide (CPS) for plate-grown cells and lipopolysaccharide (LPS) for broth-grown cells (Sherwood *et al.* 1984; Hrabak, Urbano & Dazzo, 1981). High-pressure liquid chromatography and nuclear magnetic resonance spectroscopy indicate that the non-carbohydrate substitutions (acetate, pyruvate, and betahydroxybutyrate) are the components of the trifoliin A-binding CPS which change with culture age (Hollingsworth, Abe, Sherwood & Dazzo, 1984; Sherwood *et al.* 1984; Abe, Sherwood, Hollingsworth & Dazzo, 1984). Unique determinants in the LPS are responsible for transient binding of trifoliin A to broth-grown cells (Hrabak *et al.* 1981) (Fig. 11.4). Gas chromatography and combined gas chromatography–mass spectrometry showed increases in several glycosyl components of the LPS with culture age. Quinovosamine (2-amino-2,6-dideoxyglucose) in the β-anomeric configuration was one of the components which increased in the LPS as cells entered stationary phase, and was an effective hapten inhibitor of the binding of trifoliin A.

There is a transient appearance and disappearance on *R. japonicum* of the receptor that specifically binds soybean lectin (Bhuvaneswari, Pueppke & Bauer, 1977; Vasse, Dazzo & Truchet, 1984). Most strains of *R. japonicum* have the highest percentage of soybean lectin-binding cells and the greatest number of soybean lectin-binding sites per cell in the early and mid-log phases of growth. The proportion of galactose residues in the capsular polysaccharide is high at a culture age when the cells bind the galactose-reversible soybean lectin (Mort & Bauer, 1980). A decline in lectin-binding activity of cells accompanying culture ageing is concurrent with a decline in galactose content and a rise in 4-O-methyl galactose residues in the capsular polysaccharides. The latter

methylated sugar has low affinity for the galactose-binding soybean lectin. These results suggest that the galactose residues in the capsular
polysaccharide become methylated, and as a consequence the cell loses
its ability to combine specifically with the soybean lectin. Shedding of
soybean lectin-binding capsular polysaccharides from the cells (Vasse
et al., 1984) explains why the broth culture as a whole continues to
bind soybean lectin (Tsien & Schmidt, 1980).

The profound influence of the growth phase on the composition of
lectin-binding polysaccharides of *Rhizobium* may be a major underlying
cause of conflicting data among laboratories testing the lectin-recognition
hypothesis. Furthermore, the growth-phase dependent modification of
the lectin-binding polysaccharides of rhizobia (Mort & Bauer, 1980;
Hrabak *et al.* 1981; Sherwood *et al.* 1984; Abe *et al.* 1984) may reflect
mechanisms which regulate cellular recognition in the *Rhizobium–*
legume symbiosis.

Fig. 11.4. Effect of culture age on the binding of antibody specific
for unique determinants in lipopolysaccharide of *Rhizobium trifolii*
0403 in early stationary phase. Cells were grown in a chemically
defined medium, and monitored for cell density with a Klett-
Summerson colorimeter (red filter). Samples were adjusted to 10^7
cells, and assayed by ELISA. (From Hrabak, Urbano & Dazzo, 1981,
and courtesy of the American Society for Microbiology.)

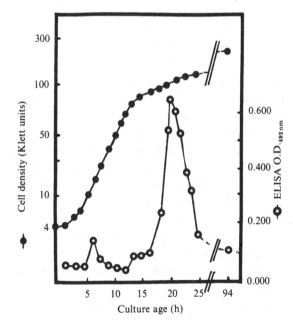

11.3.3 Genetic regulation by the bacterium

Evidence is accumulating that genetic elements important to surface polysaccharides and symbiotic recognition are encoded on very large, transmissible plasmids of *Rhizobium* (Johnston *et al*. 1978). For instance, the genes responsible for the 2-deoxy-D-glucose inhibitable attachment of *R. trifolii* to clover root hairs are encoded on the large nodulation plasmid designated pWZ2 (Zurkowski, 1980), and incorporation of the trifoliin A-binding sugar, quinovosamine into the LPS of *R. trifolii* is controlled by the clover nodulation plasmid (Russa, Urbanik, Kowalczuk & Lorkiewicz, 1982). Conjugal transfer of the nodulation plasmid from *R. trifolii* to *Agrobacterium tumefaciens* results in a hybrid which nodulates clover roots (Hooykaas *et al*. 1981), and which binds trifoliin A (Dazzo *et al*. manuscript in preparation).

11.3.4 Regulation by the plant host

An important regulation of lectin receptors occurs when the bacterium is in the root environment. For instance, Bhuvaneswari & Bauer (1978) showed that some strains of *R. japonicum* bind soybean lectin in the root environment but not in pure culture. Other strains of *R. japonicum* could bind soybean lectin better when grown in soil extract than in standard bacteriological media (Shantharam & Bal, 1981). Thus, an understanding of the biochemical basis of *Rhizobium*–legume interactions will require detailed studies of the microorganism in the rhizosphere of the host root as the normal case.

The first clue that the clover-root environment altered the lectin receptors on *R. trifolii* 0403 came from detailed studies on the orientation of attachment of these bacteria to clover root hairs (Dazzo *et al*. 1976, 1981, 1982; Dazzo & Brill, 1977; Dazzo *et al*. 1984). If fully encapsulated cells (which bind the lectin uniformly around the cell) were incubated for 15 min with seedlings, cells attached with no preferred orientation to root hair tips. However, after 4 h of incubation, additional cells began to attach along the sides of the root hair in a distinct polar orientation. This delay in polar attachment was due to an alteration of the lectin-binding capsule by enzymes released from axenically growing roots into the surrounding environment (Dazzo *et al*. 1982). The disorganization of the acidic polymers of the capsule began in the equatorial centre of the rod-shaped cell and then progressed towards the poles at unequal rates. At the same time, cells synthesized new capsular material at one cell pole (Sherwood *et al*. 1984), resulting in the *in situ* binding of trifoliin A to one pole of cells in the root environment (Dazzo *et al*. 1982). The kinetics of the capsule alteration and development of trifoliin A-binding capsules at one cell pole in the root environment matched the delay in polar attachment of the bacteria to the root hairs (Dazzo *et al.* 1982; Dazzo *et al.* 1984).

Five lines of soybean lack the 120 000 dalton soybean seed lectin but produce roots which nodulate normally (Pull, Pueppke, Hymowitz & Orf, 1978). When the molecular biology of these lines was examined, it was found that soybean has two genes for this lectin: one is expressed during seed development and the other during root development (Goldberg, Hoschek & Vodkin, 1983). The seed lectin gene is mutated and inactive (Goldberg *et al.* 1983) whereas the root lectin gene is expressed in these mutant lines of soybean (Sengupta-Gopalan, Pitas & Hall, 1984; Goldberg, personal communication).

11.4   Concluding remarks

The *Rhizobium*–legume symbiosis is a marvellous balance of many cell–cell communications which, in coordination, culminate in the formation of root nodules that fix N_2 into ammonia fertilizer for the plant symbiont. The lectin-recognition hypothesis has withstood a decade of rigorous testing as the basis for host specificity in the *Rhizobium*–legume symbiosis, and although there is substantial evidence supporting this concept, there remains no incontrovertible proof of its validity. It seems unlikely that acceptance or rejection of the lectin-recognition hypothesis will ever become universal until the recognition code is deciphered and mutant strains of both symbionts – the bacterium and the plant – with altered lectin/lectin receptors and no pleiotropic effects are available for analysis.

Portions of this work were supported by Grants 78-59-2261-0-1-050-2 and 82-CRCR-1-1040 from the Competitive Research Grant Program of the United States Department of Agriculture, Grant 80-21906 from the National Science Foundation, Grant GM-34331-01 from The National Institutes of Health, and the Michigan Agricultural Experiment Station (article No. 10815).

11.5   References

Abe, M., Sherwood, J., Hollingsworth, R. & Dazzo, F. (1984). Stimulation of clover root hair infection by lectin-binding oligosaccharide from the capsular and extracellular polysaccharides of *Rhizobium trifolii. J. Bacteriol.* **160**: 517–20.

Bhuvaneswari, T. V. & Bauer, W. D. (1978). The role of lectins in plant-microorganism interactions. III. Influence of rhizosphere/rhizoplane culture conditions on the soybean lectin-binding properties of rhizobia. *Plant Physiol.* **62**: 71–74.

Bhuvaneswari, T. V., Bhagwat, A. A. & Bauer, W. D. (1981). Transient susceptibility of root cells in four common legumes to nodulation by rhizobia. *Plant Physiol.* **68**: 1144–9.

Bhuvaneswari, T. V., Pueppke, S. G. & Bauer, W. D. (1977). Role of lectins in plant-

microorganism interactions. I. Binding of soybean lectin to rhizobia. *Plant Physiol.*
60: 486–91.

Bhuvaneswari, T. V., Turgeon, B. & Bauer, W. D. (1980). Early events in the infection of soybean (*Glycine max* L. Merr.) by *Rhizobium japonicum.* I. Location of infectible root cells. *Plant Physiol.* **66:** 1027–31.

Bishop, P. E., Dazzo, F. B., Applebaum, E. R., Maier, R. J. & Brill, W. J. (1977). Intergeneric transformation of genes involved in the *Rhizobium*–legume symbiosis. *Science* **198:** 938–9.

Bohlool, B. B. & Schmidt, E. L. (1974). Lectins: a possible basis for specificity in the *Rhizobium*–legume root nodule symbiosis. *Science* **185:** 269–71.

Burnet, F. (1971). Self-recognition in colonial marine forms and flowering plants in relation to evaluation of immunity. *Nature* **232:** 230–5.

Chen, A. P. & Phillips, D. A. (1976). Attachment of *Rhizobium* to legume roots as the basis for specific associations. *Physiol. Plant.* **38:** 83–8.

Dazzo, F. B. & Brill, W. J. (1977). Receptor site on clover and alfalfa roots for *Rhizobium. Appl. Environ. Microbiol.* **33:** 132–6.

Dazzo, F. B. & Brill, W. J. (1978). Regulation by fixed nitrogen of host–symbiont recognition in the *Rhizobium*-clover symbiosis. *Plant Physiol.* **62:** 18–21.

Dazzo, F. B. & Brill, W. J. (1979). Bacterial polysaccharide which binds *Rhizobium trifolii* to clover root hairs. *J. Bacteriol.* **137:** 1362–73.

Dazzo, F. B. & Hrabak, E. M. (1981). Presence of trifoliin A, a *Rhizobium*-binding lectin, in clover root exudate. *J. Supramol. Struct. Cell. Biochem.* **16:** 133–8.

Dazzo, F. B. & Hrabak, E. M. (1982). Lack of a direct nitrate–trifoliin A interaction in the *Rhizobium trifolii*–clover symbiosis. *Plant Soil* **69:** 259–64.

Dazzo, F. B., Hrabak, E. M., Urbano, M. R., Sherwood, J. E. & Truchet, G. L. (1981). Regulation of recognition in the *Rhizobium*–clover symbiosis. In *Current Perspectives in Nitrogen Fixation*, ed. H. Gibson & W. E. Newton, pp. 292–5. Canberra: Australian Academy of Science.

Dazzo, F. B. & Hubbell, D. H. (1975). Cross-reactive antigens and lectin as determinants of symbiotic specificity in the *Rhizobium*–clover association. *Appl. Microbiol.* **30:** 1017–33.

Dazzo, F. B., Napoli, C. A. & Hubbell, D. H. (1976). Adsorption of bacteria to roots as related to host specificity in the *Rhizobium*–clover association. *Appl. Environ. Microbiol.* **32:** 168–71.

Dazzo, F. B. & Truchet, G. L. (1983). Interactions of lectins and their saccharide receptors in the *Rhizobium*–legume symbiosis. *J. Membrane Biol.* **73:** 1–16.

Dazzo, F. B., Truchet, G. L. & Kijne, J. W. (1982). Lectin involvement in root hair tip adhesions as related to the *Rhizobium*–clover symbiosis. *Physiol. Plant.* **56:** 143–7.

Dazzo, F. B., Truchet, G. L., Sherwood, J. E., Hrabak, E. M., Abe, M. & Pankratz, H. S. (1984). Specific phases of root hair attachment in the *Rhizobium trifolii* – clover symbiosis. *Appl. Environ. Microbiol.* **48:** 1140–50.

Dazzo, F. B., Truchet, G. L., Sherwood, J. E., Hrabak, E. M. & Gardiol, A. E. (1982). Alteration of the trifoliin A-binding capsule of *Rhizobium trifolii* 0403 by enzymes released from clover roots. *Appl. Environ. Microbiol.* **44:** 478–90.

Dazzo, F. B., Urbano, M. R. & Brill, W. J. (1979). Transient appearance of lectin receptors on *Rhizobium trifolii. Curr. Microbiol.* **2:** 15–20.

Dazzo, F. B., Yanke, W. E. & Brill, W. J. (1978). Trifoliin: a *Rhizobium* recognition protein from white clover. *Biochim. Biophys. Acta* **536:** 276–86.

Deinema, M & Zevenhuizen, L. P. T. (1971). Formation of cellulose fibrils by gram negative bacteria and their role in bacterial flocculation. *Arch. Mikrobiol.* **78:** 42–57.

Diaz, C., Kijne, J. W. & Quispel, A. (1981). Influence of nitrate on pea root cell wall composition. In *Current Perspectives in Nitrogen Fixation*, ed. A. H. Gibson and W. E. Newton, p. 426. Canberra: Australian Academy of Science.

Gade, W., Jack, M. A., Dahl, J. B., Schmidt, E. L. & Wolf, F. (1981). The isolation and characterization of a root lectin from soybean (*Glycine max* L.) cultivar Chippewa. *J. Biol. Chem.* **256**: 12905–10.

Gatehouse, J. A. & Boulter, D. (1980). Isolation and properties of a lectin from the roots of *Pisum sativum*. *Physiol. Plant.* **49**: 437–42.

Goldberg, R. B., Hoschek, G. & Vodkin, L. O. (1983). An insertion sequence blocks the expression of a soybean lectin gene. *Cell* **33**: 465–75.

Hamblin, J. & Kent, S. P. (1973). Possible role of phytohaemagglutinin in *Phaseolus vulgaris* L. *Nature, New Biol.* **245**: 28–9.

Higashi, S. & Abe, M. (1980). Promotion of infection thread formation by substances from *Rhizobium*. *Appl. Environ. Microbiol.* **39**: 297–301.

Hollingsworth, R. I., Abe, M. & Dazzo, F. B. (1984). Identification of ether-linked 3-hydroxybutanoic acid in the extracellular polysaccharide of *Rhizobium trifolii* 0403. *Carbohydr. Res.* **133**: C7–C11.

Hooykaas, P. J., van Brussell, A. A. N., den Hulk-Ras, H., van Slogteren, G. M. & Schilperoort, R. A. (1981). Sym plasmid of *Rhizobium trifolii* expressed in different rhizobia species and *Agrobacterium tumefaciens*. *Nature* **291**: 351–3.

Hrabak, E. M., Urbano, M. R. & Dazzo, F. B. (1981). Growth-phase dependent immunodeterminants of *Rhizobium trifolii* lipopolysaccharide which bind trifoliin A, a white clover lectin. *J. Bacteriol.* **148**: 697–711.

Hubbel, D. H., Morales, V. M. & Umali-Garcia, M. (1978). Pectolytic enzymes in *Rhizobium*. *Appl. Environ. Microbiol.* **35**: 210–13.

Hughes, T. A. & Elkan, G. H. (1981). Study of the *Rhizobium japonicum*–soybean symbiosis. *Plant Soil* **61**: 87–91.

Hughes, T. A., Leece, J. G. & Elkan, G. H. (1979). Modified fluorescent technique using rhodamine for studies of *Rhizobium japonicum* soybean symbiosis. *Appl. Environ. Microbiol.* **37**: 1243–4.

Johnston, A. W. B., Beyon, I. L., Buchanan-Wollaston, A. V., Setchell, S. M., Hirsch, P. R. & Beringer, J. E. (1978). High frequency transfer of nodulating ability between species and strains of *Rhizobium*. *Nature* **276**: 635–8.

Kamberger, W. (1979). Role of cell surface polysaccharides in the *Rhizobium*–pea symbiosis. *FEMS Microbiol. Lett.* **6**: 361–5.

Kato, G., Maruyama, Y. & Nakamura, M. (1980). Role of bacterial polysaccharides in the adsorption process of the *Rhizobium*–pea symbiosis. *Agric. Biol. Chem.* **44**: 2843–55.

Kato, G., Maruyama, Y. & Nakamura, M. (1981). Involvement of lectins in *Rhizobium*–pea recognition. *Plant Cell. Physiol.* **22**: 759–71.

Kijne, J. W., van der Schaal, I. A. M. & de Vries, G. E. (1980). Pea lectins and the recognition of *Rhizobium leguminosarum*. *Plant Sci. Lett.* **18**: 65–74.

Kijne, J. W., van der Schaal, I. A. M., Diaz, C. L. & van Iren, F. (1982). Mannose-specific lectins and the recognition of pea roots by *Rhizobium leguminosarum*. *Lectins*, vol. 3, ed. T. C. Bog-Hansen & G. A. Spengler, pp. 521–9. Berlin: W. de Gruyter.

Kumarasinghe, M. K. & Nutman, P. S. (1977). *Rhizobium*-stimulated callose formation in clover root hairs and its relation to infection. *J. Exp. Bot.* **28**: 961–76.

Ljunggren, H. & Fahraeus, G. (1961). Role of polygalacturonase in root hair invasion by nodule bacteria. *J. Gen. Microbiol.* **26**: 521–8.

Martinez-Molina, E., Morales, V. M. & Hubbell, D. H. (1979). Hydrolytic enzyme production by *Rhizobium*. *Appl. Environ. Microbiol.* **38**: 1186–8.

Mort, A. J. & Bauer, W. D. (1980). Composition of the capsular and extracellular polysaccharides of *Rhizobium japonicum*: changes with culture age and correlations with binding of soybean seed lectin to the bacteria. *Plant Physiol.* **66**: 158–63.

Napoli, C. A. & Albersheim, P. (1980). *Rhizobium leguminosarum* mutants incapable of normal extracellular polysaccharide production. *J. Bacteriol.* **141**: 1454–6.

334   F. B. Dazzo

Napoli, C. A., Dazzo, F. B. & Hubbell, D. H. (1975). Production of cellulose microfibrils
by *Rhizobium. Appl. Microbiol.* **30:** 123–31.
Napoli, C. A. & Hubbell, D. H. (1975). Ultrastructure of *Rhizobium*-induced infection
threads in clover root hairs. *Appl. Microbiol.* **30:** 1003–9.
Paau, A. S., Leps, W. T. & Brill, W. J. (1981). Agglutinin from alfalfa necessary for
binding and nodulation by *Rhizobium meliloti. Science* **213:** 1513–15.
Palomares, A., Montega, E. & Olivares, J. (1978). Induction of polygalacturonase
production in legume roots as a consequence of extrachromosomal DNA carried by
Rhizobium meliloti. Microbios **21:** 33–9.
Pull, S. P., Pueppke, S. G., Hymowitz, T. & Orf, J. H. (1978). Soybean lines lacking
the 120 000 dalton seed lectin. *Science* **200;** 1277–9.
Reeke, J. N., Becker, J. W., Cunningham, B. A., Wang, J. C., Yahara, I. & Edelman,
G. M. (1975). Structure and function of concanavalin A. *Adv. Expt. Med. Biol.* **55:**
13–33.
Reporter, M., Raveed, D. & Norris, G. (1975). Binding of *Rhizobium japonicum* to
cultured soybean root cells: morphological evidence. *Plant Sci. Lett.* **5:** 73–6.
Russa, R., Urbanik, T., Kowalczuk, E. & Lorkiewicz, Z. (1982). Correlation between
occurrence of plasmid pUCS202 and lipopolysaccharide alterations in *Rhizobium.*
FEMS Microbiol. Lett. **13:** 161–5.
Saunders, R. E., Carlson, R. W. & Albersheim, P. (1978). A *Rhizobium* mutant
incapable of nodulation and normal polysaccharide secretion. *Nature* **271:** 240–2.
Sengupta-Gopalan, C., Pitas, J. W. & Hall, T. C. (1984). Reexamination of the role
of lectin in the *Rhizobium–Glycine* symbiosis. In *Advances in Nitrogen Fixation
Research*, ed. C. Veeger & W. E. Newton, p. 427. The Hague: Nijhoff/Junk Publishers.
Shantharam, S. & Bal, A. K. (1981). The effect of growth medium on lectin binding
by *Rhizobium japonicum. Plant Soil* **62:** 327–30.
Sherwood, J. E., Truchet, G. L., Hrabak, E. & Dazzo, F. (1984). Effect of nitrate supply
on the in vivo synthesis and distribution of trifoliin A, a *Rhizobium trifolii*-binding
lectin, in *Trifolium repens* seedlings, *Planta* **162:** 540–7.
Sherwood, J. E., Vasse, J., Dazzo, F. B. & Truchet, G. L. (1984). Development and
trifoliin A-binding properties of the capsule of *Rhizobium trifolii. J. Bacteriol.* **159:**
145–52.
Solheim, B. (1975). Possible role of lectin in the infection of legumes by *Rhizobium
trifolii* and a model of the recognition reaction between *Rhizobium trifolii* and *Trifolium
repens. NATO Advanced Study Institute Symposium on Specificity in Plant Diseases,*
Sardinia.
Stacey, G., Paau, A. S. & Brill, W. J. (1980). Host recognition in the *Rhizobium*–soybean
symbiosis. *Plant Physiol.* **66:** 609–14.
Truchet, G. L. & Dazzo, F. B. (1982). Morphogenesis of lucerne root nodules incited
by *Rhizobium meliloti* in the presence of combined nitrogen. *Planta* **154:** 352–60.
Tsien, H. C. & Schmidt, E. L. (1980). Accumulation of soybean lectin-binding
polysaccharide during growth of *Rhizobium japonicum* as determined by
hemagglutination inhibition assay. *Appl. Environ. Microbiol.* **39:** 1100–4.
van der Schaal, I. A. M. & Kijne, J. W. (1981). Pea lectins and surface carbohydrates
of *Rhizobium leguminosarum.* In *Current Perspectives in Nitrogen Fixation*, ed. A. H.
Gibson & W. E. Newton, p. 425. Canberra: Australian Academy of Science.
Vasse, J., Dazzo, F. & Truchet, G. L. (1984). Reexamination of capsule development
and lectin-binding site on *Rhizobium japonicum* 3I1b110 by the
glutaraldehyde/ruthenium red/uranyl acetate staining method. *J. Gen. Microbiol.* **130:**
3037–47.
Yao, P. Y. & Vincent, J. M. (1976). Factors responsible for the curling and branching
of clover root hairs by *Rhizobium. Appl. Environ. Microbiol.* **45:** 1–16.
Zurkowski, W. (1980). Specific adsorption of bacteria to clover root hairs, related to
the presence of the plasmid pWZ2 in cells of *Rhizobium trifolii. Microbios.* **27:** 27–30.

12

Possible functions of plant lectins

Mark A. Holden and Michael M. Yeoman

12.1   Introduction

The purpose of this article is to review and discuss the properties, structure and location of plant lectins and their 'receptors' in an attempt to establish their *in vitro* role(s). The review also attempts to make a critical appraisal of the increasing body of experimental evidence linking these sugar-binding proteins and glycoproteins with functions within the plant, especially those involving cellular recognition. However, before a start can be made it is essential to define the term 'lectin'.

12.2   Definition of the term 'lectin'

Definition of the term 'lectin' [from the Latin legere, to choose] (Boyd & Shapleigh, 1954) presents a difficult problem and the adoption of one specific set of criteria necessitates including some but not other molecules which have at one time been considered 'lectins'. Thus any discussion of the role of lectins in plants obviously depends on which definitons is used.

Classically, the term 'lectin', as employed by Boyd & Shapleigh (1954), refers to proteins, or more frequently, glycoproteins of plant origin with the ability to agglutinate red blood cells. These lectins are not necessarily blood-type specific but the term 'lectin' was adopted in order to call attention to the serological specificity of certain seed agglutinins. That is, some lectins only agglutinate erythrocytes of one particular blood group.

The discovery of agglutinins in such distantly related life forms as bacteria (Gilboa-Garber, Mizrahi & Garber, 1977) and mammals (de Waard, Hickman & Kornfield, 1976; Den & Malinzak, 1977) has led to a broadening of the term to include these non-plant lectins.

This definition however excludes molecules such as ricin and abrin which are considered by most workers to be lectins. These toxins bind to the surfaces of erythrocytes, but being monomeric and possessing only one binding site they cannot bind cells together.

A broader definition for the term has been suggested by Goldstein, Hughes, Monsigny & Sharon (1980). They suggest that a lectin is 'a sugar-binding protein or glycoprotein of non-immune origin which agglutinates cells and/or precipitates glycoconjugates'. Their definition allows the inclusion of non-agglutinating 'lectins' but removes the distinction between lectins and enzymes of sugar metabolism. Any enzyme with more than one binding site and a specificity for particular carbohydrates can act as a haemagglutinin under certain circumstances. Thus sugar kinases, glycosidases and glycosyltransferases can all act as lectins within the confines of this definition.

Another class of proteins with carbohydrate-binding properties has been demonstrated by quite different techniques. Jermyn & Yeow (1975) found that seed extracts from 94 out of 104 plant families formed insoluble complexes with artificially synthesized 'Yariv' antigens. The monovalent glycoproteins responsible were termed 'all β lectins' since the specificity of the reaction lay in the β-glycoside linkage and not in the particular sugar. Such 'specificity defined' lectins could be identical to classically defined lectins isolated from the same sources, but where comparisons have been made they have little in common in either composition or specificity (Jermyn & Yeow, 1975).

For the purposes of this article we will favour the wider definition of Goldstein *et al.* (1980) in order to include a discussion of all 'lectin-like' molecules.

12.3   Detection and assay of lectins

Study of the location of lectins within plants is an important part of any discussion of function, since any proposed role must be consistent with their spatial and temporal location in the plant. In this section the methods used for their detection and location are discussed.

The standard assay for a lectin consists of the titration of serial dilutions of the sample with red blood cells, thus determining the minimum amount of sample needed to agglutinate the cells.

There are several problems with this assay. Red blood cells may not contain the specific 'receptor' for the lectin. It has been shown that a seed extract from *Nicandra physaloides* had barely detectable lectin activity towards human erthrocytes. However, *Nicandra* lectin could be detected by double diffusion against the antiserum to *Datura stramonium* lectin at a concentration lower than the minimum required for erythrocyte agglutination. In contrast, tomato fruit juice, potato tuber, or potato fruit lectins formed precipitation lines only when used at a concentration 30-fold, or greater, than the minimum required to cause agglutination (Kilpatrick, Jeffree, Lockhart & Yeoman, 1980). The sugar specificity for *Nicandra* seed lectin is not known, but inhibition studies using *N*-acetylglucosamine oligomers have shown it to be different from *Datura* seed lectin.

The *Nicandra* lectin, therefore, although closely related to the *Datura* lectin has only a very low haemagglutinin activity. Therefore, when dealing with data concerning specific activity of a lectin it must always be borne in mind that relative lectin activity between different sources need not necessarily be an indication of the relative amount of total lectin present.

Another problem with the agglutination assay is that under most circumstances monovalent 'lectins' do bind to cell-surface saccharides. This can be dealt with, perhaps somewhat unsatisfactorily, by excluding such toxins from the defintion of lectins as is suggested by Goldstein and co-workers (1980). An alarming possibility associated with use of this assay is that during extraction, usually by grinding and homogenization of the tissue, the lectin may bind to subcellular fractions with which it is not normally associated. Thus Kohle & Kauss (1979) reported that the supposed binding of *Ricinus communis* agglutinin to the inner mitochondrial membrane was simply an artefact of the preparation procedure.

Obviously other more reliable assays for lectins are needed, and indeed, the basis for such a procedure has been established. This involves binding of the lectin to its specific antiserum which can then be assayed using an enzyme-linked immuno-sorbent procedure (ELISA). The distribution of a lectin in tissues of *Phaseolus vulgaris* has recently been determined accurately by use of such a technique (Borrebaeck & Mattiasson, 1983).

Lectins may also be detected by the precipitates they form with polysaccharides or glycoproteins, in either liquid or semi-solid (agar) media, (Goldstein, 1972). These reactions also provide information on lectin specificity as well as on the constituent sugars of the polysaccharide or glycoprotein precipitated.

Microscopy can be used to localize lectins *in situ*. Antibodies to highly purified samples of lectins can be linked to fluorescein (FITC) (Jeffree, Yeoman & Kilpatrick, 1982), ferritin or gold (Horisberger & Vonlanthen, 1980) and visualized using light or transmission electron microscopy. Immunofluorescent localization of lectins in plants has recently been reviewed by Jeffree *et al.* (1982). These techniques allow the sub-cellular sites occupied by lectins to be confirmed.

12.4   Structure of lectins

Lectins appear to be a highly heterogeneous group of proteins and glycoproteins ranging in molecular weight from 17 000 to 400 000 (Goldstein & Hayes, 1978). The sugar content can range from 0 to 20% and the number of subunits present varies between two and four in most reported studies.

Despite the heterogeneity of lectins within the plant kingdom, recent data suggest a considerable degree of homology between lectins of related species. Using amino acid sequence data a large measure of homology between several lectins from leguminous plants has been shown (Richardson, Behnke, Freisheim & Blumenthal, 1978). For example the first 25 amino acids of the β chain of the D-mannose-binding lectins from pean and lentil differ by only two amino acids (Foriers, Wuilmart, Sharon & Strosberg, 1977, and Forers *et al.* 1977). The soybean and peanut agglutinins are identical in 11 of the first 25 *N*-terminal amino acids and the β chain of the lentil lectin also showed sequence homologies with the soybean agglutinin. Shannon & Hankins (1981) have recently demonstrated that the *N*-terminal amino acid sequences from *Bauhinia, Caragana, Sophora* and *Ulex* lectins all show extensive sequence homology with one another and with pea, lentil and *Dolichos biflorus* lectins.

It has also been found that the α chain of the lectin from broad bean is homologous to a large region in the middle of the Concanavalin A (Con A) sequence, and that the β chain is composed of homologous sequences to two discrete segments of Con A. Thus the two lectins comprise circular permutations of extensive homologous sequences (Foriers, de Neve, Kanarek & Strosberg, 1978; Cunningham, Hemperly, Hopp & Edelman, 1979).

The distinctive subunits of many lectins have also been shown to possess extensive amino acid sequence homology with each other. The *N*-terminal amino acids of lectin subunits have been shown to be similar in soybean (Lotan *et al.* 1975), *Dolichos biflorus* (Etzler, Talbot & Ziaya, 1977) and *Phaseolus vulgaris* (Miller, Hsu, Heinrikson & Yachnin, 1975).

Using immunological techniques, Kilpatrick *et al.* (1980) demonstrated structural similarity among lectins from species of the *Solanaceae* but found that the highly conserved region of the molecule did not correspond. Interestingly, the carbohydrate-binding site of several legume lectins did not show the same degree of conservation of amino acid sequence as the rest of the molecule (Strosberg, Lauwereys & Foriers, 1983). Lectins of the Gramineae are also a closely related class of proteins. The lectins from rye, barley and wheat (WGA) are almost identical in physical and biological properties (Peumans & Stinissen, 1983).

The 'all β lectins' from plants as distantly related as gymnosperms and angiosperms also all showed similar protein and carbohydrate compositions (Jermyn & Yeow, 1975).

Thus, despite the heterogeneity of lectins within the plant kingdom, within particular taxonomic groups the lectins are closely related structurally. For example, solanaceous lectins are characterized by their carbohydrate content and the presence of hydroxyproline residues. Many legume lectins are rich in aspartate and serine and are low in methionine or cysteine, whereas the monocot lectins are basic proteins with a large number of cysteine residues.

The highly conserved nature of lectins from related species suggests that they may play an essential role in plant survival.

12.5 Location of lectins

12.5.1 Distribution within the plant kingdom

In a survey of extracts of 2663 different plant species haemagglutinin activity was detected in over 800 (Allen & Brilliantine, 1969). Toms & Western (1971) showed that 54% of higher plant families contained haemagglutinins and they included over 600 species and varieties of the *Leguminosae* in their data. This was substantiated by Lee, Tan & Liew (1977), who found that 79% of the 125 species and varieties of tropical *Leguminosae* they tested showed lectin activity. Several legume species have been shown to be lacking in proteins with haemagglutinin activity. However, without exception, these plants contained cross-reacting material which was immunologically closely related to one of the haemagglutinins (Shannon, 1983). All β lectins also show a very wide distribution within the plant kingdom, and 94 out of 104 families tested contain this type of lectin (Jermyn & Yeow, 1975). Again it must be stressed here that the failure to detect lectins in some genera, species and varieties does not necessarily prove that they are absent. It may be that the *in vivo* binding of lectins to endogenous sugars prevents them from acting as haemagglutinins.

12.5.2 At the organ level

The majority of lectins detected so far has been isolated from plant seeds where they are present in a more or less soluble state. However, lectins are not always confined to a single part of the plant, and the haemagglutinin content is not always highest in the seed (see Table 12.1). For example, a high level of soluble lectin was detected in the phloem exudate of *Cucurbita* species, but was entirely absent from the seeds (Sabnis & Hart, 1978).

In contrast, most membrane-associated lectins so far studied are present at high levels in the non-seed tissues of the plant. In soybean and peanut high levels of membrane-bound lectins are maintained in leaves, shoots and roots throughout plant development (Bowles, Lis & Sharon, 1979). None of these lectins have been purified and characterized but in some plants such as soybean (Bowles *et al.* 1979), wheat (Yoshida, 1978*a*,*b*) and *Datura stramonium* (Kilpatrick *et al.* 1980) it appears that they differ in sugar specificity from the soluble seed lectins isolated from the same plant. It would therefore seem that the soluble seed lectins and their membrane-associated counterparts have different binding sites and possibly even different functions.

12.5.3 At the tissue level

Jeffree & Yeoman (1981) have employed indirect immunofluorescence with antibodies raised against highly purified lectin from *Datura stramonium* seeds in order to study intercellular and intracellular localization of the lectin within the plant. They found that, in stems, the richest sources of lectin are the epidermis, endodermis and the phloem. By comparison, Sabnis & Hart (1978) found a lectin located in the phloem, and reported that 15–20% of phloem exudate from *Cucurbita* species consisted of lectin.

The majority of reports on plant lectins are concerned with seed lectins and their location within the seed. High levels of lectin have been detected both in the cotyledons and embryonic axis of lentil (Howard, Sage & Horton, 1972; Rougé 1974*a*,*b*), red kidney bean (Mialonier, Privat, Monsigny, Kalem & Durand, 1973) and pea (Rougé, 1975) but there is little lectin in the seed coat. On the other hand, lectin activity has been detected in the testa as well as the cotyledons and embryonic axis of soybean (Pueppke, Bauer, Keegstra & Ferguson, 1978). The relevance of this distribution of lectins is discussed later.

12.5.4 At the subcellular level

The studies of Jeffree & Yeoman (1981) using immunofluorescence have shown lectins associated with the plasmalemma and endomembranes of *Datura* stem and callus cells. However, cell walls appeared to be free from lectin. Bowles & Kauss (1975) also demonstrated that carbohydrate-binding proteins could be extracted from the cellular mem-

Table 12.1 *Localization of lectins at the tissue level*

Plant	Location	Reference
Arachis hypogaea (peanut)	Roots (up to 18 days from planting), soluble lectin	Schulz & Pueppke 1978
	Roots, leaves, shoots, mature plant, membrane-bound lectin	Bowles *et al.* 1979
Cucumis melo (melon)	Phloem mature plant	Sabnis & Hart, 1978
Cucumis sativus (cucumber)		
Cucurbita maxima (pumpkin)		
Datura stramonium	Phloem of stem	Jeffree & Yeoman, 1981
Dolichos biflorus (horsegram)	High level in seeds	Talbot & Etzler, 1978a,b
	Low levels in stems and leaves	
*Fomens fomentarious	Fruiting bodies	Hořejši & Kocourek, 1978
Glycine max (soybean)	Seeds, roots, and leaves of young plants	Howard *et al.* 1972
Lens culinaris (lentil)		Rougé, 1974a,b
*Marasmus oreades	Fruiting bodies	Hořejši & Kocourek, 1978
*Ononis hircina	Roots	Hořejši *et al.* 1979
*Ononis spinosa	Roots	Hořejši & Kocourek, 1978
Phaseolus vulgaris (red kidney bean)	Seeds, roots, stems and leaves of young seedlings	Mialonier *et al.* 1973
Pisum sativum (pea)	Cotyledons and embryo axis	Rougé, 1975
Ricinus communis (castor bean)	Seeds, roots, stems, leaves of young seedlings	Rougé 1974b; Youle & Huang 1976
		Hořejši *et al.* 1979
*Robinia pseudoacacia	Bark	Allen & Neuberger 1973; Allen, Neuberger & Sharon, 1973
*Solanum tuberosum (potato)	Tuber	
*Trifolium repens	Roots	Dazzo *et al.* 1978

*Indicates lectin has been purified.

branes (dictyosomes, endoplasmic reticulum, plasmalemma and mitochondria) of mung bean hypocotyls. Further, soybean agglutinin (SBA)
has been detected in protein bodies of the coyledons of *Glycine max*.
In this latter case, histochemical techniques suggested that the lectin
was associated with membrane around the protein bodies (Horisberger
& Vonlanthen, 1980). Other studies have shown lectins associated with
cellular and subcellular membranes of various plants (Kauss & Glaser,
1974; Bowles & Kauss, 1976; Bowles, Schnarrenberger & Kauss, 1976;
Yoshida, 1978*a,b*).

Lectins have also been found in cell walls of plant tissues including mung
bean hypocotyls (Kauss & Glaser, 1974) and the roots of various legumes
(Dazzo, Yanke & Brill, 1978; Gatehouse & Boulter, 1980). Clarke,
Knox & Jermyn (1975) have demonstrated that 'all-β lectins' are located
in the cell walls of cotyledon parenchyma cells of Jack and red kidney
beans whereas Con A and phytohaemagglutinin are found in cytoplasmic
sites.

12.6   Lectin receptors in plants

Most of the studies of lectin 'receptors' in plants have been
performed with puried lectins conjugated with gold, fluorescein or ferritin, and subsequently analysed by light and electron microscopy.

The term 'receptor' must be used with caution as it has associations
with work on animal hormones, and drug receptors which interact with
external molecules in a highly specific way, thereby transmitting signals
from the environment to the interior of the cell. Interactions of lectins
with plant cells will presumably be less specific because the lectin will
react with any cell surface sugar residue that is complementary to its
binding site. Thus the lectin will bind not only to its true 'receptor'
(in animal cells a rather complex carbohydrate structure, dependent
partly on the presence of specific sugars but also on monomeric forms,
spatial arrangement and distribution of the monomeric units, as well
as the nature of the polypeptide chain by which carbohydrate groups
are bound to the cell surface) but also to any cell-surface sugar with
a specificity for the binding site (the hapten).

As some lectins are known for which simple sugar inhibitors are as
yet undiscovered (Sharon & Lis, 1972), it is possible that their receptors
also involve a highly specific interaction, and not merely binding to a
single sugar hapten. The problem of analysis may therefore be resolved
by the use of procedures which selectively remove lectin bound only
to simple sugar residues and which leave lectin bound to receptors of
higher specificity. Thus, the binding of the hapten to the lectin is much

weaker than that of the lectin to the erythrocyte receptor, so that selective removal of the lectin may be possible.

However, the analogy of lectin to erythrocyte binding may be invalid. Indeed, if simple oligosaccharide residues project from cell surfaces and a lectin of appropriate specificity is a component of the cytoplasm, the lectin will bind to it, with simple sugars thus binding lectins '*in vivo*'. The distinction made between true receptors and lectin-reactive membrane glycoproteins (Goldstein & Hayes, 1978) that applies in animal cell–lectin interactions may not therefore apply in the interaction of lectins with plant cells.

For the purpose of this review the term 'lectin-binding site' will be used in a wide sense to describe plant cell components that bind specifically to lectins of that plant (or to lectins of another plant with the same sugar specificity).

12.6.1 Lectin-binding sites on the plasmalemma

Much work has been carried out on the lectin receptors of animal cells (for review see Nicolson, 1978). The discovery that incubation of erythrocytes with Con A led to the formation of patches and caps of lectin and receptor (Smith & Hollers, 1970) led to several studies of lectin–plant protoplast interactions.

Protoplasts of *Daucus carota* are agglutinated by Con A, indicating that residues related to glucose or mannose are located on the plasma membrane surface (Glimelius, Wallin & Eriksson, 1974). Whether these residues are integral parts of the membrane or associated with cell wall material is not known, but phosphotungstic acid/chrome acid and periodate/schiff's reagent (both specific for carbohydrate) each stain the plasma membrane (Roland, Lembi & Morré, 1972).

In protoplasts allowed to agglutinate, Con A binding was densest at the cell junction suggesting that, as for mammalian cells (Walther, 1976), a clustering of cell surface bound Con A is necessary for Con A induced agglutination.

Williamson, Fowke, Constabel & Gamborg (1976) have also obtained an agglutination of soybean protoplasts by treatment with Con A. Further, lectin-binding sites have been localized on the plasmalemma by an electron-microscope analysis of thin sections stained with haemocyanin bound to Con A (Williamson *et al.* 1976). Protoplasts prefixed before Con A treatment exhibited an even distribution of haemocyanin. In contrast, protoplasts treated with Con A 45 min before fixation showed a clustered distribution of the stain. If protoplasts were incubated in Con A for a further 16 h prior to fixation the haemocyanin appeared in tighter clusters. These results were interpreted as suggesting that in the living membrane there prevails an even distribution of Con A binding sites and that these sites become cross-linked by tetravalent Con A molecules, leading to patch formation. This is similar to the binding of Con A to erythrocytes (Nicolson, 1974) but unlike lectins bound to red blood cells the clusters on plant protoplasts do not aggregate and form a cap.

Similar results have been obtained of binding of Con A to lectin-binding sites in plasma membranes of *Allium porrum* L. (leek) protoplasts (Williamson, 1979). Moreover, gold linked to Con A can be visualized with the scanning electron microscope, and using this technique Williamson (1979) has demonstrated binding of Con A to tobacco leaf protoplasts. At 5°C the particles of gold–Con A were found to be distributed randomly over the surface of the protoplast, but raising of the temperature led to an association of particles into clusters. Glutaraldehyde fixation after low-temperature labelling prevented cluster formation; thus Con A appears to cause clustering of binding sites on the plant protoplast membrane.

It has been noted (Glimelius, Wallin & Eriksson, 1978) that the number of binding sites for Con A on isolated protoplasts is dependent on the enzyme preparation used to remove the cell wall. This is explained as being due to enzymic removal of binding sites by the contaminating hydrolases which are present to varying extents in different enzyme preparations. On the other hand, it is interesting to note that erythrocytes are frequently more susceptible to agglutination by lectins if they have been mildly treated with proteolytic enzymes such as trypsin or pronase (Pardoe & Uhlenbruck, 1970).

The above evidence suggests that binding sites are exposed, or indeed formed, during protoplast isolation. Further support is provided by studies on the agglutination of plastids by Con A. Thus, Shepard & Moore (1978) note that plastids isolated mechanically from the fruits of cucumber (*Cucumis sativus* L.) and pear (*Pyrus domestica* Medik.) and the leaves of pea (*Pisum sativum* L.) are not agglutinated by Con A. In contrast, similar plastids obtained enzymatically were agglutinated by Con A and the conclusion was drawn that the sugars present on the plastid surfaces to which the lectins bound 'are not necessarily indigenous but may become attached as a result of the extraction procedure'. Certainly, plastids released from protoplasts obtained by pectinase treatment alone were not agglutinated by Con A, which suggests that a cellulolytic release of material containing terminal D-glucosyl/mannosyl residues is necessary before the plastids become susceptible to agglutination. Similarly, the lectin-binding sites on the surfaces of isolated protoplasts may also represent the products of enzymatic breakdown of cell walls.

It is noteworthy that Burgess & Linstead (1976) reported that Con A agglutinates both tobacco (*Nicotiana tabacum* L.) leaf protoplasts and those prepared from a suspension culture of grape vine (*Vitis vinifera* L.). They found, however, that Con A conjugated to ferritin or to colloidal gold binds differently to the protoplasts of these two species. The plasmalemma of tobacco protoplasts were uniformly labelled while the outer surface of vine protoplasts showed a patchy distribution of label. This difference was explained by the authors, not as a cross-linking of Con A binding sites by the multivalent Con A in vine protoplasts, but as a reflection of the relative ability of both types of protoplasts to regenerate a cell wall. Tobacco protoplasts regenerate a cell wall rapidly (Burgess & Fleming, 1974) whereas vine protoplasts, under identical conditions, regenerate an atypical wall only slowly (Burgess & Linstead, 1976). As gold–Con A and ferritin–Con A bind readily to the fragments of wall remaining at the cell surface after digestion, it was assumed that labelling of the surface of freshly prepared protoplasts was due to binding of Con A to nascent wall material. The results also suggested that Con A binding to the outer surfaces of protoplasts could be due to the method of protoplast preparation rather than to binding by 'true' lectin 'receptors'. Nevertheless, a later paper suggested that 'indigenous' lectin-binding sites might be present on some protoplast surfaces (Burgess & Linstead, 1977).

12.6.2 Lectin-binding sites on organellar membranes
It has been demonstrated that the various cellular membranes of animal cells show differences in the sugar moieties exposed at their surfaces (Lis & Sharon, 1973; Mellor, Krusius & Lord, 1980). Further, the carbohydrate specificities of lectins obtained from different membrane fractions (mitochondria, plasma membrane, Golgi apparatus and endoplasmic reticulum) of higher plant cells also differ from one fraction to another (Bowles & Kauss, 1975). It is interesting then, to speculate that membranes from various organelles possess lectins with specificities for different sugar receptors, which may be present on these membranes, or those of other organelles.

Mellor *et al.* (1980), besides noting differences in the organellar distribution of peripheral sugars such as fucose, arabinose and xylose, also showed that sugar residues were present only on the luminal (non-cytoplasmic) surfaces of membranes. These results are consistent with studies on organelles of the spinach leaf (Frederick, Nies & Gruber, 1981) and

346   M. A. Holden and M. M. Yeoman

with those from experiments on lectin binding to the membranes of organelles of animal cells (Rodriguez-Boulan, Kreibich & Sabatini, 1978; Virtanen & Wariovaara, 1976; Yokoyama, Nishiyama, Kawai & Hirano, 1980). The relevance of the location of these lectin-binding sites is discussed later.

12.7   Possible functions of plant lectins
The very widespread distribution of lectins in the plant kingdom and their ability to discriminate between different saccharides on cell surfaces has prompted speculation on their role(s) *in vivo*. It must be stated at the outset that none of the following suggestions are incontrovertible; even in the widely studied case of *Rhizobium* legume symbiosis certain vexing inconsistencies are apparent (see Chapter 6). The evidence concerning the physiological role of lectins is far from complete and many of the key questions remain to be answered.

12.7.1   Regulation of plant cell wall extension
Kauss & Glaser (1974) demonstrated the presence of lectin in cell walls of mung bean hypocotyls; furthermore they detected a slight difference in the lectin content of growing and non-growing cell walls. They suggested that lectins in some way regulate cell wall extension. The extracts from the mung bean hypocotyls had a high hydroxyproline content as did a lectin extracted from the potato by Allen & Neuberger (1973). The latter was thought to be identical to the hydroxyproline-rich extensin reported by Lamport (1973). However, Kauss (1977) showed that most of the hydroxyproline could be separated from the lectin without impairing its haemagglutinating ability. Nevertheless, Kauss (1977) suggested that lectins could govern extension growth in three possible ways. 'Firstly, new polysaccharide molecules could have a lectin handle during their transport and introduction into the wall. Secondly, lectins in the wall could allow a sliding of wall polysaccharides relative to each other, regulated by protons and/or divalent ions. Thirdly, lectins could reside at the wall surface and establish wall–membrane contacts.'

12.7.2   Involvement in the storage, mobilization and transport of sugars
Krüpe (1956) first suggested that lectins might function as 'Kohlenhydraftfixieren' (carbohydrate catchers) and the specific binding of lectins to sugars since has suggested to many workers a role for lectins in sugar metabolism.

(1) Lectins in seed storage
In many cases, 1–3% of total seed protein is lectin (Callow, 1975). An increase in lectin content takes place as the seed matures and this is followed by a decrease during germination. The suggestion has therefore arisen that lectins are involved in processes connected with storage of food reserves or utilization of these reserves on germination (Clarke *et al.* 1975; Kilpatrick, Yeoman & Gould, 1979). Also, the uniform distribution of soybean agglutinin in plant protein bodies has been held as supporting evidence that lectins are involved in the binding of glycoprotein enzymes and in the utilization of storage products (Horisberger & Vonlanthen, 1980).

However, the fact that lectins are easily extracted from seeds with phosphate-buffered saline (PBS) suggests that they are not bound to their specific receptors (Kauss, 1977). Indeed this may be taken as in direct evidence against the involvement of lectin in sugar immobilization. The fact that some seeds contain no detectable lectin while lectin activity can be detected in various parts of the mature plant (Sabnis & Hart, 1978) and that some seed lectins are also found in older tissues (Howard *et al.* 1972; Mialonier *et al.* 1973; Rougé, 1974*b*; Pueppke *et al.* 1978) suggests that a role in storage of seed reserves is certainly not a universal function of plant lectins.

In some species the progressive decrease in lectin content on germination parallels the decrease in reserve protein, suggesting that lectins may act as storage proteins (Lis & Sharon, 1981). The plants in which this has been reported are the lentil (Howard *et al.* 1972; Rougé, 1974*a,b*), red kidney bean (Mialonier *et al.* 1973) and *Dolichos biflorus* (Talbot & Etzler, 1978*b*). In castor bean the disappearance of reserve protein is much faster than that of the lectin (Youle & Huang, 1976), suggesting that if some lectins play a role as seed storage proteins, that role is not shared by all haemagglutinins. Once again the evidence relating to the *in vivo* functions of lectins is contradictory and incomplete.

(2) Lectins in phloem transport
Kauss & Ziegler (1974) in *Robinia pseudoacacia*, Sabnis & Hart (1978) in three *Cucurbita* species and Jeffree & Yeoman (1981) in *Datura stramonium* have each reported the presence of lectin in phloem. Kauss & Ziegler (1974) found that the lectin from the sap of *Robina* sieve tubes is specific for N-acetyl-D-galactosamine and glycosides containing galactose. These sugars are not transported in *Robinia* sieve tubes. From this evidence Kauss & Ziegler (1974) suggested that the lectin is probably not directly involved in sugar transport. Such a suggestion is consistent

with an hypothesis proposed by Sabnis & Hart (1978). The latter workers reported that 15–20% of *Cucurbita* phloem exudate is a lectin, with a specificity directed towards a disaccharide not present in plant extracts in an unconjugated form. They suggested that *in vivo*, fraction 1 protein (in filamentous form) and fraction 2 protein (the lectin) are linked by disulphide bridges and that the lectin serves to anchor the P-protein filaments to glycoprotein or glycolipid components of the sieve element plasma membrane. (The necessity for the P-protein filaments to be anchored during normal sugar translocation in the sieve tube is generally accepted (Cronshaw, 1975).) This would indicate an indirect function of lectin in sugar transport in *Cucurbita* species.

(3) Lectins in intracellular transport

Mellor *et al.* (1980) and Frederick *et al.* (1981) showed that lectin-binding sites are constitutents of most membranes and are exposed on the luminal (extracytoplasmic) surface of organelles. Subsequently, Frederick *et al.* (1981) suggested that these moieties aid in sequestering various components into cellular compartments and/or the direction of them to their final destination. Unfortunately no mechanism for this function was proposed and lectins located in the cell walls (Kauss & Glaser, 1974) are unlikely to be involved in this way.

12.7.3   Involvement in membrane fusion and structure

Membrane-bound organelles may show close and characteristic associations with each other. For example, the surfaces of chloroplasts and microbodies (peroxisomes) appear contiguous over considerable distances in leaves and cotyledons (Gruber, Trelease, Becker & Newcomb, 1970; Schopfer, Bajracharya, Bergfeld & Falk, 1976). It would appear that recognition between membranes of organelles may be important in controlling intracellular organization (Frederick *et al.* 1981). Likewise, membrane fusion is highly specific and shows a high degree of spatial and temporal organization (Poste & Allison, 1973). Lectins and their specific interactions with 'receptors' would appear to be possible candidates to fulfill this recognition role. Bowles & Kauss (1975) detected lectins on membrane fractions enriched in plasma membrane, Golgi apparatus and endoplasmic reticulum. Furthermore, the sugar specificities of these lectins differed in various organelles. On this basis they suggested a role for lectins in membrane contact and fusion during membrane flow. It is also noteworthy that Schneider & Sievers (1981) suggest a role for lectins in the proposed communication mechanism between the endoplasmic reticulum (ER) and the amyloplast envelope which is considered to be necessary for graviperception.

Evidence against a role for lectins in intracellular organelle communication comes from the reported localization of lectin-binding sites on the extracytoplasmic surfaces of organelle membranes where they would presumably be unavailable for contact with a membrane-bound lectin from another organelle (Frederick *et al.* 1981).

Further possibilities concerning lectin function in membranes have been proposed by Bowles & Kauss (1975), and other workers. Bowles & Kauss (1975) suggest that lectins on membranes may participate in the maintenance of structural integrity within membranes, for example by binding together multi-enzyme complexes. Yoshida (1978*a*), on the other hand, has detected novel lectins in the endoplasmic reticulum of wheat germ, while Howard & Schnebli (1977) and Yoshida (1978*b*) have detected carbohydrate moieties in eukaryotic ribosomes. The suggestion has therefore arisen that lectins may serve to bind ribosomes to the endoplasmic reticulum and thus control the extent to which endoplasmic reticulum is 'rough' or 'smooth' (Yoshida, 1978*b*).

12.7.4   Lectins as enzymes

It is apparent that many enzymes of carbohydrate metabolism which possess multiple combining sites, for example, sugar kinases, glycosidases and glycosyl transferases could under some circumstances act as lectins. Thus glutaraldehyde treatment of lysozyme cross-links the enzyme which will then agglutinate human erythrocytes (Hořejší & Kocourek, 1974). Galactose oxidase will agglutinate sialidase-treated human erythrocytes if maintained at 0°C. However, the agglutinate disperses if the temperature rises (Hořejší, 1979).

The parallel between sugars that inhibit Con A and the monosaccharides that act as a substrate for yeast hexokinase suggested an enzymatic function for the lectin (Goldstein, Reichert & Misaski, 1974). However, the latter authors detected no hexokinase, phosphatase or phosphorylase activity in their lectin preparations. In contrast a highly purified lectin from mung bean, specific for α-galactosides, also possessed α-galactosidase activity (Hankins & Shannon, 1978). It has been suggested that lectins *in situ* possess highly labile enzyme catalytic sites which are destroyed during purification and this might explain Goldstein's lack of success in detecting enzymatic activity in Con A (Lis & Sharon, 1981). Hankins & Shannon (1978) further speculated that lectins in dry seeds are enzymatically inactive proteins which acquire catalytic properties during germination.

The most intensively studied enzymes with lectin-like properties are the α-galactosidases isolated from several species of legume seeds. Six out of twenty five legume species tested possessed haemagglutinins with α-galactosidase activity (Shannon & Hankins, 1981). The agglutinins extracted from soybean and lima beans are distinct from the well characterized N-acetylgalactosamine lectins previously detected in these beans (Hankins, Kindinger & Shannon, 1980). Furthermore, non-agglutinating α-galactosidases are closely related to the α-galactosidase haemagglutinins. Antisera to several 'classic' lectins also contain cross-reacting material (c.r.m.) to the mung bean haemagglutinin (with enzymatic activity) indicating that these two classes of proteins are evolutionarily related. Indeed legume species with no haemagglutinin activity cross-react with at least one of the three N-acetylgalactosamine-specific haemagglutinins. This suggests that there are related proteins in all species of legumes tested, with the same function, differing in their ability to agglutinate erythrocytes. Further studies have suggested that the N-acetylgalactosamine lectins are closely related to β-hexosaminidases (Shannon & Hankins, 1981). A further haemagglutinin has been extracted from *Phaseolus vulgaris* which displays α-mannosidase activity (Paus & Steen, 1978).

The very close similarity between agglutinins with enzyme-like activity and enzymes with the same sugar specificity suggests that ability to agglutinate erythrocytes may be incidental to the lectins' role *in vivo*. Shannon & Hankins (1981) suggest that all legumes contain functional homologues of the α-galactosamine-specific lectins, with or without the ability to cause erythrocyte agglutination.

More lectins may possess enzymatic activity than presently recognized but it is unlikely that this is a universal role. Other lectins may perform totally different *in vivo* role(s) which may or may not be a reflection of the ability to agglutinate red blood cells.

12.7.5   Protection from herbivores

While the role of various somatic tissue and seed toxins as a defence against herbivore predators is well documented (Smith, 1976), any such role for lectins has received little attention.

Lectins such as ricin and abrin are extremely toxic and it seems possible that these lectins as well as others in seeds of the Leguminosae may play a role in protection against herbivores. Ingestion of *Phaseolus* vulgaris phytohaemagglutinin kills the larvae of the bruchid beetle which, on the other hand, may eat phytohaemagglutinin-free *Vigna unquiculata* seeds (Janzen, Juster & Liener, 1976). These authors concluded that 'a major part of the adaptive significance of photohaemagglutinins in black bean and other legume seeds is to protect from insect predators'. Similarly Janzen (1981) suggested that the lethal effect on the mouse *Liomys salvini* of ingestion of *Phaseolus vulgaris* beans is due to the seed lectin. In contrast, however, Goldstein & Hayes (1978) and Liener (1976) have reported that most lectins are only slightly toxic to animals.

12.7.6 Protection from plant pathogens

In contrast to the paucity of evidence supporting the suggestion that lectins are involved in the protection of plants against attack by animals, there are several well documented instances of lectins as recognition molecules in the interactions of plants and microorganisms. Most of the evidence relating lectins to specificity in host–parasite interactions arises from work with bacterial pathogens (reviewed by Sequeira, 1978). The role of lectins in the legume–*Rhizobium* symbiosis is discussed elsewhere in this book.

(1) Interactions between lectins and bacterial pathogens

Agglutination reactions have been observed between plant lectins and avirulent cells of phytopathogenic bacteria. For example, Goodman, Huang & White (1976) reported that avirulent cells of *Erwinia amylovora* are agglutinated by lectin-like 'granules' present in the xylem fluid of apple petioles. Similar agglutinins have also been detected in rice leaves inoculated with incompatible strains of *Xanthomonas oryzae* (Horino, 1976) and in fluid from tobacco leaves previously infected with *Pseudomonas pisi* (Goodman *et al.* 1976). In the latter instance the incompatible bacteria became immobilized within tobacco leaf tissue and spread of the pathogen was prevented. Unfortunately, there is no evidence concerning the nature of the lectins involved or about the bacterial cell wall components that bound these lectins.

Inoculation of tobacco leaves with a virulent strain of *Pseudomonas solanacearum* leads to multiplication of the bacteria which then spread to adjoining tissues (Sequeira & Graham, 1977). In contrast, if leaves are inoculated with an avirulent strain, the bacteria rapidly become attached to plant cell walls and subsequently become enveloped. The cells to which bacteria become attached then collapse and the spread of bacteria is prevented. Thus attachment of the bacteria to the cell walls which results in this hypersensitivity response is the recognition even between host and pathogen. Since potato is also a host plant for *P. solanacearum* and the potato lectin is well characterized (Allen, Desai, Neuberger & Creeth, 1978) interactions between potato lectin and various strains of bacteria were tested (Sequeira & Graham, 1977). Fifty-five virulent and thirty-four avirulent strains of *P. solanacearum*, from different geographic regions and representing all major races and biotypes, were tested; all avirulent isolates bound the lectin and were strongly agglutinated, whereas virulent strains were not. The binding of lectin to the bacteria could be inhibited by chitin oligomers. The virulent bacteria possessed an extracellular polysaccharide (EPS) not formed by avirulent cells, which, when removed by washing, left the virulent strains susceptible to agglutination by the lectin. These facts suggest strongly that the EPS masks the lectin-binding sites on the bacteria which are present in the underlying lipololysaccharide (LPS) (Sequeira & Graham, 1977). This contrasts with the situation in *Rhizobium* – legume symbiosis (see Chapter 6) where binding to host cell walls appears to be necessary for infection.

Similarly, inoculation of red kidney beans with *Pseudomonas putida*, a saprophytic bacterium, leads to immobilization and encapsulation of the bacteria in the plant cell walls. The bacteria *Pseudomonas phaseolicola* and *Pseudomonas tomato* do not bind to the plant cell walls and therefore do not become encapsulated (Sing & Schroth, 1977). As only *P. putida* was agglutinated by Phytohaemagglutinin (PHA) it was concluded that plant lectins were involved in immobilization and encapsulation.

Recently, a negative relationship has been demonstrated between virulence of strains of *Erwinia stewartis* (Stewart's wilt of corn) and their susceptibility to agglutination by an agglutinin isolated from ground corn (Bradshaw-Rouse *et al.* 1981). The avirulent strains produced only small amounts of EPS and the cells lacked capsules; the virulent strains on the other hand released large amounts of EPS and the cells were capsulated. The parallels with the potato – *P. solanacearum* interactions are obvious although no direct involvement of lectins has yet been demonstrated.

The inverse relationship between virulence of any one strain and its suceptibility to agglutination by the lectin of the host does not hold true for all species of bacterial pathogens. Thus the ability of *Erwinia* strains to agglutinate in the presence of potato lectin is apparently unrelated to their pathogenicity to the crop (Ghanekar & Pérombelon, 1980). The ability of these 'opportunist' pathogens to invade potato tubers is probably due to the production of large quantities of pectic enzymes rather than to any specific host–pathogen interaction (Pérombelon, Gullings-Handly & Kelman, 1979).

Similarly, Ohyama, Pelcher, Shaefer and Fowke (1979) have reported

no effect of Con A or soybean agglutinin (SBA) on binding of *Agrobacter-*
ium cells to suspension culture cells of *Datura innoxia* and other workers
have demonstrated that potato lectin plays no role in binding *Agrobacter-*
ium cells to potato tuber discs (Pueppke, Kluepfel & Anand, 1982).
It is thought more likely that attachment of *A. tumefaciens* to plant cell
walls is due to an interaction between galacturonic acid residues in plant
cell wall pectins, and some carbohydrate component of the bacterial
cell wall (Lippincott & Lippincott, 1977).

A quite different and more direct interaction between lectins and
microorganisms has been proposed by Jones (1964) who sought to explain
the characteristic localization of lectins in seeds at such higher concentrations. Leguminous seeds are known to be relatively impermeable to
water, and in nature germination is probably accelerated by degradation
of the seed coat caused by microorganisms. Jones suggested that as some
seeds have lectins in the cotyledons and embryonic axis but none in
the testa (see section 12.5.3), lectins may act to prevent further degradation of the seed once the testa has been broken down. One particular
soil bacterium, shown to accelerate germination of *Vicia cracca* seeds,
was inhibited by semi-purified *Vicia* seed lectin (Callow, 1977). Howard
et al (1972) and Mialonier *et al.* (1973) have shown that lectin is absent
from seed coats but present in the cotyledons of *Phaseolus vulgaris* and
lentil respectively; however *Glycine max* seeds also possess lectin in
the testa (Pueppke *et al.* 1978). Thus, it is unlikely that this proposed
role for these lectins is of universal application.

(2) Interactions between lectins and fungi
The specificity of many lectins for *N*-acetylglucosamine and oligomers
of chitin, a major fungal wall component, suggests a possible role for
some lectins in protection against fungal attack.

Mirelman, Galun, Sharon & Lotan (1975) have demonstrated a direct
effect of certain lectins on the extension of fungal hyphae. Wheat germ
agglutinin (GA) and potato lectin will inhibit hyphal extension and spore
germination in *Trichoderma viride* and *Botrytis cinerea*. It is believed
that these lectins cross-link chitin microfibrils at the fungal apex and
thus inhibit fungal growth.

Marcan, Jarvis & Friend (1979) and Marcan & Friend (1979) have
demonstrated that binding of *Phytopthora infestans* (potato blight)
mycelia from an avirulent race to potato lectin initiates a hypersensitive
response and thus initiates host resistance. Other studies indicate that spores
of strains of *Ceratocystis fimbriata* that are not compatible with sweet
potato are agglutinated by a high-molecular-weight fraction from the

host (presumably a lectin). Spores from compatible strains are not agglutinated to the same extent by the 'lectin' (Kojima & Uritani, 1974).
Therefore, some lectins may play a role in protection of the plant from
fungal pathogens by binding to components of hyphal walls.

(3) Interactions between lectins and mycoplasma
Membranes of many mycoplasmas bind Con A, *Ricinus communis* lectin,
and to a lesser extent WGA, suggesting a role in recognition between
mycoplasmas and host cells (Kahane & Tully, 1976). However, more
work is necessary before any definitive conclusions can be drawn on
lectin–mycoplasma interactions.

12.7.7   *Regulation of cell division*

Nowell's (1960) discovery that extracts of red kidney bean stimulate lymphocyte division *in vitro* led some workers to search for a similar
activity with plant cells. Howard & Schnebli (1977) seeking to explain
the apparent high lectin content of many seeds, suggested that they
might play a role in differentiation and development of the embryo by
control of cell division. The addition of SBA to soybean callus led to
a slight increase in cell number, cell weight and [^{3}H]thymidine incorporation into DNA; this effect was partially prevented by N-acetylglucosa-
mine, the inhibitory sugar for the lectin.

A regulatory role for graminaceous lectins in embryogenesis of cereals
and related species has also been proposed by Peumans & Stinissen
(1983). They suggest that the lectins keep the developing embryonic
axes of seeds in a resting state, possibly by binding to receptors at the
outer side of the plasmalemma and inhibiting cell division. These authors
suggest that the synthesis of lectins during early embryogenesis is consistent with this proposal. The reversibility of the binding of lectins and
receptors is also presented as being compatible with a role in the control
of cell division. Peumans & Stinissen (1983) argue that the proposed
role of WGA in protecting plants from fungal attack (Mirelman *et al.*
1975) is inconsistent with the timing of lectin synthesis during early
embryogenesis and not during germination. Furthermore they point out
that wild relatives of modern cereals are more resistant to certain fungal
pathogens than the cultivated varieties, but contain much less lectin.

Chin & Scott (1979) reported an increase in DNA, RNA and protein
content on addition of Con A, PHA or WGA to protoplasts of both
cereals and tobacco cells. A slight transient increase in growth of seedlings of *Allium* and *Phaseolus* species caused by PHA has also been
reported (Nagl, 1972*a,b*). However, Vasil & Hubbell (1977) observed
no increase in the extent or rate of cell division when tobacco pith seg-

ments and soybean roots were incubated with SBA or PHA. They suggested that the effect of PHA on seedling growth of *Allium* and *Phaseolus* reported (Chin & Scott, 1979) may have been due to contaminants in the lectin preparations used (possibly plant growth substances).

12.7.8 Lectins in the establishment of symbiotic associations

The evidence for lectin involvement in the legume–*Rhizobium* symbiosis is discussed in Chapter 6. There is also some evidence for the involvement of lectins in establishing other symbiotic relationships.

The fact that isolated lichen fungi bind fluorescein-conjugated WGA (Galun, braun, Frensdorf & Galun, 1976) as well as the fact that certain lichens produce lectins of fungal origin suggest the involvement of lectins at some stage in establishing the symbiosis (Lockhart, Rowell & Stewart, 1978). Weis (1978, 1979) has shown a correlation of sugar release and Con A agglutinability with the ability of *Paramecium bursaria* to infest asymbiotic *P. bursaria*.

Studies on the distribution of lectin-binding sites on algal/cell surfaces also imply that these glycoconjugates are involved in interactions between algae and fungi (Sengbusch & Müller, 1983). Marx & Peveling (1983) have recently suggested that the frequency with which certain algae are found in symbiotic association with fungi in lichens is related to the number of sugar groups located at the cell surfaces.

The recently reported observation that *Glomus fasciculatus* (a vesicular arbuscular fungus) can organize chitin microfibrils in the extraradical phase of its cycle, but that during the most frequent intraradical phase (intercellular hyphae and arbuscules) the chitin polymerization is not accomplished, may have some relevance to the establishment of symbiosis (Bonfante-Fasolo, 1982). Herth (1980) has observed that there is a time gap between chitin subunit formation and polymerization into fibrils. It is interesting to speculate that plant lectin binding to short *N*-acetylglucosamine chains prevents the formation of chitin. The amorphous fungal wall may allow close contact between the fungus and host and thus facilitate establishment of symbiosis.

12.7.9 Possible involvement of lectins in grafting

Apart from the interactions between plants and pathogens only one other phenomenon has been studied in which there is recognition involving somatic cells of plants. Yeoman, Kilpatrick, Miedzybrodzka & Gould (1978) have suggested that recognition occurs between the cells of stock and scion before the formation of a graft in plants. More recent data (Parkinson & Yeoman, 1982) on the grafting of excised internodes of solanaceous species in culture supports the view that a recognition response occurs in both graft acceptance and graft rejection. Under identical conditions, autografts of *Nicandra physaloides* and *Lycopersicon esculentum* can be established successfully (re-attainment of vascular connection across the union) while *Nicandra/Lycopersicon* heterografts are unsuccessful (no vascular connections across the union).

Lectins would seem to be one of the obvious candidates for a recognition role in the formation of grafts. Certain circumstantial evidence supports this possibility. Saline extracts of *Lycopersicon* seeds agglutinate protoplasts prepared from the leaves of *Lycopersicon* or *Datura* (compatible graft partners) at a lower concentration than the minimum required to agglutinate protoplasts from *Nicandra* leaves (incompatible graft partners) (Yeoman *et al.* 1978). There is therefore an apparent correlation between compatibility in a grafting context and agglutination.

It has been suggested that either the lectin could be a soluble messenger which travels across the graft (Yeoman & Brown, 1976) or the lectin, which is a component of the plasmalemma (Jeffree & Yeoman, 1981), is able to recognize and specifically bind a diffusible carbohydrate-containing messenger. The latter hypothesis is consistent with the suggestion of Yeoman (1984) that cell wall fragments (formed during enzymatic degradation of the cell walls to form pit fields) are the determinants of the recognition response. It is also probable that these cell wall fragments contain exposed saccharide moieties that could interact with membrane-bound lectins of the plasmalemma. Presumably the lectin, once bound to the cell wall fragment, could initiate a chain of events which lead to either graft acceptance or rejection.

In this respect it is interesting to note that Ryan *et al.* (1981) have reported that a polysaccharide, which resembles pectin or other cell wall components, can induce a protection response in tomato leaves. This saccharide (PIIF) induces synthesis of two protease inhibitors several centimetres from the wound site. They claim that cleavage of cell walls by wounding or microbial attack leads to the formation of PIIF which in turn stimulates protease-inhibitor formation. PIIF may therefore play a role in protection of the plant from parasites (Ryan, 1981). Whether these large polysaccharides (mol. wt 5000–10000 in tomato and mol. wt 200000 in sycamore) are transported from the site of initial attack is not known. It may be that a smaller more mobile fraction is transported, or that the action of PIIF may be localized, producing an amplified second message that is transported (Ryan *et al.* 1981).

Irrespective of the mode of transport, it would seem that these large cell wall fragments must in some way induce the accumulation of protease inhibitors within plant vacuoles (Ryan *et al.* 1981). It is unlikely that such large fragments could pass through cell membranes to induce the proteases within the cell. It is possible that the wall fragments could be introduced into the cytoplasm by pinocytosis, but they would consequently be membrane bound and not in direct contact with any cytoplasmic component. A more attractive proposal is that cell wall fragments interact with membrane-bound 'receptors' (possibly lectins) which act as 'transducers' and convey information to the cell cytoplasm. Although entirely hypothetical, a role for lectins as 'membrane modulators', conveying information from the environment of the cell to internal organelles, is consistent with the results on lectin binding to erythrocytes (Nicolson, 1978).

12.7.10 No function

The discovery that certain strains of soybean exist which lack detectable lectin and have apparently unaltered phenotypes (Pull & Pueppke, 1978) has led Vodkin (1983) to suggest that the lectin has no critical function in the plant. Roots of such lines can be modulated by *Rhizobium japonicum* indicating that this seed lectin is not necessary for initiation of soybean–Rhizobium symbiosis. Results of large-scale screening in *Canavalis, Dolichos, Glycine* and *Pisum* indicate that lectin content may vary considerably between cultivars. The difficulty in assigning any particular function to this group of proteins and glycoproteins however does not mean that they have no function in plants. The susceptibility of lectinless plants to pathogenic invasion was not tested nor was growth under different environmental conditions. It would be surprising if some plants synthesized, in such relatively large quantities, molecules which were totally redundant under all conditions of plant growth. The metabolic expense of producing products with no function would presumably place these plants at a selective disadvantage.

The highly conserved nature of lectins of related species also suggests an important role for these molecules. The paradox however remains that certain plants can lack lectins with no obvious phenotypic effect, and plants of closely related cultivars can vary in lectin content, but the molecules are very highly conserved and present in a large number of plant groups.

12.8   Concluding remarks

The very widespread distribution of lectins in the plant kingdom as well as their highly conserved structure suggests an important role for these molecules in plants. The ability of these proteins to discriminate

358   *M. A. Holden and M. M. Yeoman*

between different saccharides suggests a function of lectins as 'recognition molecules'. Liener (1976) has suggested that the biological properties of lectins, as observed in the laboratory, bear no relation to their function in nature. This, however, seems unlikely as all plant cells contain membranes with exposed saccharide residues and it would be expected that binding of lectins to these residues would have some effect on the membranes

The discovery that a lectin from mung beans possesses galactosidase activity has prompted speculation that other lectins may possess enzymatic activity *in vivo*. Indeed it has been suggested that, either lectins in dry seeds are inactive enzymes which acquire catalytic properties during germination, or that the lectins *in situ* possess highly labile enzymatic activities which are lost during purification. The location of lectins without enzymatic activity in non-seed tissue of plants suggests that the former possibility does not apply to all lectins. Although it seems likely that more lectins with enzymatic activity will be discovered it is improbable that lectins as a group are a class of enzymes that possess particularly labile catalytic sites. However, there seems to be no distinction between 'classical' lectins and those which have been found to function as enzymes; rather they are part of a group of glycoproteins and proteins with the ability specifically to bind carbohydrate-containing molecules.

There is also evidence which suggests that in a variety of host–parasite systems, specificity involves recognition by complementary surface components. compelling evidence, reviewed here, suggests that some lectins may play a role in these interactions. Lectins in cell walls and on plant membranes appear to function by binding to carbohydrate constituents of plant pathogens. The vast informational potential of glycoproteins that contain even a limited number of saccharide residues may be capable of explaining the large number of discriminations between susceptible and resistant strains of phytopathogens.

Other suggested roles for lectins also depend on their ability specifically to bind saccharide residues, whether in maintaining multi-enzyme systems at the appropriate membrane site, binding ribosomes to endoplasmic reticulum or sequestering and channelling glycoproteins destined for secretion. The majority of roles suggested for lectins involve the binding of sugar residues to membrane-bound lectins. It seems possible that in plant cells, as in animal cells, some lectins serve as 'transducers' of information either from the exterior of the cell to the cytoplasm or across internal membranes.

The authors gratefully acknowledge Mrs E. Raeburn and Miss J. Wood for typing the manuscript.

12.9   References

Allen, H. J. & Brilliantine, L. (1969). A survey of haemagglutinins in various seeds. *J. Immunol.* **102:** 1295–9.

Allen, A. K., Desai, N. N., Neuberger, A. & Creeth, J. M. (1978). Properties of potato lectin and the nature of its glycoprotein linkages. *Biochem. J.* **171:** 665–74.

Allen, A. K. & Neuberger, A. (1973). The purification and properties of the lectin from potato tubers, a hydroxyproline-containing glycoprotein. *Biochem. J.* **135:** 307–14.

Allen, A. K., Neuberger, A. & Sharon, N. (1973). The purification, composition and specificity of wheat germ agglutinin. *Biochem. J.* **131:** 155–62.

Anand, V. K., Pueppke, S. G. & Heberlein, G. T. (1977). The effect of lectins on *Agrobacterium tumefaciens* caused by crown gall tumour induction. *Pl. Physiol.* **59:** 5–109.

Barkai-Golan, R., Mirelman, D. & Sharon, N. (1978). Studies on growth inhibition by lectins on *Penicillia* and *Apergilli*. *Arch. Microbiol.* **116:** 119–24.

Barrett, J. T. & Howe, M. L. (1968). Hemagglutination and hemolysis by lichen extracts. *Appl. Microbiol.* **16:** 1137–9.

Baumann, C., Rüdiger, H. & Strosberg, A. D. (1979). A comparison of the two lectins from *Vicia cracca*. *FEBS Lett.* **102:** 216–18.

Beyenbach, J., Weber, C. & Kleinig, H. (1974). Sieve tube proteins from *Cucurbita maxima*. *Planta* **119:** 113–24.

Bonfante-Fasolo, P. (1982). Cell wall architecture in a mycorrhizal association as revealed by cryoultramicrotomy. *Protoplasma* **111:** 113–21.

Borrebaeck, C. A. K. & Mattiasson, B. (1983). Distribution of a lectin in tissues of *Phaseolus vulgaris*. *Physiol. Plant* **58:** 29–32.

Bowles, D. J. & Kauss, H. (1975). Carbohydrate-binding proteins from cellular membranes of plant tissue. *Pl. Sci. Lett.* **4:** 411–18.

Bowles, D. J. & Kauss, H. (1976). Charaterisation, enzymatic and lectin properties of isolated membranes from *Phaseolus aureus*. *Biochim. Biophys. Acta* **443:** 360–74.

Bowles, D. J., Lis, H. & Sharon, N. (1979). Distribution of lectins in membranes of soybean and peanut plants. 1. General distribution in root, shoot and leaf tissue at different stages of growth. *Planta* **145:** 193–8.

Bowles, D. J., Schnarrenberger, C. & Kauss, H. (1976). Lectins as membrane components of mitochondria from *Ricinus communis*. *Biochem. J.* **160:** 375–82.

Boyd, W. C. & Shapleigh, E. (1954). Specific precipitating activity of plant agglutinins (lectins). *Science* **119:** 419.

Bradshaw-Rouse, J. J., Whately, M. H., Coplin, D. L., Woods, A. Sequeira, C. & Kelman, A. (1981). Agglutination of *Erwinia stewartii* strains with a corn agglutinin: correlation with extracellular polysaccharide production and pathogenicity. *App. & Environ. Microbiol.* **42:** 344–50.

Burgess, J. & Fleming, E. N. (1974). Tobacco protoplasts rapidly regenerate cell wall. *J. Cell Sci.* **14:** 439–49.

Burgess, J. & Linstead, P. J. (1976). Ultrastructural studies of the binding of concanavalin A to the plasmalemma of higher plant protoplasts. *Planta* **130:** 73–9.

Burgess, J. & Linstead, P. J. (1977). Membrane mobility and the concanavalin A binding system of the plasmalemma of higher plant protoplasts. *Planta* **136:** 253–9.

Burke, D., Kaufman, P., McNeil, M. & Albershein, P. (1974). The structure of plant cell walls VI. A survey of the walls of suspension-cultured monocots. *Plant Physiol.* **54:** 109–15.

Callow, J. A. (1975). Plant lectins. *Curr. Adv. Pl. Sci.* **18:** 181–93.

Callow, J. A. (1977). Recognition, resistance and the role of plant lectins in host–parasite interactions. *Adv. Bot. Res.* **4:** 1–49.

Chang, C.-M., Rosen, S. D. & Barondes, S. H. (1977). Cell surface location of an endogenous lectin and its receptor in *Polysphondylium Pallidam*. *Exp. Cell Research* **104:** 101–9.

Chin, J. C. & Scott, K. J. (1979). Effect of phytolectins on isolated protoplasts from plants. *Ann. Bot.* **43:** 33–44.

Clarke, A. E., Anderson, R. L. & Stone, B. A. (1979). form and function of arabinogalactans and arabinogalactan-proteins. *Phytochem.* **18:** 521–40.

Clarke, A. E., Knox, R. B. & Jermyn, M. A. (1975). Localisation of lectins in legume cotyledons. *J. Cell Sci.* **19:** 157–67.

Cronshaw, J. (1975). P-proteins. In *Phloem Transport*, ed. S. Aronoff, J. Dainty, P. R. Gorham, L. M. Srivastava & C. A. Swanson, pp. 74–115. New York, London: Plenum Press.

Cunningham, B. A., Hemperly, J. J., Hopp, T. P. & Edelman, G. M. (1979). Favin versus concanavalin A: circularly permuted amino acid sequences. *Proc. Natl. Acad. Sci. USA* **76:** 3218–22.

Dazzo, F. B., Truchet, G. L., Sherwood, J., Hrabale, M., Abe, M. & Pankratz, H. S. (1984). Specific phases of root hair attachment in *Rhizobium trifolii*. – clover symbiosis. *App. Environ. Microbiol.* **48:** 1140–50.

Dazzo, F. B., Yanke, W. E. & Brill, W. J. (1978). Trifoliin: a *Rhizobium* recognition protein from white clover. *Biochim. Biophys. Acta* **539:** 276–86.

Den, H. & Malinzak, D. A. (1977). Isolation and properties of β-D-galactoside-specific lectin from chick embryo thigh muscle. *J. Biol. Chem.* **252:** 5444–8.

Drysdale, R. G., Herrick, P. R. & Franks, H. (1968). The specificity of the haemagglutinin of the castor bean *Ricinus communis*. *Vox Sang.* **15:** 194–202.

Dulaney, J. T. (1979). Binding interactions of glycoproteins with lectins. *Mol. & Cell. Biochem.* **22:** 43–63.

Ensgraber, A. (1958). Die Phythamagglutine und ihre Funktion in der Pflanze als Kohlenhydrat-transtortsubstanzen. *Ber. dt. bot. Gesn.* **71:** 349–61.

Etzler, M. E., Talbot, C. F. & Ziaya, P. R. (1977). NH_2-Terminal sequences of the subunits of *Dolichos biflorus* lectin. *FEBS Lett.* **82:** 39–41.

Fenton, C. A. L. & Labavitch, J. M. (1980). Lectin-mediated agglutination of plant protoplasts. *Physiol. Plant.* **49:** 393–7.

Foriers, A., de Neve, R., Kanarek, L. & Strosberg, A. D. (1978). Common ancestor for concanavalin A and lentil lectin? *Proc. Natl. Acad. Sci. USA* **75:** 1136–9.

Foriers, A., Van Driessche, E., de Neve, R., Kanarek, L., Strosberg, A. D. & Wuilmart, C. (1977). The subunit structure and *N*-terminal sequences of the α- and β-subunits of the lentil lectin (*Lens culinaris*). *FEBS Lett.* **75:** 237–40.

Foriers, A., Wuilmart, C., Sharon, N. & Strosberg, A. D. (1977). Extensive sequence homologies among lectins from leguminous plants. *Biochem. Biophys. Res. Commun.* **75:** 980–6.

Frederick, S. E. & Newcomb, E. H. (1969). Microbody like organelles in leaf cells. *Science* **163:** 1353–5.

Frederick, S. E., Nies, B. & Gruber, P. J. (1981). An ultrastructural search for lectin-binding sites on surfaces of spinach leaf organelles. *Planta* **152:** 145–52.

Galbraith, W., & Goldstein, I. J. (1970). Phytohaemagglutinins: a new class of metalloproteins. Isolation, purification and some properties of the lectin from *Phaseolus lunatus*. *FEBS Letts.* **9:** 197–201.

Galun, M., Braun, A., Frensdorf, E. & Galun, E. (1976). Hyphal walls of isolated lichen fungi. Autoradiographic localisation of precursor incorporation and binding of fluorescein-conjugated lectins. *Arch. Microbiol.* **108:** 9–16.

Gaspari-Campani, A., Barbieri, L., Lorenzoni, E. & Stripe, F. (1977). Inhibition of protein synthesis by seed-extracts. *FEBS Lett.* **76:** 173–6.

Gatehouse, J. A. & Boulter, D. (1980). Isolation and properties of a lectin from the roots of *Pisum sativum* (garden pea). *Physiol. Plant.* **49:** 437–42.

Ghanekar, A. & Pderombelon, M. C. M. (1980). Interactions between potato lectin and some phytobacteria in relation to potato tuber decay caused by *Erwinia carotovora*. *Phytopath. Z.* **98:** 137–49.

Gilboa-Garber, L., Mizrahi, L. & Garber, N. (1977). Mannose-binding haemagglutinins in extracts of *Pseudomonas aerwyinosa*. *Can J. Biochem.* **55:** 975–81.

Glimelius, K., Wallin, A. & Eriksson, T. (1974). Agglutinating effects of concanavalin A on isolated protplasts of *Daucus carota*. *Physiol. Plant.* **31:** 225–30.

Glimelius, K., Wallin, A. & Eriksson, T. (1978). Ultrastructural visualisation of sites binding concanavalin A on the cell membrane of *Daucus carota*. *Protoplasma* **97:** 291–300.

Goldstein, I. J. (1972). Use of concanavalin A for structural studies. In *Methods of Carbohydrate Chemistry*, ed. R. L. Whistler & J. M. Be Millar, vol. 6, p. 106–19. New York: Academic Press.

Goldstein, I. J. & Hayes, C. E. (1978). The lectins: carbohydrate binding proteins of plants and animals. *Adv. Carbohyd. Chem. Biochem.* **35:** 127–340.

Goldstein, I. J., Hughes, R. C., Monsigny, M., Osawa, T. & Sharon, N. (1980). What should be called a lectin? *Nature* **285:** 66.

Goldstein, I. J., Reichert, L. M. & Misaki, A. (1974). Interaction of concanavalin A with a model substrate. *Ann. New York Acad. Sci.* **234:** 283–96.

Goldstein, I. J. & So, L. L. (1965). Protein–carbohydrate interaction. III. Agar gel-diffusion studies on the interaction of Concanavalin A, a lectin isolated from Jack Bean with polysaccharides. *Arch. Biochem. Biophys.* **111:** 407–14.

Goodman, R. N., Huang, P. Y. & White, J. A. (1976). Ultrastructural evidence for immobilisation of an incompatible bacterium, (*Pseudomonas pisi* in tobacco leaf tissue. *Phytopathol.* **66:** 754–64.

Graham, T. L., Sequeira, L. & Huang, T. R. (1977). Bacterial lipopolysaccharides as inducers of disease resistance in tobacco. *Appl. & Environ. Microbiol.* **34:** 424–32.

Green, T. R. & Ryan, C. A. (1973). Wound-induced proteinase inhibitor in tomato leaves. *Pl. Physiol.* **51:** 19–21.

Gruber, P. J. Trelease, R. N., Becker, W. M. & Newcomb, E. H. (1970). A correlative ultrastructural and enzymatic study of cotyledonary microbodies following germination of fat-storing seeds. *Planta* **93:** 269–88.

Hakomori, S-I. (1975). Structures and organisation of cell surface glycolipids dependency on cell growth and malignant transformation. *Biochim. Biophys. Acta* **417:** 55–89.

Hankins, C. N., Kindinger, J. I. & Shannon, L. M. (1979). 1. Immunological cross-reactions between the enzymic lectin from Mung beans and other well characterised lectins. *Plant Physiol.* **64:** 104–7.

Hankins, C. N., Kindinger, J. I. & Shannon, L. M. (1980). Legume α-galactosidases which have haemagglutinin properties. *Plant Physiol.* **65:** 618–22.

Hankins, C. N. & Shannon, L. M. (1978). The physical and enzymatic properties of a phytohaemagglutinin from Mung beans. *J. Biol. Chem.* **253:** 7791–7.

Hawkes, J. G., Lester, R. N. & Skelding, A. D. (eds). (1979). The biology and Taxonomy of the Solanaceae. London and New York: Academic Press.

Hepler, P. K. (1982). Endoplasmic reticulum in the formation of the cell plate and plasmodesmata. *Protoplasma* **111:** 121–33.

Herth, W. (1980). Calcofluor white and congo red inhibit chitin microfibril assembly of *Poterioochromonas:* evidence for a gap between polymerisation and microfibril formation. *J. Cell. Biol.* **87:** 442–50.

Hořejší, V. (1979). Galactose oxidase – an enzyme with lectin properties. *Biochim. Biophys. Acta* **577:** 383–8.

Hořejší, V. & Kocourek, J. (1974). Studies on phytohaemagglutinins XXI. The covalent oligomers of lysozyme – first case of semisynthetic haemagglutinins. *Experientia* **30:** 1348–9.

Hořejší, V. & Kocourek, J. (1978). Studies on lectins. XXXVII. Isolation and characterisation of the lectin from Jimson-weed seeds (*Datura stramonium* L.) *Biochim. Biophys. Acta* **532:** 92–7.

Hořejší, V., Ticha, M. & Kocourek, J. (1979). Affinity Electrophoresis. *Trends Biochem. Sci.* **4:** N6–N7.

Horino, O. (1976). Induction of bacterial leaf blight resistance by incompatible strains of *Xanthomonas oryzae*. In *Biochemistry and Cytology of Plant Parasite Interactions*, ed. K. Tomiyama *et al.*, pp. 43–55. Tokyo: Kodansha.

Horisberger, M. & Vonlanthen, M. (1980). Ultrastructural localisation of soybean agglutinin on thin sections of *Glycine max* (Soybean) var. Altona by the gold method. *Histochemistry* **65:** 181–6.

Howard, I. K., Sage, A. J. & Horton, C. B. (1972). Studies on the appearance and location of haemagglutinins from a common lentil during the life cycle of the plant. *Archs Biochem. Biophys.* **149:** 323–6.

Howard, G. A. & Schnebli, H. P. (1977). Eukaryote ribosomes possess a binding site for concanavalin A. *Proc. Natl. Acad. Sci. USA* **74:** 818–21.

Howe, M. L. & Barrett, J. T. (1970). Studies on a haemagglutinin from the lichen *Parmelia michantiana*. *Biochim. Byophys. Acta* **215:** 97–104.

Jaffe, C. L., Ehrlich-Rogozinski, S., Lis, H. & Sharon, N. (197). Transition metal requirements of soybean agglutinin. *FEBS Lett.* **82:** 191–6.

Janzen, D. H. (1981). Lectins and plant–herbivore interactions. In *Recent Advances in Phytochemistry*, vol. 15, *The Phytochemistry of Cell Recognition and Cell Surface Introductions*, ed. F. A. Loewus & C. A. Ryan, pp. 241–58. New York: Plenum Press.

Janzen, D. H., Juster, H. B. & Liener, I. E. (1976). Insecticidal action of the phytohaemagglutinin in black beans on a Bruchid Beetle. *Science* **192:** 795–6.

Jeffree, C. E. & Yeoman, M. M. (1981). A study of the intracellular and intercellular distribution of the *Datura stramonium* lectin using an immunofluorescent technique. *New Phytol.* **87:** 463–71.

Jeffree, C. E. & Yeoman, M. M. & Kilpatrick, D. C. (1982). Immunofluorescence studies on plant cells. *Int. Rev. Cyt.* **80:** 231–65.

Jermyn, M. A. & Yeow, Y. M. (1975). A class of lectins present in the tissue of seed plants. *Aust. J. Pl. Physiol.* **2:** 501–31.

Jones, D. A. (1964). The lectin in the seeds of *Vicia cracca* L. II. A population study and possible function for the lectin. *Heredity* **19:** 459–69.

Kabat, E. A. (1978). Dimensions and specificities of recognition sites on lectins and antibodies. *J. Supramolecular Structure* **8:** 79–88.

Kahane, I. & Tully, J. G. (1976). Binding of plant lectins to mycoplasma cells and membranes. *J. Bacteriology* **128:** 1–7.

Kauss, H. (1977). The possible physiological role of lectins. In *Cell Wall Biochemistry Related to Specificity in Host Plant Pathogen Interactions*, ed. B. Solheim & J. Raa, pp. 347–58. Universitetsforlaget Tromsø-Oslo-Bergen.

Kauss, H. & Glaser, C. (1974). Carbohydrate binding proteins from plant cell walls and their possible involvement in extension growth. *FEBS Lett.* **45:** 304–7.

Kauss, H. & Ziegler, H. (1974). Carbohydrate-binding proteins from the sieve-tube sap of *Robinia pseudoacacia*. *Planta* **121:** 197–200.

Kilpatrick, D. C. (1980). Purification and some properties of lectin from the fruit juice of tomato (*Lycopersicon esculentum*). *Biochem. J.* **185:** 269–72.

Kilpatrick, D. C., Jeffree, C. E., Lockhart, C. M. & Yeoman, M. M. (1980). Immunological evidence for structural similarity among lectins from species of the Solanaceae. *FEBS Lett.* **113:** 129–33.

Kilpatrick, D. C. & Yeoman, M. M. (1978). Purification of the lectin from *Datura stramonium*. *Biochem. J.* **175:** 1151–3.

Kilpatrick, D. C., Yeoman, M. M. & Gould, A. R. (1979). Tissue and subcellular

Functions of plant lectins 363

distribution of the lectin from *Datura stramonium* (Thorn Apple). *Biochem. J.* **184**: 215–19.

Köhle, H. & Kauss, H. (1979). binding of *Ricinus communis* agglutinin to the mitochondrial inner membrane as an artefact during preparation. *Biochem. J.* **184**: 721–3.

Kojima, M. & Uritani, I. (1974). The possible involvement of a spore agglutinating factor(s) in various plants in establishing host specificity by various strains of black rot fungus, *Ceratocystis fimbiata. Plant Cell Physiol.* **15**: 733–7.

Krüpe, M. (1956). Blutgruppenspezifische Pflanzliche Eiweiskorper (Phytoagglutinine), pp. 1–131. Stuttgart: Enke-Verlag.

Lamport, D. T. A. (1973). The glycopeptide linkages of extensin: O-D-galactosy serine and O-L-arabinosyl hydroxyproline. In *Biogenesis of Plant Cell Wall Polysaccharides*, ed. F. A. Loewus, pp. 149–65. London: Academic Press.

Lamport, D. T. A. (1980). Structure and function of plant glycoproteins. In *The Biochemistry of Plants*, ed. P. K. Stumpf & E. E. Conn, pp. 501–36. New York, London, Toronto, Sydney, San Francisco: Academic Press.

Larkin, P. J. (1978). Plant protoplast agglutination by lectins. *Pl. Physiol.* **61**: 626–9.

Lee, D. W., Tan, G. S. & Liew, F. Y. (1977). A survey of lectins in South-east Asian Leguminosae. *Planta Medica* **31**: 83–93.

Li, E. & Kornfeld, S. (1977). Effects of wheat germ agglutinin on membrane transport. *Biochim. Biophys. Acta* **469**: 202–10.

Liener, I. E. (1976). Phytohaemagglutinins (phytolectins). *Ann. Rev. Pl. Physiol.* **27**: 291–319.

Lippincott, J. A. & Lippincott, B. B. (1977). Nature and specificity of the bacterium–host attachment in *Agrobacterium* infection. In *Cell Wall Biochemistry related to Specificity in Host–Plant Pathogen Interactions*, ed. B. Solheim & J. Raa, pp. 439–51. Oslo: Universitetshorlaget.

Lis, H., Sela, B-A. Sachs, L. & Sharon, N. (1970). Specific inhibition by N-acetyl-D-galactosamine of the interaction between soybean agglutinin and animal cell surfaces. *Biochim. Biophys. Acta* **211**: 582–5.

Lis, H. & Sharon, N. (1973). The biochemistry of plant lectins (phytohemagglutinins). *Ann. Rev. Biochem.* **42**: 541–74.

Lis, H. & Sharon, N. (1981). Lectins in higher plants. In *The Biochemistry of Plants*, ed. P. K. Stumpf & E. E. Conn, pp. 371–447. New York, London, Toronto, Sydney, San Francisco: Academic Press.

Lockhart, C. M., Rowell, P. & Stewart, W. D. P. (1978). Phytohaemagglutinins from the nitrogen fixing lichens *Peltigera canina* and *P. polydactyla. F.E.M.S. Microbiology Letters* **3**, 127–30.

Lotan, R., Cacan, R., Cacan, M., Debray, H., Carter, W. G. & Sharon, N. (1975). On the presence of two types of subunit in soybean agglutinin. *FEBS Lett.* **57**: 100–3.

Lotan, R., Galun, E., Sharon, N. & Mirelman, D. (1975). Interaction of wheat germ agglutinin with microorganisms. Abstract 1001 of *10th FFBS Meeting*, Paris.

Marcan, H. & Friend, J. (1979). Lectins in potato tuber tissues and relevance to host–parasite interaction. *Arabinogalactan protein News* 2.

Marcan, H., Jarvis, M. L. & Friend, J. (1979). Effect of methyl glycosides and oligosaccharides on cell death and browning of potato tuber discs induced by mycelial components of *Phytophthora infestums. Physiol. Plant Pathol.* **14**: 1–9.

Marx, M. & Peveling, E. (1983). Surface receptors in lichen symbionts visualized by fluorescence microscopy after use of lectins. *Protoplasma* **114**: 52–61.

Mellor, R. B., Krusius, T. & Lord, M. J. (1980). Analysis of glycoconjugated saccharides in organelles isolated from castor bean endoplasm. *Pl. Physiol.* **65**: 1073–5.

Mellor, R. B., Roberts, L. M. & Lord, J. M. (1980). N-Acetylglucosamine transfer

364   M. A. Holden and M. M. Yeoman

reactions and glycoprotein biosynthesis in castor bean endosperm. *J. exp. Bot.* **31:** 993–1003.

Mialonier, G., Privat, J. P., Monsigny, M., Kahlem, G. & Durand, R. (1973). Isolemenet, propriétés physico-chimiques et localisation *in vivo* d'une phytohémagglutine (lectine) de *Phaseolus vulgaris* L. (var. rouge). *Physiol. Vég.* **11:** 519–37.

Miller, J. B., Hsu, R., Heinrikson, R. & Yachnin, S. (1975). Extensive homology between the subunits of the phythaemagglutinin mitogenic proteins derived from *Phaseolus vulgaris*. *Proc. Natl. Acad. Sci. USA* **72:** 1388–91.

Mirelman, D., Galun, D., Sharon, N. & Lotan, R. (1975). Inhibition of fungal growth by wheat germ agglutinin. *Nature* **256:** 414–16.

Monsigny, M., Kieda, C. & Roche, A. C. (1979). Membrane lectins. *Biol. Cellulaire* **36:** 289–300.

Morimoto, N., Shimizu, S. & Yamada, K. (1978). Capping of saccharides on the plasma membrane of lymphocytes as studied by fluorescein-labelled lectins. *Histochem. J.* **10:** 223–8.

Nagl, W. (1972*a*). Phytohaemagglutinin: transitory enhancement of growth in *Phaseolus* and *Allium*. *Planta* **106:** 269–72.

Nagl, W. (1972*b*). Phytohämagglutinin, temporäre erhöhung der mitotischen ativität bei *Allium*, und partieller autagonismus gegenüber Colchicin. *Exp. Cell Res.* **74:** 599–602.

Nicolson, G. L. (1974). The interaction of lectins with animal cell surfaces. *Int. Rev. Cytol.* **39:** 89–109.

Nicolson, G. L. (1978). Ultrastructural localization of lectin receptors. In *Advanced Techniques in Biological Electron Microscopy*, ed. J. K. Koehler, vol. 2, pp. 1–38. Berlin and New York: Springer-Verlag.

Nowell, P. C. (1960). Phytohaemagglutinin: an initiator of mitosis in cultures of normal human leukocytes. *Cancer Res.* **20:** 462–6.

Ohyama, K., Pelcher, L. E., Schaefer, A. & Fowke, L. C. (1979). *In vitro* binding of *Agrobacterium tumefaciens* to plant cells from suspension culture. *Pl. Physiol.* **63:** 382–7.

Paau, A. S., Leps, W. T. & Brill, W. J. (1981). Agglutinin from Alfalfa necessary for binding and nodulation by *Rhizobium meliloti*. *Science* **213:** 1513–15.

Pardoe, G. I. & Uhlenbruck, G. (1970). Characteristics of antigenic determinants of intact cell surfaces. *J. med. Lab. Technol.* **27:** 249–63.

Parkinson, M. & Yeoman, M. M. (1982). Graft formation in cultured explanted internodes. *New Phytol.* **91:** 711–19.

Paulova, M., Ticha, M., Entlicher, G., Kostir, J. V. & Kocourek, J. (1971). Studies on phytohaemagglutininfrom the lentil (*Lens esculentum*). *Biochim. Biophys. Acta* **252:** 388–95.

Paus, E. & Steen, H. B. (1978). Mitogenic effect of α-mannosidase on lymphocytes. *Nature* **272:** 452–4.

Perombelon, M. C. M., Gullings-Handly, J. & Kelman, A. (1979). Population dynamics of *Erwinia caratovora* and pectolytic *Clostridium* spp. in relation to decay of potatotes. *Phytopathol.* **69:** 167–73.

Peumans, W. J. & Stinissen, H. M. (1983). Gramineae lectins: occurrence, molecular biology and phsyiological function. In *Chemical Taxonomy, Molecular Biology and Function of Plant Lectins*, ed. I. J. Goldstein & M. E. Etzler, New York: Alan R. Liss Inc.

Peumans, W. J., Stinissen, H. M., Tierens, M. & Carlier, A. R. (1982). *In vitro* synthesis of lectins in cell-free extracts from dry wheat and rye embryos. *Pl. Cell Reports.* **1:** 212–16.

Poste, G. & Allison, A. C. (1973). Membrane fusion. *Biochim. Biophys. Acta* **300:** 421–65.

Pueppke, S. G., Bauer, W. D., Keegstra, K. & Ferguson, A. L. (1978). Role of lectins in plant microorganisms interactions. II. Distribution of soybean lectin in tissues of *Glycine max* (L.) Merr. *Pl. Physiol.* **61**: 779–84.

Pueppke, S. G., Kluepfel, D. A. & Anand, V. K. (1982). Interaction of *Agrabacterium* with potato lectin and Concanavalin-A and its effect on tumor induction in potato. *Physiol. Plant Pathol.* **20**: 35–42.

Pull, S. P. & Puppke, S. G. (1978). Soybean lines lacking the 120 000-Dalton seed lectin. *Science* **200**: 1277–9.

Reisfeld, R. A., Borjeson, J., Chessin, L. N. & Small, P. A. (1967). Isolation and characterisation of a mitogen from pokeweek (*Phytolacca americana*). *Proc. Natl. Acad. Sci. USA* **58**: 2020–7.

Reitherman, R. W., Rosen, S. D., Frazier, W. A. & Barondes, S. H. (1975). Cell surface species-specific high affinity receptors for discoidin: Developmental regulation in *Dictyostelium discoideum*. *Proc. Natl. Acad. Sci. USA* **72**: 3541–5.

Richardson, C., Behnke, W. D., Freishman, J. H. & Blumenthal, K. M. (1978). The complete amino acid sequence of the α-subunit of pea lectin, *Pisum sativum*. *Biochim. Biophys. Acta* **537**: 310–19.

Rodriguez-Boulan, E., Kreibich, G. & Sabatini, D. D. (1978). Spatial orientation of glycoproteins in membranes of rat liver microsomes. I. Localisation of lectin-binding sites in microsomal membranes. *J. Cell Biol.* **78**: 874–93.

Roland, J. C., Lembi, C. A. & Morré, D. J. (1972). Phosphotungstic acid-chromic acid as a selective stain for plasma membranes of plant cells. *Stain Technol.* **47**: 195–200.

Rougé, P. (1974*a*). Physiologie végétale: étude de la phytohémagglutinine des graines des Lentille au cours de la germination et des premiers stades du development de la plante. Evolution dans les cotylédons. *Comptes Rendus de L'Academie des Sciences, Series D* **278**: 449–52.

Rougé, P. (1974*b*). Physiologie végétale: étude de la phytohémagglutinine des graines de Lentille au cours de la germination et des premiers stades du development de la plante. Evolution dans les racines, les tiges et les feuilles. *C. R. Acad. Sci. Paris Ser. D* **278**: 3083–6.

Rougé, P. (1975). Physiologie végétale: devenir des phytohémagglutinines provenant des diverses parties de la graine dans les jeunes germinations du Pois. *C. R. Acad. Sci. Paris Ser. D* **280**: 2105–8.

Ryan, C. A. (1981). Proteinase inhibitors. In *The Biochemistry of Plants*, ed. P. K. Stumpf & E. E. Conn, pp. 351–70. New York, London, Toronto, Sydney, San Francisco: Academic Press.

Ryan, C. A., Bishop, P., Pearce, G., Darvill, A. G., McNeil, M. & Albersheim, P. (1981). A sycamore cell wall polysaccharide and a chemically related tomato leaf polysaccharide possess similar proteinase inhibitor-inducing activities. *Pl. Physiol.* **68**: 616–18.

Sabnis, D. D. & Hart, J. W. (1978). The isolation and some properties of a lectin (haemagglutinin) from *Cucurbita* phloem exudates. *Planta* **142**: 97–101.

Schneider, E. & Sievers, A. (1981). Concanavalin A binds to the endoplasmic reticulum and the starch grain surface of root statocytes. *Planta* **152**: 177–80.

Schopfer, P., Bajracharya, D., Bergfeld, R. & Falk, H. (1976). Phytochrome-mediated transformation of glyoxysomes into peroxisomes in the cotyledons of mustard (*Sinapis alba* L.) seedlings. *Planta* **133**: 73–80.

Schulz, B. C. & Pueppke, S. G. (1978). Peanut lectin interaction with Rhizobia and distribution within *Arachis hypogaea*. *Pl. Physiol.* **61**: Suppl. 328.

Sengbusch, P. V. & Müller, U. (1983). Distribution of glycoconjugates at algal cell surfaces as monitored by FITC-conjugated lectins. Studies on selected species from *Cyanophyta, Pyrrhophyta, Raphidophyta, Euglenophyta, chromophyta* and *Chlorophyta*. *Protoplasma* **114**: 103–13.

Sequeira, L. (1978). Lectins and their role in host–pathogen specificity. *Ann. Rev. Phytopathol.* **16**: 453–81.

Sequeira, L. & Graham, T. L. (1977). Agglutination of avirulent strains of *Pseudomonas solanacearum* by potato lectin. *Physiol. Plant Pathol.* **11**: 43–54.

Shannon, L. M. (1983). Structural properties of legume lectins. In *Chemical Taxonomy, Molecular Biology and Function of Plant Lectins*, ed. I. S. Goldstein & M. E. Etzler. New York: Alan R. Liss.

Shannon, L. M. & Hankins, C. N. (1981). Enzymatic properties of phytohemagglutinins. In *Recent Advances in Phytochemistry*, vol. **15**: *The Phytochemistry of Cell Recognition and Cell Surface Interactions*, ed. F. A. Loewns & G. A. Ryan. New York and London: Plenum Press.

Sharon, N. (1977). Lectins. *Sci Am.* **236**: 108–19.

Sharon, N. & Lis, H. (1972). Lectins: cell agglutinating and sugar-specific proteins. *Science* **177**: 949–59.

Shepard, D. V. & Moore, K. G. (1978). Concanavalin A-mediated agglutination of plant plastids. *Planta* **138**: 35–9.

Sing, V. O. & Schroth, M. N. (1977). Bacteria-plant cell surface interactions: active immobilization of saprophytic bacteria in plant leaves. *Science* **197**: 759–61.

Smith, C. W. & Hollers, J. C. (1970). The pattern of binding of fluorescein-labelled concanavalin A to the motile lymphocyte. *J. Reticuloendothel. Soc.* **8**: 458–64.

Smith, P. M. (1976). *The Chemotaxonomy of Plants*. London: Edward Arnold.

Strosberg, A. D., Lauwerys, M. & Foriers, A. (1983). Molecular evolution of legume lectins. In *Chemical Taxonomy, Molecular Biology and Function of Plant Lectins*, ed. I. J. Goldstein & M. E. Etzler. New York: Alan R. Liss.

Talbot, C. F. & Etzler, M. E. (1978*a*). Isolation and characterization of a protein from leaves and stems of *Dolichos biflorus* that cross reacts with antibodies to the seed lectin. *Biochemistry* **17**: 1474–9.

Talbot, C. F. & Etzler, M. E. (1978*b*). Development and distribution of *Dolichos biflorus* lectin as measured by radioimmunoassay. *Plant Physiol.* **61**: 847–50.

Toms, G. c. & Western, A. (1971). Phytohaemagglutinins. In *Chemotaxonomy of the Leguminosae*, ed. J. B. Harborne, D. Boulter & L. Turner, pp. 367–462. New York: Academic Press.

Vasil, I. H. & Hubbell, D. H. (1977). The effect of lectins on cell division in tissue cultures of soybean and tobacco. In *Cell wall biochemistry related to specificity in host-plant pathogen interactions*, ed. B. Solheim & J. Raa, pp. 361–7. Tromsø, Oslo, bergen: Universitetsforlaget.

Virtanen, I. & Wartiovaara, J. (1976). Lectin receptor sites on rat liver cell nuclear membranes. *J. Cell Sci.* **22**: 335–44.

Vodkin, L. O. (1983). Structure and Expression of Soybean Lectin Gene. In *Chemical Taxonomy, Molecular Biology and Function of Plant Lectins*, ed. I. J. Goldstein & M. E. Etzler. New York: Alan R. Liss.

Walther, B. J. (1976). Mechanism of cell agglutination by Concanavalin A. In *Concanavalin A as a tool*, ed. H. Bittinger & H. P. Schnebli, pp. 231–48. London, New York, Sydney, Toronto: John Wiley & Son.

de Waard, A., Hickman, S. & Kornfeld, S. (1976). Isolation and properties of β-galactoside binding lectins of calf heart and lung. *J. Biol. Chem.* **251**: 7581–7.

Weber, C., Franke, W. W. & Kartenbeck, J. (1974). Structure and biochemistry of phloem-proteins isolated from *Cucurbita maxima*. *Exp. Cell Res.* **87**: 79–106.

Weis, D. S. (1978). Correlation of infectivity and Con A. Agglutinability of algae exymbiotic from *Paramecium busaria*. *J. Protozool.* **25**: 366–70.

Weis, D. S. (1979). Correlation of sugar release and Concanavalin A agglutinability with infectivity of symbiotic algae from *Paramecium bursaria*. *J. Protozool.* **26**: 117–19.

Williamson, F. A. (1979). Concanavalin A binding sites on the plasma membrane of leek stem protoplasts. *Planta* **144:** 209–15.

Williamson, F. A., Fowke, L. C., Constable, F. C. & Gamborg, O. L. (1976) Labelling of concanavalin A sites on the plasma membrane of soybean protoplasts. *Protoplasma* **89:** 305–16.

Yeoman, M. M. (1984). Cellular recognition systems in grafting. In *Encyclopedia of Plant Physiology*, ed. H. F. Linskens & J. Heslop-Harrison, pp. 453–70. Heidelberg, New York, London: Springer-Verlag.

Yeoman, M. M. & Brown, R. (1976). Implications of the formation of the graft union for organisation in the intact plant. *Annals of Botany* **40:** 1265–76.

Yeoman, M. M. Kilpatrick, D. C., Miedzybrodzka, M. B. & Gould , A. R. (1978). Cellular interactions during graft formation in plants, a recognition phenomenon? pp. 139–160. In *Cell-Cell Recognition*, ed. A. S. G. Curtis, pp. 139–60. S. E. B. Symposium XXXII, pp. 139–60. Cambridge:Cambridge University Press.

Yokoyama, M., Nishiyama, F., Kawai, N. & Hirano, H. (1980). The staining of golgi membranes with *Ricinus communis agglutinin–horseradish peroxidase conjugate in mice tissue cells. Exp. Cell Res.* **125:** 47–53.

Yoshida, K. (1978a). Novel lectins in the endoplasmic reticulum of wheat germ and their possible role. *Plant Cell Physiol.* **19:** 1301–5.

Yoshida, K. (1978b). The presence of ribosomal glycoproteins; agglutination of free and membrane bound ribosomes from wheat germ by concanavalin A. *J. Biochem.* **83:** 1609–14.

Youle, R. J. & Huang, H. C. (1976). Protein bodies from the endosperm of castor bean. *Pl. Physiol.* **58:** 703–9.

INDEX

Helianthus tuberosus, cont'd
cyclic AMP-binding proteins 135
Helminthosporum victoriae, Avena
  susceptibility 309
Hordeum,
  adenylate cyclase 132
  cyclic AMP,
    effects 140
    interaction with gibberellins 137
    levels 123
  cyclic AMP-binding proteins 134–5
  lectins 339
  resistance genes 306
  response to *Erysiphe graminis*
    infection 287
Hordeum vulgare, cyclic AMP-binding
  proteins 134
host–pathogen interactions 281–310,
  319–31, 351–5
  fungal infections 283–91
  adhesion of pathogen to host 295
  host response 285–91
  hypersensitive response 287–9
  penetration of host 295–305
  recognition 294–305
  resistance, genetics 305–10
  role of phytoalexins 288–9, 290
  host specificity 281–3
  Rhizobium–legume attachment 319–31
  role of lectins 351–5

IAA (indole-3-acetic acid),
  effects on cyclic AMP biosynthesis 133
  structure 2
IBA (indole-3-butyric acid), structure 2
Ipomoea,
  effects of cyclic AMP 141
  lectin 353–4

Kalanchoe, cyclic AMP levels 123
Kluyveromyces lactis, cell wall
  components 244

Lactuca,
  cyclic AMP,
    interaction with auxins 137
    interaction with gibberellins 138
    levels 123
  lectin-binding sites 342–6
  Allium porrum 344
  Daucus carota 343
  Glycine max 342
  Nicotiana tabaccum 344
lectins,
  definition 335–6
  detection 336–8
  distribution 339–42
  enzymes 349–50
  growth regulation 346

protection from pathogens 351–4
protection from predation 350–1
regulation of cell division 354–5
role in grafting 355–7
role in membrane fusion/structure 348–9
role in *Rhizobium*–legume
  attachment 319–31, 351
role in sugar metabolism 346–8
role in symbiosis 355
structure 338–9
legumes,
  endocytosis of *Rhizobium* 225
  Rhizobium attachment 319–31, 352
Lemna, effects of cyclic AMP 141
Lemna gibba, cyclic AMP levels 123
Lens culinaris, lectins 340, 341, 347, 388
Lolium, cyclic AMP levels 123
Lycopersicon esculentum, lectin 356

Malus pumila, lectin-like granules 351
Marasmas oreades, lectins 341
Medicago sativa, *Rhizobium*
  attachment 321, 322, 323
mucor, cyclic AMP-binding protein 118

NAA (naphthalene-1-acetic acid),
  structure 2
Nicandra physaloides, lectins 337–8, 356
Nicotiana tabaccum,
  auxin-binding sites 5, 14–16, 33, 35–8,
    46–7
  cyclic AMP,
    effects 141
    levels 123
  cyclic GMP 145
  ethylene-binding sites 70
  fusicoccin-binding sites 24
  lectin-binding sites 344, 345
  lectins 351
  protoplasts,
    membrane ultrastructure 232, 234
    surface charge 219

Ononis hircina, lectins 341
Ononis spinosa, lectins 341
Oryza sativa,
  agglutinins 351
  cyclic AMP–gibberellins interaction 137

pallidin, *Polysphondylium pallidum* 203
pallidin receptors, *Polysphondylium
  pallidum* 204
Phacelia, cyclic AMP–gibberellins
  interaction 137
Phaseolus,
  adenylate cyclase 132
  auxin-binding sites 5, 36–7, 38–9
  cyclic AMP,
    interaction with gibberellins 137